AF889158

Advancing Leadership and Collaboration in Project Management

emerald PUBLISHING
ice Publishing

Advancing Leadership and Collaboration in Project Management

Culture, Knowledge and Team Dynamics

Mohammed Reyasudin Basir Khan
PhD, CEng, MIET

Published by Emerald Publishing Limited,
Floor 5, Northspring, 21-23 Wellington Street, Leeds LS1 4DL.

ICE Publishing is an imprint of Emerald Publishing Limited

Other ICE Publishing titles:
Advancing Project Management: Frameworks, Digital Transformation, Sustainability and Governance
Khan, Mohammed Reyasudin Basir ISBN 978-1-83742-799-4
Sustainability and Digital Engineering Management: African Case Studies in Sustainable Digital Transformation
Telukdarie, Arnesh ISBN 978-1-83662-645-9
Essentials of Construction Planning and Scheduling
Robertson, Graham ISBN 978-0-7277-6592-5

A catalogue record for this book is available from the British Library

ISBN 978-1-83742-803-8

Copyright 2026 © Mohammed Reyasudin Basir Khan.
Published under exclusive license by Emerald Publishing Limited

Permission to use the ICE Publishing logo and ICE name is granted under licence to Emerald from the Institution of Civil Engineers. The Institution of Civil Engineers has not approved or endorsed any of the content herein.

All rights, including translation, reserved. Except as permitted by the Copyright, Designs and Patents Act 1988, no part of this publication may be reproduced, stored in a retrieval system or transmitted in any form or by any means, electronic, mechanical, photocopying or otherwise, without the prior written permission of the publisher, Emerald Publishing Limited, Floor 5, Northspring, 21-23 Wellington Street, Leeds LS1 4DL.

This book is published on the understanding that the author is solely responsible for the statements made and opinions expressed in it and that its publication does not necessarily imply that such statements and/or opinions are or reflect the views or opinions of the publisher. While every effort has been made to ensure that the statements made and the opinions expressed in this publication provide a safe and accurate guide, no liability or responsibility can be accepted in this respect by the author or publisher.

While every reasonable effort has been undertaken by the author and the publisher to acknowledge copyright on material reproduced, if there has been an oversight please contact the publisher and we will endeavour to correct this upon a reprint.

Cover photo: shutterstock.com/g/worawee

Commissioning Editor: Michael Fenton
Content Development Editor: Ryan Molyneux
Books Production Lead: Emma Sudderick
Project Manager: Rajachitra S

Typeset by: KnowledgeWorks Global Limited

Printed and bound by CPI Group (UK) Ltd, Croydon, CR0 4YY

Contents

Preface

Project success has traditionally been viewed through the lens of technical methodologies and process-driven frameworks. However, as organisations navigate an era defined by digital disruption, global collaboration and increasing sociopolitical complexity, it is increasingly evident that methodologies alone cannot guarantee success. *Advancing Leadership and Collaboration in Project Management: Culture, Knowledge and Team Dynamics* was developed to explore the human and organisational factors that underpin resilience and innovation in modern project environments. As a companion to *Advancing Project Management: Frameworks, Digital Transformation, Sustainability, and Governance*, this volume shifts the focus from technical frameworks to the people-centred elements that transform project strategies into measurable outcomes.

The insights presented in this book are grounded in rigorous empirical research based on data from over 2000 professionals across sectors including logistics, IT, construction, telecommunications and energy. While these studies are rooted in Malaysian industry cases, a context that reflects the unique challenges of emerging and transitional economies, the findings resonate globally. Malaysia serves as a significant laboratory for assessing how leadership styles, cultural context and organisational learning can be adapted to deliver projects with agility and sustainability in resource-constrained or rapidly changing environments.

The development of this work was motivated by the widening gap between traditional, linear planning and the complex realities of contemporary projects. Projects often fail not because of poor technical scoping but because of misaligned leadership, communication breakdowns or a lack of cultural and stakeholder integration. This book argues that, in the modern era, for a project to succeed, leadership must evolve from control-based models to adaptive, empowering approaches that foster distributed collaboration and continuous learning.

The book is organised into four major parts:

Part 1: Leadership examines how different leadership styles, such as transformational, transactional and servant leadership, influence work performance and team productivity in sectors like construction and telecommunications. These chapters highlight the importance of adaptive strategies tailored to specific project requirements and team dynamics.

Part 2: Team dynamics investigates the challenges of remote and hybrid work, the success factors for industrial engineering projects and the integration of renewable technologies within high-stakes frameworks like oil and gas. This section emphasises that teams must operate as dynamic ecosystems where trust and collaboration are essential.

Part 3: Knowledge and learning focuses on the adoption of Agile practices and their impact on workplace effectiveness. It explores the barriers to implementation in IT organisations and demonstrates how sustainability-driven project management can be embedded through internal stakeholder roles and accountability structures.

Part 4: Communication and culture addresses the critical role of communication competence and cultural intelligence in managing distributed or virtual teams. The chapters provide insights into risk management amid political instability and the success factors for managing global virtual engineering teams across different time zones.

This book is intended for a broad audience of academic researchers, postgraduate students, practitioners and policy makers.
For scholars, it provides fresh empirical evidence that extends established theories, such as the resource-based view and social exchange theory, into new industrial contexts. For practitioners, it offers actionable guidance for building high-performing, adaptive teams in an era of rapid change.

By bridging the gap between scholarly knowledge and practical application, this book seeks to position resilience and leadership as strategic capabilities. It is my hope that these insights will encourage project leaders to look beyond processes and embrace the human and organisational dimensions needed to navigate the next era of project management innovation.

Acknowledgement

The completion of this second volume represents both a continuation and an evolution of the journey that began with the first book. While earlier work explored innovation, tools, sustainability and governance in project environments, this volume turns to the human and organisational dimensions that ultimately determine project success. This shift reflects not only an academic progression but also the lived realities of practitioners navigating complexity, digital transformation and sustainability imperatives.

I would like to extend my sincere appreciation to the contributors of this volume. Each chapter reflects rigorous empirical work, thoughtful analysis and a shared commitment to advancing project leadership beyond traditional process-driven perspectives. Their dedication, discipline and scholarly integrity have made this collection both cohesive and impactful. It has been a privilege to guide and collaborate with researchers who are deeply engaged in strengthening the human side of project management.

My gratitude also goes to Universiti Tun Abdul Razak and the Tun Razak Graduate School for providing an environment that supports research excellence, intellectual discourse and industry engagement. The encouragement to bridge theory and practice has shaped this book's direction and reinforced its applied orientation.

I am equally thankful to industry practitioners, project managers, engineers and organisational leaders who participated in the various empirical studies featured in this volume. Their insights, experience and openness to share real-world challenges have enriched the research and ensured that the findings remain grounded in practice.

To my academic colleagues, mentors and collaborators, thank you for the ongoing conversations that challenge assumptions and inspire deeper inquiry. To my family, whose patience and understanding sustain every scholarly endeavour, your support remains the foundation of all achievements.

This book stands as a reflection of collective effort. It affirms the belief that leadership, collaboration, culture and knowledge are not peripheral to project management but central forces that shape resilience, innovation and long-term organisational success.

About the editor

Mohammed Reyasudin Basir Khan, Tun Razak Graduate School, Universiti Tun Abdul Razak, Kuala Lumpur, Malaysia

Mohammed Reyasudin Basir Khan is a Professor of Artificial Intelligence and a Project Management specialist with over 14 years of experience at the intersection of academia, research and industry. While his technical expertise spans artificial intelligence (AI), data analytics, computing and engineering, his career is equally distinguished by his work in project management, organisational performance and leadership of large-scale initiatives. He has successfully directed complex, multidisciplinary projects across domains such as healthcare, energy analytics, environmental, social and governance (ESG) and business transformation, combining technology adoption with proven project management methodologies.

As Deputy Vice Chancellor (Research) at Universiti Tun Abdul Razak (UNIRAZAK), Prof. Reyasudin leads the strategic enhancement of the university's research and project delivery ecosystem. His responsibilities include building research capacity, increasing the number of funded projects and publications, and cultivating strong industry partnerships. He also serves as Senior Director of URIMPACT, a business unit that leverages AI and project management frameworks to execute AI and digital transformation initiatives. In this role, he oversees programmes in research, innovation, digital apps, business analytics, and stakeholder engagement, ensuring effective delivery through governance structures and project management office (PMO)-style practices.

At the Tun Razak Graduate School, he teaches and supervises in AI, data analytics, business analytics and project management, equipping postgraduate students and executives with both technical and managerial skills. His earlier roles included academic leadership and programme development, where he contributed to curriculum design, accreditation, and compliance with bodies such as the Malaysian Qualification Agency (MQA), the Engineering Accreditation Council (EAC), the Engineering Technology Accreditation Council (ETAC) and the Technology and Technical Accreditation Council (TTAC).

Prof. Reyasudin holds a BEng and PhD in Electrical and Computer Engineering from Universiti Tenaga Nasional (UNITEN). He is a Professional Engineer (Board of Engineers, Malaysia), Chartered

Engineer (Engineering Council, UK), Chartered Scientist (Science Council, UK), European Engineer (FEANI) and Professional Technologist (Malaysian Board of Technologists). His professional credentials also include Registered Electrical Energy Manager (Energy Commission, Malaysia), Certified Energy Manager (ASEAN Centre of Energy), Professional Data Analytics (Google) and Data Science Specialist (IBM).

He has published widely in international journals and at conferences, and has served as an editor, reviewer and chair for academic and professional events. With deep experience in project governance, accreditation management and delivery of high-impact initiatives, he combines the rigour of project management with the innovation of emerging technologies to drive measurable outcomes in both academia and industry.

List of contributors

Sim Tek Chiang: Tun Razak Graduate School, Universiti Tun Abdul Razak (UNIRAZAK), Kuala Lumpur, Malaysia.

Syuhada Assila Tamrin: Tun Razak Graduate School, Universiti Tun Abdul Razak (UNIRAZAK), Kuala Lumpur, Malaysia.

Shatana Manoharan: Tun Razak Graduate School, Universiti Tun Abdul Razak (UNIRAZAK), Kuala Lumpur, Malaysia.

Kuah Shen Hsing: Tun Razak Graduate School, Universiti Tun Abdul Razak (UNIRAZAK), Kuala Lumpur, Malaysia.

Abdullah Haniff Mohd: Tun Razak Graduate School, Universiti Tun Abdul Razak (UNIRAZAK), Kuala Lumpur, Malaysia.

Previndrarasan Krishnamutty: Tun Razak Graduate School, Universiti Tun Abdul Razak (UNIRAZAK), Kuala Lumpur, Malaysia.

Ahmad Farid Azmi: Tun Razak Graduate School, Universiti Tun Abdul Razak (UNIRAZAK), Kuala Lumpur, Malaysia.

Huzaifah Azman Mohamad: Tun Razak Graduate School, Universiti Tun Abdul Razak (UNIRAZAK), Kuala Lumpur, Malaysia.

Nur Aisya Khairul Nizar: Tun Razak Graduate School, Universiti Tun Abdul Razak (UNIRAZAK), Kuala Lumpur, Malaysia.

Chong Peng Lean: MILA University, Nilai, Negeri Sembilan, Malaysia.

Ahmad Anwar Zainuddin: Kulliyyah of Information and Communication Technology, IIUM, Gombak, Malaysia.

Nurul Hafeezah Mohd Zaharin: Tun Razak Graduate School, Universiti Tun Abdul Razak (UNIRAZAK), Kuala Lumpur, Malaysia.

Tanaraj Muniandy: Tun Razak Graduate School, Universiti Tun Abdul Razak (UNIRAZAK), Kuala Lumpur, Malaysia.

Hussain Ibrahim Saleh Saad: Tun Razak Graduate School, Universiti Tun Abdul Razak (UNIRAZAK), Kuala Lumpur, Malaysia.

Ahmad Fairuz Mohd Amin: Tun Razak Graduate School, Universiti Tun Abdul Razak (UNIRAZAK), Kuala Lumpur, Malaysia.

List of abbreviations

ADKAR	awareness, desire, knowledge, ability and reinforcement
AI	artificial intelligence
APM	Agile project management
APR	Agile project readiness
ASEAN	Association of Southeast Asian Nations
AVE	average variance extracted
AWS	Amazon Web Services
CI	cultural intelligence
CSQ	communication satisfaction questionnaire
DOSM	Department of Statistics Malaysia
DV	dependent variable
EPCO	engineering, procurement, construction and operations
ESG	environmental, social and governance
FDI	foreign direct investment
GSCM	green supply chain management
GVET	global virtual engineering team
HR	human resource
HTMT	Heterotrait–Monotrait ratio
IoT	Internet of Things
IPMA	International Project Management Association
ISR	internal stakeholder role
IT	information technology
IV	independent variable
KMO	Kaiser-Meyer-Olkin
MLQ	multifactor leadership questionnaire
MNC	multinational corporation
MV	mediating variable
OPM3	organisational project management maturity model
PLA	planet (sustainability dimension)
PLS	partial least squares
PMBOK	project management body of knowledge

PMI	Project Management Institute
PMMM	project management maturity model
PPL	people
PRO	profit
RBV	resource-based view
RE	renewable energy
SD	standard deviation
SDG	sustainable development goal
SEM	structural equation modelling
SLS	servant leadership scale
SME	small and medium-sized enterprise
SPM	sustainability project management
SPS	sustainability project success
TAM	technology acceptance model
TBL	triple bottom line
TQM	total quality management
VIF	variance inflation factor

Advancing Leadership and Collaboration in Project Management: Culture, Knowledge and Team Dynamics

emerald PUBLISHING ice Publishing

Mohammed Reyasudin Basir Khan
ISBN 978-1-83742-803-8
https://doi.org/10.1108/978-1-83742-802-120261002
Emerald Publishing Limited: All rights reserved

Chapter 1
Reframing project management in the era of complexity

Mohammed Reyasudin Basir Khan

Tun Razak Graduate School, Universiti Tun Abdul Razak (UNIRAZAK), Kuala Lumpur, Malaysia

1.1. Introduction

Project management has traditionally been conceptualised as a discipline of planning, coordination and control. Foundational standards, such as the project management body of knowledge (PMBOK) framework, formalise processes for scope definition, schedule control, cost monitoring and risk mitigation (PMI, 2025). These models emerged from engineering and industrial management traditions that prioritised predictability and standardisation. However, the assumptions underpinning these models have been increasingly challenged. Globalisation has intensified interdependence across supply chains and stakeholder ecosystems (Flyvbjerg, 2014). Digital transformation has compressed innovation cycles and altered coordination mechanisms (Kerzner, 2025). Sustainability imperatives and environmental, social and governance (ESG) accountability have expanded the definition of project success beyond the traditional iron triangle of cost, time and scope (Silvius and Schipper, 2014). Simultaneously, sociopolitical volatility and regulatory complexity have introduced institutional uncertainty into project environments (Too and Weaver, 2014).

Scholars have increasingly recognised that project management is no longer merely technical but also sociotechnical (Bredillet, 2010). Projects are embedded within complex adaptive systems shaped by human relationships, cultural norms and governance structures (Maylor *et al.*, 2008). The discipline, therefore, stands at an epistemological crossroads.

This chapter argues that contemporary project management must be reframed around the integration of capabilities. In this, leadership, team dynamics, knowledge systems and communication culture operate as interdependent drivers of resilience and innovation.

1.1.1 Background and context

Project management has undergone considerable change over the last 50 years. Originally rooted in the engineering, defence and construction sectors, the discipline focused on systematic planning, resource allocation and performance oversight within well-defined boundaries. Classical models such as the Waterfall approach and later formalised frameworks such as the *PMBOK Guide* (see PMI, 2025) established best practices focused on scope, cost, time and quality control. These models were created for environments where predictability was reasonably expected and uncertainty could be managed through thorough upfront planning.

However, the modern project landscape is significantly different. Globalisation has increased supply chain interdependencies, exposing projects to systemic risks beyond organisational control. Digital transformation has accelerated innovation cycles, shortened product lifecycles and reshaped collaboration mechanisms. Sustainability imperatives, including environmental accountability and social governance standards, have broadened stakeholder expectations beyond economic performance alone. Furthermore, geopolitical instability and regulatory complexity introduce additional layers of institutional uncertainty, challenging traditional governance models.

In this changing landscape, projects are no longer standalone technical tasks but sociotechnical systems situated within fluid ecosystems. The growing use of virtual teams, cross-cultural collaboration and digital platforms further complicates coordination and leadership. These changes require a rethinking of basic assumptions about how projects are managed, governed and assessed.

The background to this book, therefore, lies in the recognition that project management is undergoing not merely incremental change but structural and epistemological transformation. The discipline must evolve from a control-oriented paradigm toward a capability-based and resilience-driven framework suited to complexity.

1.1.2 Purpose of the book

The primary purpose of this book is to reconceptualise project management through the lens of integrated organisational capabilities. Rather than treating leadership, team behaviour, learning processes and communication systems as secondary considerations, this book positions them as central determinants of project success in complex environments.

The book advances three overarching objectives. First, it seeks to reposition project management as a sociotechnical discipline grounded in systems thinking and awareness of complexity. Interdependent organisational, technological and institutional forces influence projects. Understanding these interconnections is essential for sustainable performance.

Second, the book aims to foreground human and organisational capabilities as strategic assets. Leadership adaptability, collaborative team dynamics, institutional learning and culturally intelligent communication are not peripheral soft skills but structural enablers of resilience and innovation. By integrating insights from strategic management, organisational behaviour and governance theory, the book offers a holistic perspective on project capability development.

Third, the book aims to bridge the gap between theoretical advancement and practical application. While academic research increasingly recognises complexity and sustainability dimensions, practitioners often operate within frameworks that remain procedurally focused. This book provides conceptual clarity alongside applied insight, encouraging alignment between scholarly discourse and industry realities.

Ultimately, the purpose of the book is to challenge readers, such as scholars, practitioners and policy makers alike, to rethink project management as an evolving capability system capable of navigating uncertainty, fostering innovation and sustaining long-term value creation.

1.1.3 Structure of the chapter

This chapter establishes the conceptual and intellectual foundation for the book. It begins by situating project management within the broader transformation of organisational and institutional environments, highlighting the limitations of traditional, linear, control-based paradigms.

Following this contextual framing, the chapter consolidates key theoretical perspectives, including systems thinking, complexity theory and capability-based views, to justify a shift toward integrated organisational capabilities. These perspectives provide the analytical grounding for the subsequent discussion.

The chapter is structured around four interdependent capability domains: leadership, team dynamics, knowledge and learning, and communication and culture. Each domain is examined in depth to demonstrate its role in enabling resilience, adaptability and sustainability within complex project environments. Rather than presenting these domains as isolated constructs, the chapter emphasises their systemic interdependence and mutual reinforcement.

The chapter concludes by synthesising these domains into an integrated capability framework that positions innovation, resilience and sustainability as central outcomes of contemporary project management practice. In doing so, it establishes the intellectual architecture on which the book's broader discussions are built.

1.2. Theoretical foundation: systems thinking, complexity and strategic capability

Rather than dispersing theoretical references throughout the chapter, this section consolidates the intellectual grounding for the proposed reframing of project management. The argument advanced in this book is not merely that projects are becoming more difficult to manage; rather, it is that the ontological nature of projects has shifted. Projects now operate within interconnected, adaptive and institutionally embedded environments that cannot be adequately explained through linear planning models alone.

To support this reconceptualisation, three interrelated theoretical perspectives are synthesised here: systems thinking, complexity theory and strategic capability perspectives (including the resource-based view and dynamic capabilities theory). Individually, each provides valuable insights. Collectively, they offer a robust explanatory foundation for understanding contemporary project management as an integrated sociotechnical capability system.

1.2.1 Systems thinking: projects as open and embedded systems

Systems theory conceptualises organisations as open systems that interact continuously with their environments (Bertalanffy, 1973). Unlike closed systems, open systems exchange information, resources and feedback with external actors. Regulatory constraints, stakeholder expectations, cultural norms and economic pressures shape them.

Projects, when viewed through this lens, are not isolated operational units. They are temporary systems embedded within broader organisational and institutional ecosystems. A project draws on capital, labour, technological resources, regulatory approvals and stakeholder expectations, and transforms them into outputs that generate not only deliverables but also reputational, environmental and social consequences.

This systemic view introduces two critical implications. First, interdependence becomes central. Decisions made within one functional area influence outcomes elsewhere. Schedule compression may affect quality. Budget constraints may impact stakeholder satisfaction. A communication breakdown in one department may lead to risk escalation across the system.

Second, feedback loops become fundamental to adaptation. Outputs generate reactions, regulatory scrutiny, client feedback and community responses that inform project decision-making. Failure to recognise feedback mechanisms leads to fragmentation and breakdown in coordination (Senge, 1990).

Applying systems thinking to project management, therefore, shifts the analytical focus from task execution to systemic integration. Technical performance cannot be separated from social, cultural and environmental context. Leadership, communication and governance mechanisms become system regulators rather than peripheral considerations.

However, while systems theory explains interdependence and embeddedness, it does not fully capture the unpredictable and emergent nature of contemporary project environments. This limitation leads to the relevance of complexity theory.

1.2.2 Complexity theory: nonlinearity, emergence and adaptation

Complexity theory challenges the assumption that organisational outcomes follow predictable and proportionate cause-and-effect relationships. In complex systems, outcomes emerge from dynamic interactions between multiple agents and subsystems (Stacey, 2011). Small disturbances can produce disproportionate effects and patterns of behaviour may arise without centralised control.

Project environments increasingly exhibit characteristics of complex adaptive systems (Huemann and Turner, 2024). These include

- multiple interacting stakeholders with divergent interests
- rapid technological shifts
- uncertain regulatory landscapes
- cross-cultural collaboration dynamics
- interdependent supply chain networks.

In such environments, uncertainty cannot be entirely reduced through planning. Risk registers and predictive models provide structure but they cannot eliminate emergent behaviour.

Complexity theory introduces three additional implications for project management.

- *Nonlinearity*: minor communication errors may escalate into major coordination failures. Conversely, small innovations may generate substantial performance improvements.

- *Emergence*: team dynamics, innovation outcomes and stakeholder alignment may arise from interaction patterns rather than formal structures.
- *Adaptation*: stability is not achieved through rigidity but through continuous adjustment. Adaptive capacity becomes more important than rigid compliance.

This perspective challenges traditional command-and-control leadership models. Leaders cannot anticipate every contingency; instead, they must cultivate adaptive environments that enable distributed intelligence and rapid interpretation of feedback.

Yet systems and complexity perspectives primarily describe environmental and organisational characteristics. They do not fully explain how organisations build the internal capacity to respond effectively. This is where strategic capability theories become essential.

1.2.3 Resource-based and dynamic capabilities perspectives: capability as strategic asset

The resource-based view (Barney, 1991) argues that sustainable competitive advantage arises from valuable, rare, inimitable and non-substitutable resources. In project contexts, tangible resources, such as capital equipment, technical tools or software, are often replicable. What differentiates performance is intangible capabilities embedded within organisational routines and human capital.

These include

- leadership adaptability
- team cohesion and trust
- learning systems and institutional memory
- communication transparency
- cultural intelligence.

Such capabilities are not easily imitated because they are socially complex and historically embedded.

Dynamic capabilities theory (Teece, 2007) extends this argument by emphasising the organisation's ability to sense environmental change, seize emerging opportunities and reconfigure internal resources accordingly. Projects serve as practical arenas where dynamic capabilities are enacted and tested. Under volatile conditions, organisations must

- sense shifts in stakeholder expectations or regulatory frameworks
- seize innovation opportunities through coordinated action
- reconfigure structures, processes and resource allocation.

Without dynamic capabilities, projects become rigid and vulnerable to disruption. Strategic capability perspectives, therefore, provide the missing link between environmental complexity and organisational response. They explain why some organisations adapt effectively while others struggle under similar external pressures.

1.2.4 Integrating the three perspectives

Individually, systems thinking explains interdependence, complexity theory explains nonlinearity and emergence, while capability perspectives explain strategic adaptation. Together, they form a coherent framework for reframing project management.

- Systems thinking clarifies that projects are embedded within multi-layered environments.
- Complexity theory reveals that these environments are dynamic, nonlinear and unpredictable.
- Capability theory explains how organisations can build internal strengths to navigate such environments.

The integration of these perspectives leads to a central conclusion: project success in contemporary environments depends not solely on planning precision but on the integration of adaptive organisational capabilities within interconnected systems.

Leadership becomes a capability for system integration and sense-making. Team dynamics become mechanisms for distributed adaptation. Learning systems become vehicles for institutional memory and renewal. Communication and culture become governance infrastructures that regulate feedback and coordination.

Thus, rather than viewing project management as a linear sequence of processes, these theories collectively justify understanding it as a dynamic capabilities system operating within complex sociotechnical environments.

This integrated theoretical foundation provides the conceptual basis for the capability-based framework developed in the remainder of this chapter.

1.3. Leadership in the context of complexity and sustainability

Leadership research in project management has evolved considerably over time. Early approaches emphasised trait-based and behavioural models, focusing on personal characteristics and supervisory control. Subsequent scholarship moved toward relational and transformational perspectives, recognising that leadership effectiveness depends on interaction patterns within teams and organisational systems (Müller and Lecoeuvre, 2014). More recently, in response to environmental uncertainty and institutional volatility, scholars have emphasised adaptive and distributed leadership approaches suited to complex project environments (Uhl-Bien and Arena, 2018).

This evolution reflects broader structural changes. As projects become embedded in dynamic sociotechnical ecosystems, leadership cannot be confined to planning oversight and compliance enforcement. Instead, it functions as an integrative capability that enables alignment, learning and adaptation across interdependent systems.

1.3.1 From control to adaptation

Traditional project leadership prioritised planning discipline, milestone monitoring and variance control (Kerzner, 2025). These mechanisms remain important, but they assume that uncertainty can be sufficiently reduced through forecasting and procedural standardisation. In increasingly complex environments, this assumption no longer holds.

Adaptive leadership recognises that ambiguity is structural rather than incidental. Leaders must therefore interpret shifting conditions, reconcile diverse stakeholder interests and facilitate coordinated response rather than rigid compliance.

Key elements of adaptive leadership include

- *strategic sense-making*: enabling leaders to interpret weak environmental signals and emerging risks
- *ethical judgement under uncertainty*: balancing competing stakeholder expectations
- *stakeholder negotiation competence*: aligning divergent interests without suppressing legitimate differences
- *psychological safety facilitation*: encouraging transparent communication of risks and dissent.

Rather than attempting to eliminate uncertainty, adaptive leaders cultivate organisational capacity to respond constructively. Innovation and resilience often emerge from managing variation intelligently rather than enforcing stability.

1.3.2 Ethical and sustainability-oriented leadership

Project legitimacy increasingly depends on integrating sustainability (Silvius and Schipper, 2014). Economic performance alone is insufficient to secure stakeholder trust. ESG accountability frameworks and public scrutiny require leaders to consider social and environmental implications alongside financial outcomes.

This expands leadership responsibility beyond internal coordination toward governance stewardship. Decisions must account for long-term environmental impact, regulatory compliance and reputational exposure. Ethical integrity becomes a strategic asset, particularly in digitally connected environments where transparency is amplified.

Sustainability-oriented leadership strengthens resilience by anticipating external pressures and aligning project objectives with broader societal expectations. It reframes performance as multidimensional rather than purely operational.

1.3.3 Distributed and shared leadership

Complex projects involve specialised knowledge distributed across teams and organisational boundaries. Hierarchical authority alone cannot effectively manage such diversity. Shared leadership models enhance agility by enabling expertise to influence decision-making regardless of formal position (Carson *et al.*, 2007).

Distributed leadership supports

- faster problem resolution through empowered domain experts
- improved knowledge integration across functions
- greater accountability through shared ownership.

Figure 1.1 Evolution of project leadership paradigms

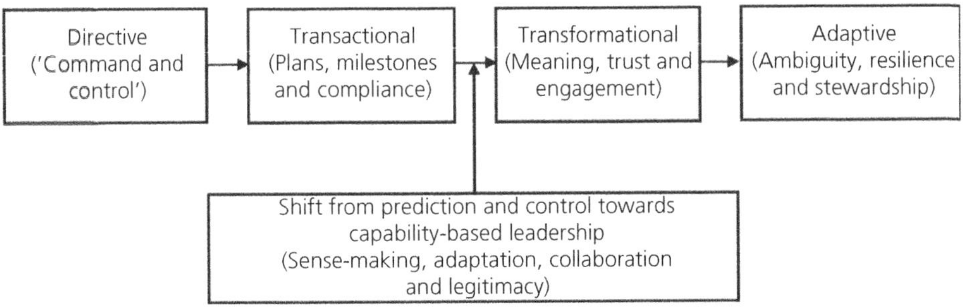

However, distributed leadership does not eliminate the need for central coordination. The formal leader remains responsible for coherence and strategic alignment. Authority becomes relational rather than purely positional, reflecting the networked nature of contemporary project environments.

1.3.4 Leadership as an integrative capability

Viewed through systems, complexity and capability perspectives, leadership emerges as a meta-capability. It integrates interdependent subsystems, enables adaptation under uncertainty and supports organisational reconfiguration in response to environmental change.

In an era defined by volatility and sustainability imperatives, effective project leadership is less about enforcing control and more about orchestrating coordinated adaptation within complex sociotechnical systems. Figure 1.1 shows the evolution of project leadership paradigms from control-oriented, hierarchical models toward adaptive, distributed and sustainability-driven approaches.

1.4. Team dynamics in complex adaptive environments

Teams are the primary execution units of projects, yet contemporary research increasingly conceptualises them not as static task groups but as dynamic and adaptive systems (Edmondson, 2018). In complex project environments characterised by uncertainty, interdependence and rapid change, team performance emerges from patterns of interaction rather than formal structure alone.

From a complexity perspective, project teams function as micro-systems embedded within broader organisational and institutional ecosystems. Relational dynamics, communication norms and shared mental models shape their behaviour. Consequently, effectiveness depends not only on technical expertise but also on the quality of internal coordination and collective sense-making.

1.4.1 Psychological safety and performance

Psychological safety – the shared belief that it is safe to speak up without fear of punishment or humiliation – has become a central construct in understanding team performance

(Edmondson, 2018). In complex projects, early disclosure of risks, ambiguities and errors is essential for adaptive response. When uncertainty signals are suppressed, minor issues can escalate into systemic failures.

High-performing project teams typically demonstrate

- *open dialogue*: where information flows transparently across hierarchical boundaries
- *constructive dissent*: allowing assumptions to be challenged without destabilising cohesion
- *shared accountability*: reinforcing collective responsibility for outcomes
- *trust density*: strengthening relational resilience under pressure.

Psychological safety does not eliminate conflict; rather, it reframes conflict as a resource for learning and innovation. In volatile environments, teams that encourage candid discussion are better positioned to identify emerging risks and generate creative solutions.

1.4.2 Digital collaboration and distributed intelligence

Digital transformation has fundamentally reshaped how project teams coordinate. Virtual platforms, real-time dashboards and collaborative software enable geographically dispersed teams to share information instantaneously (Marnewick and Marnewick, 2019). These technologies enhance transparency and accelerate decision cycles.

However, digital infrastructure alone does not guarantee effective collaboration. Technology amplifies existing relational patterns rather than replacing them. Teams with strong trust and communication norms leverage digital tools productively; teams with fragmented relationships may experience information overload or misalignment.

Collective intelligence – the capacity of a group to solve problems collaboratively – emerges from interaction quality rather than technological sophistication. Distributed intelligence depends on

- clear role clarity and boundary understanding
- effective knowledge sharing routines
- norms supporting responsiveness and accountability
- integration of diverse expertise across functions.

In complex adaptive environments, digital platforms function as enablers but relational alignment remains the foundational determinant of performance.

1.4.3 Team dynamics as adaptive capability

When viewed through systems and complexity lenses, team dynamics represent an adaptive capability within project environments. Teams regulate information flow, interpret environmental signals and coordinate response across functional boundaries.

Strong team capability enhances resilience, accelerates innovation and mitigates systemic risk. Conversely, fragmented team dynamics amplify uncertainty and erode performance stability.

Thus, effective project management requires deliberate cultivation of team relational infrastructure alongside technical coordination mechanisms. Figure 1.2 illustrates the project team

Figure 1.2 Project team ecosystem network

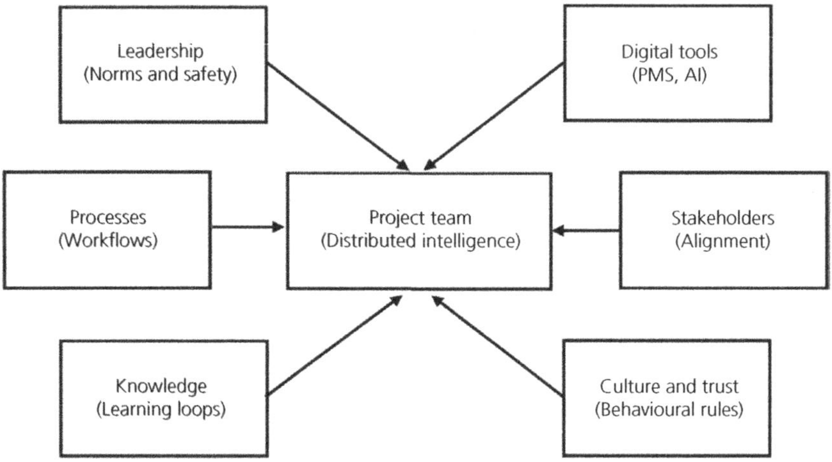

ecosystem as an interconnected network in which trust, communication, expertise and digital collaboration function as mutually reinforcing components.

1.5. Knowledge and organisational learning as resilience drivers

In complex project environments, knowledge is not merely informational content; it is a strategic capability that underpins resilience and long-term value creation. Organisational learning theory emphasises that performance improvement depends on systematic feedback, reflection and adaptation (Argyris and Schön, 1997). Similarly, knowledge-based perspectives position knowledge as a central source of competitive advantage, embedded within organisational routines and social processes (Grant, 1996).

Projects are inherently knowledge-intensive undertakings. They generate experiential insights regarding stakeholder engagement, risk management, technological integration and governance practices. However, without deliberate institutionalisation, such knowledge remains tacit and episodic. Lessons learned may be documented yet fail to influence future decision-making if not integrated into structural processes.

From a systems perspective, learning functions as a feedback mechanism that enables projects and organisations to adjust behaviour in response to environmental signals. From a complexity perspective, learning enhances adaptive capacity by improving pattern recognition and response agility.

Three levels of learning are particularly relevant in project contexts:

- *single-loop learning*, which focuses on correcting errors within existing frameworks and improving operational efficiency

- *double-loop learning*, which questions underlying assumptions, strategies or decision criteria, enabling deeper reflection and strategic adjustment
- *triple-loop learning*, which transforms governance structures, cultural norms and institutional logics, reshaping how projects are conceived and managed.

While single-loop learning improves execution, double- and triple-loop learning build long-term resilience. They allow organisations to reconfigure processes, anticipate emerging risks and align project practices with evolving sustainability expectations.

In contemporary environments marked by regulatory scrutiny and environmental accountability, learning capability becomes central to legitimacy. Organisations that embed structured reflection and knowledge-sharing mechanisms are better positioned to integrate sustainability considerations into strategic planning and risk governance.

Thus, knowledge and organisational learning operate as resilience drivers, strengthening adaptive capacity and ensuring that project experience contributes to cumulative institutional capability rather than isolated performance outcomes. Figure 1.3 presents the triple-loop learning model, demonstrating how operational improvement, strategic reassessment and governance transformation interact to enhance project resilience.

1.6. Communication and culture as structural governance mechanisms

Communication is frequently treated in project management as a soft interpersonal skill. Yet empirical evidence consistently identifies communication failure as one of the most significant contributors to project underperformance (PMI, 2025). Missed deadlines, scope creep,

Figure 1.3 Triple-loop learning in projects

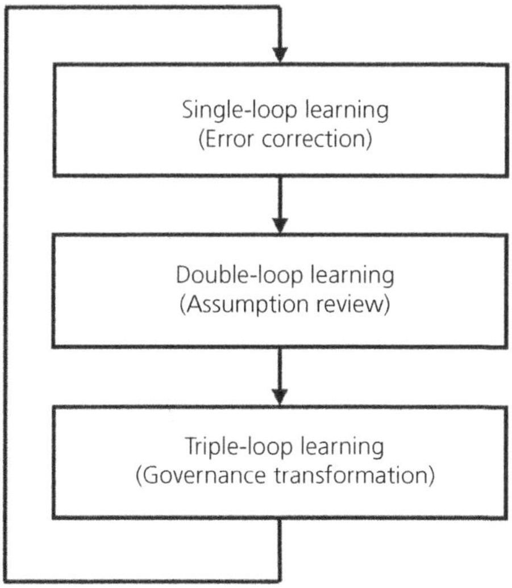

stakeholder dissatisfaction and risk escalation often stem not from technical incapacity but from breakdowns in information flow and misaligned expectations.

From a systems perspective, communication functions as the circulatory system of a project. It regulates coordination across interdependent units, enables feedback loops and sustains alignment between operational execution and strategic objectives. From a complexity perspective, communication determines how uncertainty signals are interpreted, amplified or suppressed within adaptive systems.

Importantly, communication does not operate in isolation from culture. Cultural norms shape how messages are framed, interpreted and responded to. Thus, communication and culture together form structural governance mechanisms that influence transparency, trust and institutional legitimacy.

1.6.1 Cultural intelligence in global projects

As projects increasingly span geographical, organisational and disciplinary boundaries, cultural diversity becomes a defining feature of project environments. Cultural intelligence – the capability to function effectively across diverse cultural contexts – influences negotiation styles, perceptions of authority and approaches to conflict resolution (Earley and Ang, 2003).

In multinational or cross-sector projects, differences in communication norms can significantly affect coordination. Variations in power distance may influence willingness to question authority. High-context or low-context communication styles may affect the clarity of expectations. Divergent attitudes toward uncertainty may shape risk reporting behaviour.

Without cultural sensitivity, misinterpretation can escalate into mistrust or conflict. Conversely, culturally intelligent leadership and team practices enhance relational cohesion and facilitate constructive dialogue across boundaries.

Cultural intelligence, therefore, operates as a strategic capability within complex project systems. It strengthens adaptability by improving interpretive alignment and reducing friction in diverse stakeholder environments.

1.6.2 Communication architecture and transparency

Beyond interpersonal interaction, communication also functions as organisational architecture. Formal reporting structures, escalation pathways and information-sharing protocols shape how knowledge flows through project systems. Transparent communication mechanisms strengthen the legitimacy of governance and reinforce stakeholder trust (Too and Weaver, 2014).

In contemporary environments, transparency is no longer optional. Digital platforms, real-time data analytics and public reporting standards increase scrutiny of project decisions. Sustainability reporting and ESG disclosure frameworks further institutionalise accountability for communication, extending governance expectations beyond financial metrics to environmental and social performance.

Structured communication architecture ensures that

- risks are escalated promptly
- stakeholders receive consistent and accurate information
- decision rationales are documented transparently
- accountability pathways are clearly defined.

When the communication infrastructure is weak or fragmented, governance breakdown becomes likely. Conversely, robust communication systems reinforce ethical integrity, stakeholder alignment and long-term legitimacy.

1.6.3 Communication as institutional infrastructure

Taken together, communication and culture operate as institutional infrastructure rather than merely interpersonal competencies. They regulate information flow, mediate stakeholder relationships and sustain governance coherence.

From a capability perspective, effective communication architecture enhances organisational resilience. It enables rapid feedback integration, supports adaptive decision-making and strengthens trust under uncertainty.

In complex sociotechnical environments, communication is not an auxiliary function; it is a structural determinant of coordination, legitimacy and sustainable project performance. Figure 1.4 depicts the culture alignment model, highlighting how communication norms, governance structures and stakeholder expectations intersect to shape institutional coherence.

1.7. Integrated capability framework

The preceding sections have examined leadership, team dynamics, organisational learning, and communication and culture as distinct but interdependent domains. These domains should not be understood as isolated variables. Rather, they operate as mutually reinforcing capabilities within a complex sociotechnical system.

Figure 1.4 Culture alignment model

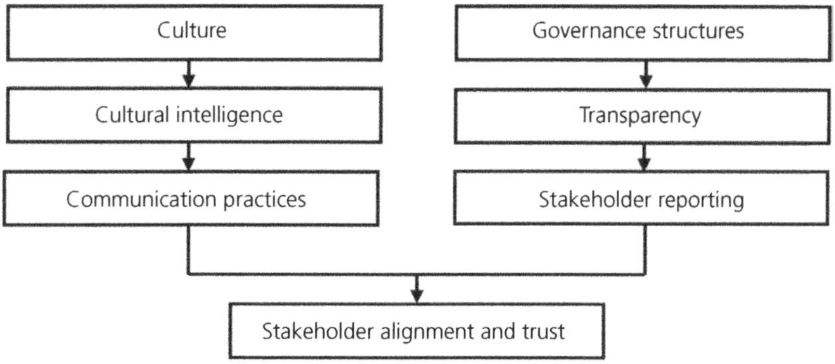

Leadership establishes direction, norms and ethical boundaries. It shapes the psychological climate within which teams operate. In turn, team dynamics determine how effectively knowledge is generated, shared and applied. Organisational learning transforms accumulated experience into institutional capability, embedding adaptation into governance structures. Communication and cultural intelligence align stakeholders, regulate transparency and sustain legitimacy across internal and external environments.

These interactions are systemic rather than linear. Leadership influences team trust and safety. Team trust enhances learning behaviour. Learning capability improves strategic decision-making. Effective communication ensures that knowledge circulates and governance mechanisms remain responsive. Feedback from stakeholders then informs leadership recalibration. The system evolves continuously.

From a systems perspective, the framework represents an interconnected network of feedback loops. From a complexity perspective, performance outcomes emerge from interaction effects rather than from single-factor optimisation. From a capability perspective, these domains collectively constitute strategic assets that are difficult to imitate and therefore central to long-term competitiveness.

When integrated effectively, the four domains generate higher-order outcomes:

- *innovation*: through psychologically safe teams and adaptive leadership
- *resilience*: through embedded learning and transparent governance
- *sustainability*: through ethical leadership and stakeholder-aligned communication
- *long-term value*: through coordinated capability deployment rather than short-term optimisation.

The integrated capability framework, therefore, reframes project success not as compliance with predefined constraints but as the dynamic alignment of leadership behaviour, relational systems, knowledge processes and governance architecture.

Project performance in complex environments is not the product of superior scheduling alone. It is the outcome of systemic capability coherence. Figure 1.5 shows the integrated

Figure 1.5 Integrated project capability framework

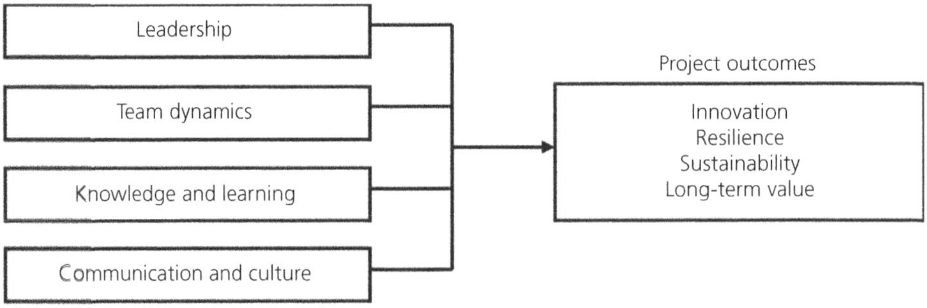

project capability framework, positioning leadership, team dynamics, learning capability and communication culture as interdependent drivers of innovation, resilience and sustainability.

1.8. Conclusion

Project management is undergoing a structural transformation. Mechanistic control models built on linear planning, rigid hierarchies and narrow performance metrics are increasingly insufficient in environments characterised by technological disruption, global interdependence, regulatory scrutiny and sustainability imperatives. The discipline must therefore evolve beyond procedural optimisation toward integrated capability systems designed for complexity.

Systems thinking clarifies that projects are not isolated execution units but open systems embedded within broader organisational and societal contexts. Interdependence, feedback loops and stakeholder interaction shape outcomes. Complexity theory further explains why predictive control alone cannot guarantee success; nonlinearity, emergence and adaptive behaviour define contemporary project environments. Resource-based and dynamic capability perspectives underscore that sustainable advantage derives not merely from tools or processes but from intangible assets such as leadership quality, team cohesion, learning capacity and communication integrity.

Leadership, teams, organisational learning, and communication and culture collectively form the structural pillars of resilience. When aligned, they enable innovation, strengthen governance legitimacy and support long-term value creation. When fragmented, they generate systemic vulnerability.

The transformation of project management is therefore not incremental but epistemological. It requires a shift from control to adaptation, from hierarchy to distributed intelligence and from short-term delivery metrics to sustainability-oriented performance. In an era defined by uncertainty and accountability, project management must become an adaptive, human-centred and strategically integrated discipline capable of navigating complexity while creating enduring value.

REFERENCES

Argyris C and Schön DA (1997) Organizational learning: A theory of action perspective. *Revista Española de Investigaciones Sociológicas* **77/78**: 345–348.

Barney J (1991) Firm resources and sustained competitive advantage. *Journal of Management* **17(1)**: 99–120.

Bertalanffy LV (1973) *General System Theory: Foundations, Development, Applications.* Braziller, New York, NY, USA.

Bredillet CN (2010) Blowing hot and cold on project management. Project Management Journal **41(3)**: 4–20.

Carson JB, Tesluk PE and Marrone JA (2007) Shared leadership in teams: An investigation of antecedent conditions and performance. *Academy of Management Journal* **50(5)**: 1217–1234.

Earley PC and Ang S (2003) *Cultural Intelligence: Individual Interactions Across Cultures.* Stanford University Press, Stanford, CA, USA.

Edmondson AC (2018) *The Fearless Organization: Creating Psychological Safety in the Workplace for Learning, Innovation, and Growth*. Wiley, Hoboken, NJ, USA.

Flyvbjerg B (2014) What you should know about megaprojects and why: an overview. *Project Management Journal* **45(2)**: 6–19.

Grant RM (1996) Toward a knowledge-based theory of the firm. *Strategic Management Journal* **17(S2)**: 109–122.

Huemann M and Turner JR (eds) (2024) *The Handbook of Project Management*. Routledge, Abingdon, UK.

Kerzner H (2025) *Project Management: A Systems Approach to Planning, Scheduling, and Controlling*, 14th edn. Wiley, Hoboken, NJ, USA.

Marnewick C and Marnewick AL (2019) The demands of Industry 4.0 on project teams. *IEEE Transactions on Engineering Management* **67(3)**: 941–949.

Maylor H, Vidgen R and Carver S (2008) Managerial complexity in project-based operations: A grounded model and its implications for practice. *Project Management Journal* **39(1_suppl)**: S15–S26.

Müller R and Lecoeuvre L (2014) Operationalizing governance categories of projects. *International Journal of Project Management* **32(8)**: 1346–1357.

PMI (Project Management Institute) (2025) *A Guide to the Project Management Body of Knowledge (PMBOK® Guide)*, 8th edn. PMI, Newtown Square, PA, USA.

Senge PM (1990) *The Fifth Discipline: The Art and Practice of the Learning Organization*. Doubleday, New York, NY, USA.

Silvius AJG and Schipper R (2014) Sustainability in project management: A literature review and impact analysis. *Social Business* **4(1)**: 63–96.

Stacey RD (2011) *Strategic Management and Organisational Dynamics: The Challenge of Complexity*. 6th edn. Pearson, Harlow, UK.

Teece DJ (2007) Explicating dynamic capabilities: The nature and microfoundations of (sustainable) enterprise performance. *Strategic Management Journal* **28(13)**: 1319–1350.

Too EG and Weaver P (2014) The management of project management: A conceptual framework for governance. *International Journal of Project Management* **32(8)**: 1382–1394.

Uhl-Bien M and Arena M (2018) Leadership for organizational adaptability: A theoretical synthesis and integrative framework. *The Leadership Quarterly* **29(1)**: 89–104.

Part I

Leadership

Mohammed Reyasudin Basir Khan
ISBN 978-1-83742-803-8
https://doi.org/10.1108/978-1-83742-802-120261003
Emerald Publishing Limited: All rights reserved

Chapter 2
The influence of leadership styles on project success in building construction: the mediating role of effective communication

Sim Tek Chiang and Mohammed Reyasudin Basir Khan

Tun Razak Graduate School, Universiti Tun Abdul Razak (UNIRAZAK), Kuala Lumpur, Malaysia

2.1. Introduction

Leadership has long been recognised as a central determinant of organisational effectiveness. Yet, its importance becomes particularly pronounced within the construction industry, where projects are characterised by complexity, uncertainty and interdependence among diverse stakeholders. Construction projects typically involve multidisciplinary teams comprising project managers, engineers, architects, contractors, subcontractors, suppliers and clients, all operating within constrained timelines and strict budgetary parameters. The dynamic, high-pressure nature of such environments places considerable demands on leadership capabilities. Effective leadership is therefore not limited to directing tasks; it encompasses motivating teams, facilitating coordination, managing conflict, shaping organisational culture and aligning stakeholder expectations toward common project objectives (Bresnen and Marshall, 2000; Jugdev and Müller, 2005).

Frequent design changes, site constraints, regulatory compliance requirements, safety risks and resource uncertainties further complicate the construction sector. These factors heighten the need for leaders who can interpret ambiguity, respond adaptively and sustain team morale under pressure. Leadership behaviours directly influence communication patterns, decision-making processes and the quality of collaboration within project teams. Where leadership is weak or inconsistent, projects are more likely to experience coordination breakdowns, interpersonal conflict, low engagement and performance inefficiencies. Conversely, effective leadership fosters trust, shared accountability and a collective commitment to project success.

Within the broader leadership literature, three dominant paradigms have received significant attention in project management research: transformational, transactional and servant leadership. Each represents a distinct theoretical orientation with differing implications for team motivation and performance.

Transformational leadership, as conceptualised by Bass (1985), is grounded in leaders' ability to inspire followers to transcend individual self-interest and commit to a shared organisational vision. This style emphasises idealised influence, inspirational motivation, intellectual stimulation and individualised consideration. In construction contexts, transformational leaders encourage innovation, collaborative problem-solving and proactive engagement – qualities that are essential when teams must adapt to unforeseen technical and environmental challenges (Guha *et al.*, 2025; Rehan *et al.*, 2024). By articulating a compelling vision and reinforcing shared purpose, transformational leaders enhance intrinsic motivation and cultivate ownership among team members. Such characteristics are particularly valuable in construction projects, where coordination across multiple stakeholders requires not only compliance but also commitment and initiative.

Transactional leadership, in contrast, is grounded in exchange theory and emphasises structured supervision, performance monitoring and contingent reward systems (Ridic *et al.*, 2025). Transactional leaders define clear roles, establish performance standards and reinforce compliance through incentives and corrective actions. In construction settings, where safety regulations, contractual obligations and quality control procedures must be strictly adhered to, transactional mechanisms can support operational discipline and short-term efficiency. However, empirical evidence suggests that while transactional leadership may secure compliance, it may not sufficiently promote creativity, innovation or deep engagement – attributes necessary for navigating complex, evolving project environments (Dvir *et al.*, 2006). Overreliance on transactional approaches may therefore limit adaptability and reduce the sustainability of long-term performance.

Servant leadership introduces a relational and ethical perspective by prioritising the development, empowerment and wellbeing of team members (Greenleaf, 2013). Servant leaders emphasise listening, empathy, stewardship and commitment to follower growth. In construction environments characterised by demanding schedules, physical strain and high stress, servant leadership can strengthen psychological safety, team cohesion and job satisfaction (Liden *et al.*, 2008). By fostering trust and mutual respect, servant leadership may reduce conflict and enhance cooperative problem-solving. Nevertheless, its effectiveness may depend on the context; in highly time-sensitive or crisis-driven scenarios, a solely supportive approach may need to be complemented by directive leadership behaviours.

Although leadership has been extensively theorised, a persistent gap remains in understanding how these leadership styles operate within the specific realities of construction project management. Many leadership studies are conducted in general organisational contexts, with limited attention to industry-specific conditions such as contractual fragmentation, temporary project structures and multi-tier subcontracting networks. Construction projects are temporary organisations, and leadership dynamics within them may differ from those in permanent organisational settings. Consequently, empirical investigation tailored to construction environments is necessary to generate contextually relevant insights.

Moreover, while leadership is acknowledged as a key driver of performance, the mechanisms by which leadership behaviours translate into project success remain unclear. One critical mechanism is effective communication. Construction projects rely heavily on information exchange across hierarchical and functional boundaries. Clear communication reduces ambiguity, prevents errors, aligns expectations and facilitates conflict resolution. Leadership behaviours shape communication climate by influencing openness, feedback quality and information transparency (Rehan *et al.*, 2024). Leaders who encourage dialogue and shared understanding create environments where coordination improves and misunderstandings decrease. Thus, communication may function as a mediating variable that explains how leadership influences measurable project outcomes.

Project success itself is multidimensional. Traditional metrics emphasise cost, time and quality; however, contemporary perspectives also consider stakeholder satisfaction, team morale and long-term organisational capability (Jugdev and Müller, 2005). In the construction industry, where project failure can result in significant financial and reputational consequences, understanding the behavioural drivers of success becomes especially important.

The study presented in this chapter examines the influence of transformational, transactional and servant leadership styles on project success in building construction, while investigating effective communication as a mediating mechanism. Drawing on survey data from 237 construction professionals, including project managers, engineers, architects and contractors, the study provides empirical evidence regarding which leadership approaches most strongly predict project outcomes.

By integrating established leadership theory with industry-specific analysis, this research contributes to both scholarly discourse and professional practice. Academically, it extends leadership research into the construction domain by empirically testing relational and motivational models within a high-complexity project environment. In practice, it offers evidence-based guidance for construction managers seeking to enhance team engagement, coordination and project performance.

Ultimately, as construction projects grow increasingly complex due to technological advancement, regulatory scrutiny and stakeholder expectations, leadership effectiveness will remain a decisive factor in achieving sustainable success. Understanding how different leadership styles shape communication and performance provides a foundation for developing more adaptive, strategically aligned leadership practices in the construction industry.

2.1.1 Conceptual framework

The conceptual framework of this study, illustrated in Figure 2.1, proposes that leadership styles significantly influence project success in building construction, both directly and indirectly, through effective communication. The framework integrates transformational, transactional and servant leadership theories to explain variations in project performance, while positioning effective communication as a mediating mechanism that translates leadership behaviours into measurable outcomes.

Transformational leadership is theorised to enhance project success by fostering trust, articulating a compelling vision, and stimulating innovation within project teams. By encouraging

Figure 2.1 Conceptual framework

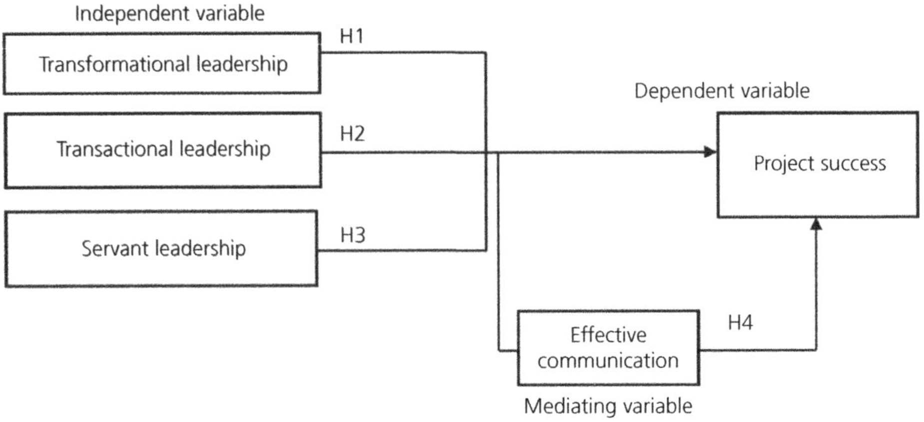

intellectual engagement and shared ownership of project objectives, transformational leaders strengthen cohesion and commitment, thereby improving productivity and overall project performance (Guha *et al.*, 2025). In construction environments characterised by uncertainty and interdependence, such motivational and relational behaviours are expected to produce superior outcomes.

Transactional leadership, grounded in exchange-based performance management, is expected to influence project performance through structured accountability, task clarity and compliance mechanisms (Dvir *et al.*, 2006). While this leadership style may enhance short-term operational efficiency and procedural adherence, particularly in safety-critical construction tasks, its capacity to foster long-term engagement, creativity and adaptive problem-solving may be limited when applied in isolation.

Servant leadership, by contrast, emphasises follower development, empowerment and relational trust (Liden *et al.*, 2008). In construction project settings where teamwork and morale significantly affect performance, servant leadership is expected to foster collaboration and psychological safety, thereby enhancing project outcomes.

Effective communication is positioned within the framework as a critical mediating variable. Leadership behaviours shape communication climate, influencing openness, clarity, feedback quality and information transparency. In turn, high-quality communication strengthens coordination, reduces conflict and aligns stakeholders toward shared objectives (Rehan *et al.*, 2024). Thus, leadership is expected to influence project success not only directly, but also indirectly through its impact on communication processes.

This framework recognises that both structural leadership behaviours and relational dynamics within teams shape project outcomes in construction. Figure 2.1 presents the conceptual model linking leadership styles, effective communication and project success.

2.1.2 Hypotheses

Based on the conceptual framework and supporting literature, the following hypotheses are proposed.

- **H1**: transformational leadership has a positive and significant influence on project success in construction project management.
- **H2**: transactional leadership has a non-significant or negative influence on project success in construction project management.
- **H3**: servant leadership has a positive and significant influence on project success in construction project management.
- **H4**: effective communication mediates the relationship between leadership styles (transformational, transactional and servant) and project success.

Collectively, these hypotheses aim to empirically test whether leadership effectiveness in construction depends not only on behavioural orientation but also on the quality of communication processes through which leadership influence is enacted.

2.2. Methodology

2.2.1 Research design

The research design for this study adopts a quantitative method approach to examine the relationship between leadership styles and project success in building construction project management. Quantitative data were collected through structured surveys to measure leadership styles, effective communication and project success.

2.2.2 Population and sampling procedures

The sampling procedure for this study used stratified random sampling to ensure a representative selection of construction project stakeholders, including project managers, team members, architects, engineers and contractors. Stratification was based on project roles so that each subgroup is proportionately represented, reducing bias and improving the reliability and validity of findings by enabling comparisons across different perspectives on leadership and work performance.

The sampling frame includes construction projects from diverse locations and project types, such as residential, commercial and infrastructure developments, to enhance the generalisability of results and capture variations in leadership practices across different contexts. The targeted sample size of 200 respondents is considered sufficient to achieve statistical significance and support robust analysis (Jimoh, 2025).

2.2.3 Data collection method

Quantitative data were gathered through a structured electronic questionnaire distributed to key stakeholders, including project managers, team members, architects and contractors. Data collection was conducted using both online questionnaires and selected in-person engagement to improve response rates. To ensure accurate and informed responses, participants must have had at least 2 years of experience in their respective construction roles, and representativeness was monitored by reviewing demographic characteristics throughout data collection and making adjustments to reflect the broader construction professional population, as necessary.

2.2.4 Operationalisation and measurement

The operationalisation and measurement of variables in this study were structured to ensure that leadership styles, team dynamics and project performance were accurately captured using established, validated instruments. Leadership styles were measured using the multifactor leadership questionnaire (MLQ) developed by Bass and Avolio (Bass, 1985) and validated by Wang *et al.* (2023), which assesses transformational and transactional leadership across key dimensions such as idealised influence, inspirational motivation, intellectual stimulation, individualised consideration, contingent reward and management by exception. Servant leadership was measured using the servant leadership scale (SLS) developed by Liden *et al.* (2008), which assesses core elements such as empathy, listening, stewardship and commitment to others' growth. Communication effectiveness was measured using the communication satisfaction questionnaire (CSQ), which evaluates clarity, openness, feedback and satisfaction with communication processes. Project success was assessed through self-reported measures of perceived productivity, deliverable quality and timeliness in meeting deadlines. All constructs were measured using a consistent five-point Likert scale format to enable statistical comparison and support the quantification of perceptions across different stakeholder groups.

2.3. Results and analysis

2.3.1 Response rate

The survey involved 237 participants representing a wide range of roles within the construction sector, providing a comprehensive view of perceptions regarding leadership dynamics in project management. The sample was deliberately structured to include key stakeholders, including project managers, engineers, architects and contractors. This broad representation enabled the study to capture diverse viewpoints and better reflect the complex, multi-stakeholder nature of the construction industry.

2.3.2 Descriptive analysis

The sample was demographically diverse, as shown by Table 2.1, where 60% (142 respondents) identified as male, reflecting the current workforce composition of the construction industry, while ensuring meaningful female participation (40%; 95 respondents) to support a more inclusive understanding of leadership and its effects across genders. Participants represented a wide age range, with the largest segment falling within the 26 to 35 age group (45.6%), followed by those aged 36 to 45 (28.6%) and smaller proportions aged 46 and above, allowing the study to capture different generational perspectives on leadership practices.

The data reveal that a significant majority of the 237 valid respondents are highly educated, with Bachelor's degree holders constituting the largest segment at 57.2% (135 respondents). This is followed by those with a Diploma (17.1%), high school qualification (14.3%) and a Master's degree (11.4%). Respondents also varied significantly in experience, averaging around 10 years in the construction sector; the sample included a balanced mix of early-career professionals and seasoned veterans with over 20 years of industry experience, which strengthened the quality of insights on leadership effectiveness in real project environments.

Table 2.1 Demographic profile

Characteristic	Category	Frequency	Percentage: %
Gender	Male	142	60.0
	Female	95	40.0
Age	16–25	20	8.6
	26–35	108	45.6
	36–45	68	28.6
	46–55	20	8.6
	Over 55	21	8.6
Qualification	High school	34	14.3
	Diploma	41	17.1
	Bachelor's degree	135	57.2
	Master's degree	27	11.4
	PhD	0	0.0
Marital status	Single	102	42.9
	Married	122	51.4
	Other/prefer not to say	13	5.7
Annual income	< $30 000	27	11.4
	$30 000–$60 000	34	14.3
	$60 001–$100 000	68	28.6
	$100 001–$150 000	27	11.4
	> $150 000	40	17.1
	Prefer not to say	41	17.1
Job title	Entry level	47	20.0
	Individual contributor	54	22.9
	Manager	75	31.4
	Director	6	2.7
	Vice-president	7	2.9
	CEO	7	2.9
	Chairperson	7	2.9
	Other professionals	34	14.3
Experience	< 5 years	41	17.1
	6–10 years	74	31.4
	11–15 years	68	28.6
	16–20 years	14	5.7
	> 20 years	40	17.1

In addition, participants were drawn from multiple geographic areas and project types, including residential, commercial and infrastructure developments, enriching the findings by incorporating diverse cultural and organisational contexts. Overall, these sample characteristics provide a strong basis for examining leadership dynamics in construction project management and support findings relevant to different stakeholder groups and work settings.

2.3.3 Reliability analysis

Reliability analysis was conducted to assess the internal consistency of the measurement instruments used in this study, using Cronbach's alpha (α) as the primary indicator. A Cronbach's alpha value of 0.70 or higher is generally accepted as evidence of satisfactory reliability in social science research (Nunnally and Bernstein, 1994). Table 2.2 presents a summary of the reliability results.

The results showed strong reliability across all study scales, indicating that the instruments consistently measured their intended constructs. Transformational leadership demonstrated excellent internal consistency, with a Cronbach's alpha of 0.89, suggesting that the items effectively captured a unified construct aligned with its motivational and developmental elements (Bass, 1985). Transactional leadership exhibited acceptable reliability, as measured by Cronbach's alpha of 0.75, indicating reliable measurement of structured leadership behaviours such as performance monitoring and compliance (Ridic *et al.*, 2025). Servant leadership also demonstrated strong reliability, with a Cronbach's alpha of 0.82, indicating consistent measurement of supportive and service-oriented leadership characteristics (Greenleaf, 2013). In addition, effective communication achieved a Cronbach's alpha of 0.85, while project success recorded a Cronbach's alpha of 0.78. Overall, these findings confirm that the instruments used in this study were reliable and provide a sound basis for further statistical analysis of leadership influences on project success in construction project management.

2.3.4 Normality analysis

Normality testing was conducted to assess whether the data satisfied the assumptions required for parametric statistical analysis. Since Pearson correlation and multiple regression analysis assume that the residuals are approximately normally distributed, preliminary diagnostic procedures were performed.

Table 2.2 Summary of reliability analysis (Cronbach's alpha)

Measurement instrument (scale)	Cronbach's alpha, α	Interpretation
Transformational leadership	0.89	Excellent
Transactional leadership	0.75	Acceptable
Servant leadership	0.82	Strong
Effective communication	0.85	Strong
Project success	0.78	Acceptable

Descriptive statistics indicated that skewness values for all principal variables ranged from –0.84 to 0.76, while kurtosis values ranged from –0.91 to 1.12. These values fall within the commonly accepted thresholds of ±1 for skewness and ±2 for kurtosis, suggesting no substantial deviation from normality.

Normality was further examined using graphical methods. Histogram outputs showed distributions approximating a bell-shaped curve. Additionally, normal Q-Q plots of the standardised residuals indicated that most data points were closely aligned with the diagonal reference line, with only minor deviations observed at the extreme tails. No systematic patterns of distortion were detected.

The Shapiro–Wilk test was also conducted. Although several variables produced statistically significant results ($p < 0.05$), this outcome is expected with larger sample sizes ($n = 237$), where the test becomes highly sensitive to small and practically negligible departures from normality. Given the acceptable skewness and kurtosis values, supportive graphical inspection and adequate sample size, the assumption of normality was considered reasonably satisfied.

Therefore, parametric analyses including Pearson correlation and multiple regression were deemed appropriate for subsequent hypothesis testing.

2.3.5 Correlation analysis

The Pearson correlation analysis presented in Table 2.3 provides critical empirical evidence regarding the associations between leadership behaviours, communication and project outcomes. The results confirm that leadership styles are significant determinants of success in the complex and high-stakes environment of construction project management.

Transformational leadership demonstrates the strongest positive correlation with project success ($r = 0.684$, $p < 0.01$), confirming it as the most influential factor in enhancing performance. This finding aligns with the theoretical framework, suggesting that leaders who inspire a shared vision and empower their teams foster the innovation and collaboration necessary for complex construction tasks. By motivating team members to move beyond self-interest and take ownership of project goals, transformational leaders directly improve overall productivity.

Servant leadership also shows a strong positive correlation with project success ($r = 0.621$, $p < 0.01$), highlighting the importance of team wellbeing and supportive practices. In an industry often marked by high stress and long hours, leaders who prioritise the growth and development of their team members enhance morale and cohesion. This supportive environment is statistically linked to improved job satisfaction and better project execution.

Effective communication is identified as a significant positive correlate ($r = 0.585$, $p < 0.01$), reinforcing its position in the conceptual framework as a key mediator. The data support the view that open, consistent communication is essential for aligning diverse stakeholders, including architects, engineers and contractors, and for ensuring that leadership intent is translated into successful deliverables.

Table 2.3 Pearson correlation matrix for leadership style and project success

		Project success	Transformational leadership	Servant leadership	Effective communication	Transactional leadership
Project success	Pearson correlation	1	0.684[a]	0.621[a]	0.585[a]	–0.142
	Sig. (1-tailed)		0.000	0.000	0.000	0.185
	n	237	237	237	237	237
Transformational leadership	Pearson correlation	0.684[a]	1	0.542[a]	0.512[a]	0.095
	Sig. (1-tailed)	0.000		0.000	0.000	0.320
	n	237	237	237	237	237
Servant leadership	Pearson correlation	0.621[a]	0.542[a]	1	0.495[a]	0.074
	Sig. (1-tailed)	0.000	0.000		0.000	0.381
	n	237	237	237	237	237
Effective communication	Pearson correlation	0.585[a]	0.512[a]	0.495[a]	1	0.110
	Sig. (1-tailed)	0.000	0.000	0.000		0.245
	n	237	237	237	237	237

Note: [a]Correlation is significant at the 0.01 level (1-tailed); $n = 237$ reflects the analysed survey sample.

Conversely, transactional leadership exhibits a weak, non-significant negative relationship with project success ($r = -0.142$, $p > 0.05$). This aligns with the study's findings that while transactional structures may ensure basic compliance and short-term target adherence, they often fail to inspire the deep commitment or creativity required for long-term engagement in high-pressure settings. These results suggest that a reliance on purely transactional methods, such as reward-and-penalty systems, may be less effective than more relational leadership approaches in driving project success.

2.3.6 Multiple regression analysis

The summary in Table 2.4 provides the overall fit for the leadership framework. The value of 0.721 indicates that leadership styles and communication explain 72.1% of the variance in project success, a strong yet standard result for organisational behaviour research.

These coefficients represent the individual impact of each variable within the adjusted model. The explanatory power of the regression model ($R^2 = 0.721$) suggests that a significant portion of project success in construction management is determined by leadership and communication. While these internal factors are dominant, the remaining 27.9% of the variance may be attributed to external contextual challenges typical of the industry, such as site constraints, resource shortages and weather disruptions. Within this framework, transformational leadership remains the primary driver of success ($\beta = 0.65$), supporting the theory that inspiring a shared vision and empowering teams are critical for navigating high-pressure environments.

Similarly, the impact of servant leadership, as evidenced by a positive correlation of 0.621 and β of 0.55, underscores the importance of team wellbeing and morale in improving project outcomes. In contrast, the analysis confirmed transactional limitations: a non-significant correlation and a negative beta ($\beta = -0.12$) indicate that while transactional structures ensure basic compliance, they do not reliably predict the innovation or long-term commitment required for complex projects. Ultimately, effective communication serves as a critical mediator ($\beta = 0.48$), with a strong correlation of 0.585 reinforcing its role as the essential link that translates leadership behaviours into measurable project results by facilitating clarity and conflict resolution among diverse stakeholders.

2.3.7 ANOVA

Analysis of variance (ANOVA) was conducted to assess the overall statistical significance and model fit of the regression model examining the effects of leadership styles on project outcomes. As shown in Table 2.5, the resulting F-value of 149.78 represents the ratio of variance explained by the model to the unexplained residual variance; this high value, reflecting

Table 2.4 Multiple regression model summary

R	R^2	Adjusted R^2	Std. error of the estimate
0.849	0.721	0.705	1.142

Note: predictors: (constant), transformational leadership, servant leadership, effective communication, transactional leadership; dependent variable: project success.

Table 2.5 ANOVA summary

Source of variation	*df*	Sum of squares	Mean square	*F*	Significance *F*
Regression	4	2570.18	642.55	149.78	<0.001
Residual	232	994.56	4.29		
Total	236	3564.74			

the model's four predictors and a sample size of 237, indicates that the integrated leadership and communication variables fit the data significantly better than a model with no predictors. Furthermore, the significance F-value (< 0.001) is well below the standard 0.05 threshold, providing strong evidence against the null hypothesis and suggesting that the predictor variables play a significant, non-random role in explaining project success variability.

With an R^2 of 0.721 (Table 2.4), the model explains approximately 72.1% of the variability in project outcomes. This confirms that while transformational and servant leadership styles, complemented by effective communication, are the primary drivers of success, a portion of the variance (27.9%) is attributable to other contextual factors common in the construction environment, such as site constraints, weather disruptions or resource shortages.

2.3.8 Hypotheses summary

Table 2.6 summarises the results of the hypotheses. The statistical analyses conducted in this study were designed to evaluate the proposed hypotheses regarding the influence of leadership styles on project outcomes in building construction project management. Through regression and mediation analyses, the results demonstrate how different leadership approaches shape team dynamics, engagement and overall project success, offering valuable insights to support more effective construction management practices.

The analysis provides strong evidence that leadership style significantly influences project outcomes. The findings support hypothesis 1, showing that transformational leadership has a significant positive effect on project success ($\beta = 0.65$, $p < 0.01$). This suggests that leaders who inspire and motivate their teams enhance collaboration, innovation and commitment, thereby driving project productivity and successful delivery. Hypothesis 2 was also supported,

Table 2.6 Summary of hypotheses

Variable	Beta, β	*p*-value	Significance	Result
Transformational leadership	0.65	<0.01	Significant	Supported (H1)
Transactional leadership	–0.12	>0.05	Not significant	Supported (H2)
Servant leadership	0.55	<0.01	Significant	Supported (H3)
Effective communication	0.48	<0.01	Significant	Supported (H4)

as transactional leadership demonstrated a weak, non-significant negative relationship with project success ($\beta = -0.12, p > 0.05$). While transactional approaches may ensure short-term compliance through structure, they appear less effective in fostering the intrinsic motivation and long-term involvement required in high-pressure construction environments. Hypothesis 3 was also supported, with servant leadership showing a significant positive impact on project success ($\beta = 0.55, p < 0.01$). This confirms that leaders who prioritise the wellbeing and development of team members strengthen morale and cohesion, which directly improves project execution.

Furthermore, the results validate hypothesis 4 by confirming that effective communication plays a significant mediating role in the relationship between leadership styles and project success ($\beta = 0.48, p < 0.01$). This indicates that leadership practices are most effective when communication channels are open and clear, serving as the essential mechanism that translates leadership into positive work outcomes. Overall, the findings emphasise that transformational and servant leadership styles are superior for improving engagement and success in construction settings. These results suggest that construction managers should prioritise developing these leadership competencies and invest in robust communication practices to navigate the demands of increasingly complex project environments.

2.4. Practical insights

The findings of this study offer several practical implications for construction project management practice. Most notably, the results demonstrate that leadership is not merely a supervisory function but a strategic driver of performance in complex construction environments. Leadership behaviours directly influence team motivation, coordination quality and ultimately project outcomes.

Transformational leadership emerged as the most influential predictor of project success. For construction organisations, this suggests that managers should prioritise developing inspirational communication, vision articulation and intellectual stimulation skills. Leaders who encourage ownership, innovation and shared accountability are better positioned to navigate uncertainty, resolve conflicts and sustain team engagement under pressure. Construction firms should therefore incorporate transformational leadership competencies into promotion criteria, succession planning and leadership development programmes.

Servant leadership also demonstrated a significant positive relationship with project success, particularly in strengthening morale, trust and collaboration. In an industry characterised by demanding schedules and high stress levels, supportive leadership behaviours contribute to stronger team cohesion and reduced interpersonal conflict. Managers should actively cultivate listening skills, empathy and developmental support to enhance psychological safety and cooperative problem-solving within project teams.

In contrast, transactional leadership showed limited predictive value. While structured performance monitoring and compliance mechanisms remain essential for safety and procedural control, over-reliance on reward–penalty systems may inhibit creativity and intrinsic motivation. Construction leaders should therefore balance procedural discipline with relational and motivational leadership behaviours.

Effective communication was confirmed as a critical mechanism through which leadership translates into performance outcomes. Open, consistent and transparent communication practices strengthen alignment with project objectives, reduce misunderstandings and improve coordination among diverse stakeholders. Organisations should invest in communication training, structured reporting systems, and collaborative platforms to reinforce leadership effectiveness.

Overall, the evidence suggests that construction firms seeking sustainable performance improvements must move beyond control-oriented leadership models toward integrated approaches that combine inspiration, support and structured communication. Such alignment enhances productivity, engagement and resilience in increasingly complex project environments.

2.5. Conclusion and recommendations

2.5.1 Implications and strategic considerations

The findings of this study highlight the critical importance of effective leadership in improving project success within the construction industry. Strong leadership enhances team cohesion, supports clear communication and contributes directly to improved project outcomes. The results suggest that construction organisations should prioritise developing transformational leadership, which showed the strongest positive influence on project success, as these approaches are most significantly associated with higher productivity, greater engagement and healthier team dynamics. In contrast, the analysis found that transactional leadership demonstrated a weak, non-significant negative relationship with project success, indicating that while it may ensure short-term compliance, it is less effective in fostering the intrinsic motivation required for long-term success. Leadership training initiatives such as mentoring, coaching and structured workshops are recommended to help managers build the skills needed to inspire teams, encourage innovation and strengthen commitment to project goals. At the same time, the study confirms that effective communication acts as a critical mediator in the relationship between leadership and success, showing that leadership practices are most effective when supported by open, supportive communication channels.

In addition, understanding how leadership shapes team dynamics provides a basis for creating a more collaborative and supportive work culture. Construction leaders are encouraged to promote teamwork through joint goal-setting, team-building activities and cross-functional collaboration to improve problem-solving and strengthen interpersonal relationships. The study also implies that leadership development and communication frameworks should be embedded into organisational policies and standard operating procedures to ensure consistent application across all project phases. By integrating transformational and servant leadership practices with robust communication and team-building strategies, construction firms can improve project execution, enhance employee wellbeing and build greater resilience in managing the demands and challenges of modern construction environments.

2.5.2 Limitations

This study provides valuable insights into the influence of leadership on job performance in building construction project management. However, several limitations should be acknowledged, as they may affect the interpretation and generalisability of the findings. Recognising these constraints is important because it clarifies the scope of the conclusions and highlights areas for future research to strengthen and extend the current results.

First, the sample size and study coverage may limit broader applicability. Although 237 participants from various professional roles in the construction industry were included, the sample was drawn from specific geographical locations and project settings. Construction practices, leadership expectations and team structures can differ significantly across regions and project types. As a result, the findings may not fully represent the diversity of leadership dynamics across the wider construction sector.

Second, the study relied heavily on self-reported data to measure leadership behaviours, team dynamics and project performance. While self-report questionnaires are commonly used in leadership studies, they can introduce response bias, particularly social desirability bias, in which participants may provide answers they believe are expected or acceptable rather than a fully accurate reflection of their experiences. This may lead to overestimating or underestimating both leadership effectiveness and performance outcomes.

Third, the study applied a cross-sectional research design, which captures perceptions and relationships at a single point in time. While this approach is useful for identifying associations between leadership styles and work performance, it does not allow for examining changes over time or determining causal relationships. Leadership influence in construction projects may evolve across phases, including initiation, execution, monitoring and completion, as teams face varying levels of complexity and pressure.

Finally, the study did not examine cultural and contextual influences in depth. This is a notable limitation because construction teams often consist of individuals from diverse cultural backgrounds, and cultural values can shape leadership expectations, communication styles and perceptions of authority and teamwork. Leadership approaches that are effective in one cultural or organisational setting may not produce the same outcomes in another.

In summary, while the study offers an important foundation for understanding the effects of leadership in construction project management, its limitations highlight the need for broader sampling, more objective measurement strategies, longitudinal designs and deeper exploration of cultural factors. Addressing these areas in future study will enhance the robustness of the findings and contribute to more accurate and practical leadership strategies to improve project outcomes and workforce performance in the construction industry.

2.5.3 Recommendations and future research

Future research should continue to examine the influence of leadership styles on job performance in the construction industry to strengthen theoretical understanding and improve practical application. In particular, further investigations should expand beyond the current scope by exploring additional leadership approaches and assessing how leadership influences project outcomes across time. Longitudinal studies would be especially valuable in capturing the evolving nature of leadership and its long-term effects on team effectiveness and project success. Additionally, greater attention should be given to the role of contextual factors, including culture and team dynamics, since these elements may shape how leadership behaviours are perceived and how effectively they translate into improved performance. Strengthening research in these areas will help develop evidence-based best practices for leadership in construction project management.

One important recommendation for future studies is to increase sample size and diversity to enhance the generalisability of findings. Because leadership practices and team expectations may differ across geographical regions, research should incorporate participants from a wider range of locations and organisational settings. Including multiple project types, such as residential, commercial and infrastructure developments, would also allow researchers to examine whether leadership styles vary with project complexity and stakeholder involvement. A more diverse sample would yield stronger comparative insights and support the development of leadership strategies more applicable across the construction sector (Cascio and Aguinis, 2008).

In addition, future studies should incorporate objective performance measures to complement self-reported survey data. While self-reported measures are useful for capturing individual perceptions and experiences, they may be influenced by biases such as social desirability, recall limitations or misinterpretation of leadership behaviours. The inclusion of objective indicators, such as project completion rates, budget performance, safety records and quality outcomes, would provide more robust evidence of leadership effectiveness and enhance the validity of findings (Robbins and Judge, 2011). Using multi-source assessments, including supervisor evaluations and peer feedback, may further strengthen the accuracy and credibility of performance measurement.

Longitudinal research designs are also strongly recommended, as they can provide a deeper understanding of how leadership influences performance over time. Construction projects progress through multiple phases and leadership demands may change depending on project stage, urgency and emerging challenges. By tracking teams throughout a project lifecycle, researchers can examine how leadership behaviours evolve, how team dynamics develop and whether leadership effects are sustained over time. Such designs are also better suited to exploring causal relationships than to relying solely on correlational evidence.

Finally, future studies should explore the influence of cultural context on leadership effectiveness in construction environments. Construction teams often comprise individuals from diverse cultural backgrounds, and cultural values can shape expectations regarding authority, communication, teamwork and motivation. Examining dimensions such as power distance, collectivism versus individualism, and leadership preferences can help researchers develop culturally responsive leadership frameworks better suited to diverse, multinational project settings. This line of inquiry may provide construction managers with practical guidance in adapting leadership strategies to fit the cultural characteristics of their teams and organisations (Hofstede, 1980).

In summary, future studies can strengthen and extend current understanding of leadership in construction project management by expanding sample diversity, incorporating objective performance indicators, adopting longitudinal designs and examining cultural influences.

These improvements will enhance the quality of the available evidence and provide construction managers with stronger guidance to improve team dynamics, work performance and overall project success.

REFERENCES

Bass BM (1985) *Leadership and Performance Beyond Expectations*. Free Press, New York, NY, USA.

Bresnen M and Marshall N (2000) Partnering in construction: a critical review of issues, problems and dilemmas. *Construction Management and Economics* **18(2)**: 229–237.

Cascio WF and Aguinis H (2008) Research in industrial and organizational psychology from 1963 to 2007: changes, choices, and trends. *Journal of Applied Psychology* **93(5)**: 1062–1081.

Dvir D, Sadeh A and Malach-Pines A (2006) Projects and project managers: the relationship between project managers' personality, project types, and project success. *Project Management Journal* **37(5)**: 36–48.

Greenleaf RK (2013) *Servant Leadership: A Journey into the Nature of Legitimate Power and Greatness*. Paulist Press, Mahwah, NJ, USA.

Guha P, Kar S and Jha KN (2025) Exploring leadership styles and their effect on the safety and quality performances in construction projects. *Journal of Construction Engineering and Management* **151(9)**: 04025120.

Hofstede G (1980) *Culture's Consequences: International Differences in Work-related Values*. Sage, Beverly Hills, CA, USA.

Jimoh AL (2025) Empowering performance through leadership: the developmental role of human capital initiatives. *Journal of Intellectual Capital* **27**: 1–16.

Jugdev K and Müller R (2005) A retrospective look at our evolving understanding of project success. *Project Management Journal* **36**(4):19–31.

Liden RC, Wayne SJ, Zhao H and Henderson D (2008) Servant leadership: development of a multidimensional measure and multilevel assessment. *The Leadership Quarterly* **19(2)**: 161–177.

Nunnally JC and Bernstein IH (1994) *Psychometric Theory*, 3rd edn. McGraw-Hill, New York, NY, USA.

Rehan A, Thorpe D and Heravi A (2024) Project success factors for leadership practices and communication: challenges in the construction sector. *International Journal of Managing Projects in Business* **17(3)**: 562–590.

Ridic O, Palalic R, Jukic T *et al.* (2025) Transactional leadership. In *International Encyclopedia of Business Management*. Elsevier, Amsterdam, the Netherlands, pp. 105–111.

Robbins SP and Judge TA (2011) *Organizational Behavior*, 14th edn. Pearson Education, Upper Saddle River, NJ, USA.

Wang JCK, Liu WC, Kee YH *et al.* (2023) Measuring educational leadership in Singapore: re-examining the psychometric properties of the Multifactor Leadership Questionnaire. *Frontiers in Psychology* **14**: 1280038.

Advancing Leadership and Collaboration in Project Management: Culture, Knowledge and Team Dynamics

emerald PUBLISHING ice Publishing

Mohammed Reyasudin Basir Khan
ISBN 978-1-83742-803-8
https://doi.org/10.1108/978-1-83742-802-120261004
Emerald Publishing Limited: All rights reserved

Chapter 3
Strategic leadership in vendor management: driving performance and innovation in telecommunications infrastructure

Syuhada Assila Tamrin and Mohammed Reyasudin Basir Khan

Tun Razak Graduate School, Universiti Tun Abdul Razak (UNIRAZAK), Kuala Lumpur, Malaysia

3.1. Introduction

Telecommunications infrastructure represents one of the most critical enablers of modern economic systems. In an increasingly digitalised world, communication networks support not only business transactions but also public services, financial systems, education platforms, healthcare delivery and government administration. High-speed connectivity, cloud computing, 5G deployment, Internet of Things (IoT) integration and cybersecurity frameworks have transformed telecommunications into a strategic national asset (Blanutsa, 2024). As digital transformation accelerates, telecommunication infrastructure projects have become more technologically sophisticated, capital-intensive and operationally complex. These projects typically involve large-scale coordination across engineering teams, regulators, suppliers, subcontractors and service providers.

Within this ecosystem, vendors occupy a pivotal role. Vendors supply essential equipment, materials, technical services and specialised expertise required for network construction, maintenance and system upgrades. Their performance directly affects installation quality, compliance with technical standards, adherence to regulatory requirements, cost control and project timelines. Weak vendor performance may lead to delays, cost overruns, rework, safety risks and service disruptions, thereby undermining organisational credibility and financial sustainability. Conversely, strong vendor performance enhances operational efficiency, improves service reliability and strengthens competitive advantage in a rapidly evolving industry.

Despite substantial investment in telecommunications infrastructure globally, many projects continue to experience implementation challenges. Studies in infrastructure and project management highlight persistent issues such as schedule slippage, budget escalation, coordination failures and technical non-compliance (Ezeigweneme *et al.*, 2023). In telecommunications construction, these challenges are often magnified by rapid technological obsolescence and stringent performance expectations. Vendors must continuously adapt to evolving standards such as 5G architecture, advanced fibre systems and digital monitoring technologies (Larsson, 2025). In this context, vendor performance cannot be understood merely as contract fulfilment; rather, it reflects broader capabilities, relational quality and evaluation systems embedded in vendor management practices.

The theoretical grounding for understanding vendor performance can be drawn from several management perspectives. The resource-based view (RBV) argues that sustained competitive advantage derives from valuable, rare, inimitable and non-substitutable resources (Huang *et al.*, 2023). In telecommunications infrastructure projects, vendor technological expertise can be viewed as an external strategic resource that complements organisational capabilities (Basiru *et al.*, 2023). Vendors possessing advanced technical knowledge, skilled personnel and the capacity for innovation contribute directly to project efficiency and technical reliability. Sposito *et al.* (2024) emphasise that expertise and experience significantly influence project outcomes, reinforcing the importance of technological capability as a driver of performance. From this perspective, telecommunications firms must evaluate vendors not only on cost criteria but also on their technical competence and capacity for continuous innovation.

Beyond capability considerations, relational dynamics also shape vendor outcomes. Social exchange theory suggests that relationships are sustained through reciprocity, trust and mutual benefit. In project environments characterised by uncertainty and interdependence, strong vendor relationship management fosters collaboration, reduces opportunistic behaviour and improves coordination. Effective communication and trust-building mechanisms enhance transparency, align expectations and support joint problem-solving. This relational perspective becomes particularly important in telecommunications infrastructure projects where unforeseen technical challenges and regulatory adjustments require adaptive coordination. Vendors that maintain constructive partnerships with client organisations are more likely to respond proactively to emerging issues, thereby improving overall performance.

Structured evaluation mechanisms further influence vendor performance. Total quality management (TQM) and performance management theories emphasise systematic monitoring, feedback loops and continuous improvement as foundations for organisational excellence (Tajouri *et al.*, 2025). Comprehensive vendor rating systems enable organisations to objectively assess quality, cost efficiency, delivery reliability and service responsiveness. By institutionalising evaluation criteria and performance metrics, organisations create accountability and encourage vendors to maintain consistent standards. Structured rating mechanisms also support data-driven decision-making in vendor selection, retention and development strategies. In supply chain sustainability research, systematic monitoring is shown to enhance transparency and resilience, particularly in complex project environments (Sarkis, 2020).

Although prior research has explored contractor and supplier performance in industries such as construction, manufacturing and energy, empirical attention has been limited to vendor performance in telecommunications infrastructure projects. Telecommunications projects differ from traditional construction contexts in several ways: rapid technological evolution, network interdependence, regulatory sensitivity and high expectations for service continuity (Vargas and Tien, 2023). Consequently, findings from other industries cannot be assumed to apply fully to the delivery of telecommunications infrastructure.

The study presented in this chapter addresses this gap by examining three key determinants of vendor performance: technological expertise, vendor relationship management and comprehensive vendor rating systems. Technological expertise captures the vendor's ability to deploy advanced systems, integrate emerging technologies and maintain compliance with evolving technical standards. Vendor relationship management reflects communication quality, trust, collaboration and long-term partnership orientation. Comprehensive vendor rating systems represent structured evaluation mechanisms that promote accountability and continuous performance enhancement.

By integrating these dimensions, the study adopts a multidimensional view of vendor performance that recognises capability, relational and governance-based influences. The research employs a quantitative approach to examine the relationships between these variables in telecommunication infrastructure projects. Through empirical analysis, the study seeks to provide evidence-based insights that inform both theory and practice.

From a practical perspective, understanding the drivers of vendor performance is essential for reducing project risk and enhancing infrastructure reliability. Telecommunications firms operate in highly competitive environments where service disruption, implementation delays or technical failures can significantly affect market reputation and stakeholder trust. Strengthening vendor technological capability, fostering collaborative partnerships and institutionalising structured evaluation systems are therefore strategic priorities rather than administrative procedures.

In summary, vendor performance in telecommunications infrastructure projects reflects the complex interplay between technical capabilities, relational dynamics and structured governance mechanisms. By investigating these factors collectively, this study contributes to a deeper understanding of how vendor management practices can support project success and long-term industry sustainability. As digital transformation continues to reshape global economies, strengthening vendor performance will remain central to delivering resilient, efficient and future-ready telecommunications infrastructure.

Based on theoretical and empirical findings, a conceptual framework is proposed in Figure 3.1 to explain vendor performance in telecommunications construction projects. The framework positions technological expertise, vendor relationship management and comprehensive vendor rating systems as key independent factors influencing vendor performance. Technological expertise supports compliance with project requirements and enables vendors to execute technical tasks effectively. Vendor relationship management strengthens collaboration and improves coordination, reducing conflict and improving project efficiency. Vendor rating systems provide accountability and continuous improvement through structured evaluation and feedback. The framework emphasises that these factors also interact, as rating

Figure 3.1 Conceptual framework

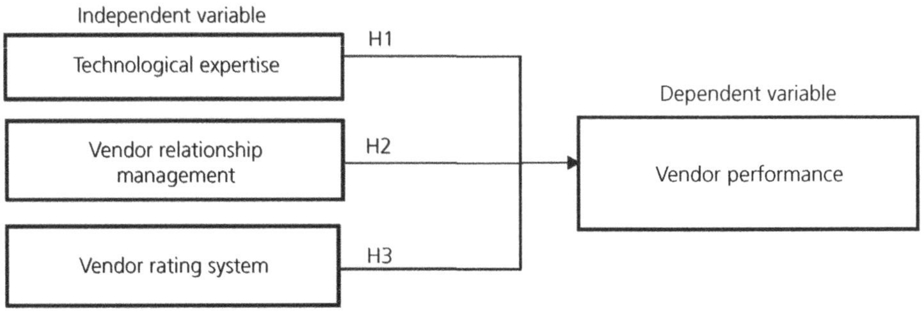

systems strengthen monitoring, relationship management improves engagement and technical expertise supports successful execution. Together, they create an integrated approach to improving vendor performance and project outcomes in telecommunications infrastructure development.

Drawing from this framework, three hypotheses are developed. The first hypothesis posits a significant positive relationship between technological expertise and vendor performance, consistent with the expectation that technically capable vendors deliver better outcomes. The second hypothesis posits that vendor relationship management is positively associated with vendor performance, emphasising the roles of trust, communication and collaboration. The third hypothesis proposes that a comprehensive vendor rating system significantly enhances vendor performance by strengthening evaluation, accountability and performance improvement. These hypotheses reflect consistent support from prior research and form the basis for further quantitative investigation in the context of telecommunications construction projects.

3.2. Methodology

3.2.1 Research design

The study presented in this chapter examines the factors influencing vendor performance in the telecommunication industry, with particular emphasis on technological expertise, vendor relationship management and comprehensive evaluation frameworks. A cross-sectional research design was adopted to collect data from vendors at a single point in time (Hajra *et al.*, 2025). This approach is suitable for describing current conditions and analysing the relationships among the study variables. By using cross-sectional data, the research efficiently captures responses from different vendor groups within the industry. The research process began with an extensive review of relevant literature to identify the key variables that shape vendor performance. Based on the findings of this review, a conceptual model was developed and corresponding hypotheses were formulated. Following this, a structured questionnaire was designed to gather data from managers and vendors involved in telecommunication construction projects. This instrument enabled the systematic collection of data on the identified variables and supports empirical testing of the proposed model.

3.2.2 Population and sampling procedures

In this research, the target population comprises individuals directly involved in vendor management, procurement and vendor performance evaluation. These participants were selected for their relevant operational knowledge and practical experience that align closely with the study objectives. The study focuses on employees from a major telecommunications infrastructure provider in Malaysia operating large-scale network development projects, as well as vendors who frequently collaborate with this organisation. This setting is appropriate because large infrastructure providers are actively engaged in vendor selection, contract administration, coordination and performance monitoring across multiple complex projects. Such an environment provides a realistic, comprehensive context for examining operational challenges and identifying key drivers of vendor performance.

In addition, the study incorporates a diverse range of vendors, including material suppliers, technical service providers and subcontractors, who contribute directly to infrastructure development. These vendors play a crucial role in meeting project specifications by delivering equipment, specialised services, and technical expertise. By including both organisational representatives and vendors, the research captures the interaction between client and supplier perspectives and examines how collaborative practices influence performance outcomes. This dual perspective provides a more holistic understanding of vendor performance, particularly regarding communication quality, trust development, coordination mechanisms and their combined impact on project execution and overall infrastructure delivery.

This study uses simple random sampling to select respondents to examine factors influencing vendor performance in the telecommunications infrastructure industry. Simple random sampling is widely used in quantitative research because it gives every individual in the population an equal chance of selection, thereby reducing bias and improving the representativeness of the sample. As a result, findings derived from the sample are more reliable and can be generalised to the wider population.

The target population comprises individuals directly involved in vendor performance activities for telecommunication infrastructure projects, including project managers, procurement officers and vendors. These groups are appropriate for the study because they are actively engaged in vendor management processes, performance monitoring and project execution. Since vendor performance can be influenced by multiple stakeholders across different roles and organisational levels, random sampling helps include varied perspectives. It reduces the risk of overrepresenting any particular group.

To implement the sampling process, the population was first defined and a sampling frame was developed using company records, industry directories and professional networks. Participants were then selected at random using unbiased methods, such as random number generation. The sample size of 209 respondents was determined to ensure adequate statistical power and to reflect the industry's stakeholder diversity. Data were collected through structured questionnaires covering technological expertise, vendor relationship management and vendor performance evaluation. The completed questionnaires were cleaned and analysed using descriptive and inferential statistics to identify patterns and relationships relevant to the research objectives. Finally, the results were interpreted in relation to the research hypotheses to produce meaningful insights and recommendations to enhance vendor performance in telecommunication infrastructure projects.

3.2.3 Data collection method

This study collected primary data using a structured questionnaire as the main research instrument. To improve efficiency and broaden participation, the questionnaire was distributed electronically through Google Forms. This method supports quick data collection and enables responses from a geographically diverse group of vendors and project managers within the telecommunications industry.

The questionnaire was developed through an extensive literature review and input from subject-matter experts to ensure its validity and alignment with the research objectives. It was designed to capture demographic details and measure the key variables related to vendor performance. The instrument is organised into four sections: vendor performance, technological expertise, vendor relationship management and comprehensive vendor rating systems.

Each section contains a set of statements evaluated using a five-point Likert scale ranging from 1 (strongly disagree) to 5 (strongly agree). This approach enables the study to assess respondents' perceptions and experiences quantitatively and to support statistical analysis. The questionnaire is structured clearly to encourage accurate and thoughtful responses, ensuring that the collected data are reliable and suitable for achieving the study's research goals.

3.2.4 Operationalisation and measurement

In this study, the independent variables consist of technological expertise, vendor relationship management and the comprehensive vendor rating system. Technological expertise is regarded as a key contributor to vendor performance in the telecommunications sector. The vendor's technological capabilities, the skill level of its personnel and the extent to which it demonstrates innovation were assessed. Technological capabilities reflect the vendor's ability to adopt and apply advanced tools and systems that support project implementation. Skill level focuses on the training and experience of vendor employees and how these competencies contribute to effective project execution. Innovation refers to the vendor's ability to introduce new ideas, adopt emerging technologies and continuously improve processes in response to a competitive and evolving industry environment.

The strength of communication, trust, collaboration and the development of long-term partnerships between vendors and project teams measures vendor relationship management. Communication assesses how clearly and consistently information is exchanged between parties during project implementation. Trust reflects the degree of reliability and confidence in the relationship, including the ability to rely on commitments and agreed-on responsibilities. Collaboration evaluates the extent to which vendors and project managers engage in cooperative planning and joint problem-solving to achieve shared goals. Long-term partnerships measure the stability and continuity of vendor relationships, as well as the willingness of both parties to maintain consistent cooperation over time.

The comprehensive vendor rating system was evaluated as a structured mechanism for monitoring and improving vendor performance, supported by clear performance criteria.

It includes assessing service quality, cost efficiency, timely delivery and customer service. Service quality focuses on compliance with quality standards and the frequency of defects or rework. Cost efficiency examines the vendor's ability to control costs and meet budget requirements while delivering expected outcomes. Timely delivery measures whether vendors consistently meet deadlines and reduce delays during execution. Customer service evaluates vendors' responsiveness, ability to address project needs, support and effective issue management.

The dependent variable in this study is vendor performance, a multidimensional construct that reflects how well vendors contribute to the success of telecommunications infrastructure projects. Vendor performance was measured by efficiency, effectiveness and client satisfaction, supported by indicators such as adherence to timelines, quality of deliverables, compliance with project requirements and cost management. Efficiency is assessed through the vendor's ability to complete tasks within scheduled timeframes and minimise disruptions. Effectiveness is evaluated by the quality and accuracy of deliverables, as well as the consistency of outputs with technical specifications and project goals. Client satisfaction measures the extent to which stakeholders are satisfied with the vendor's service delivery and overall project contribution. By integrating these indicators, the study provides a comprehensive assessment of vendor performance. It supports a clearer understanding of how technological expertise, relationship management and structured rating systems influence vendor outcomes in the telecommunications industry.

3.3. Results and analysis

3.3.1 Response rate

All 209 respondents were included in the frequency analysis, with no missing data, ensuring a complete dataset for demographic profiling.

3.3.2 Demographic profile

A frequency distribution analysis was conducted to summarise and describe the respondents' demographic profile. This analysis examined key characteristics such as gender, age and department to identify patterns within the sample and support the interpretation and generalisability of the findings. Understanding these demographic trends also helps explain how respondent backgrounds may influence their perspectives.

The demographic results given in Table 3.1 show that most respondents were mid-career professionals. The largest age group was 35 to 44 years old, representing 44.5% of the sample, followed by the 25 to 34 age group at 32.1%. Together, these categories account for 76.6% of all respondents, indicating that the findings largely reflect the perspectives of individuals who are actively involved in decision-making and operational responsibilities. Respondents in the 35 to 44 age range are likely to have substantial experience in vendor coordination, performance monitoring and contract execution, which strengthens the reliability of their evaluations. At the same time, respondents aged 25 to 34 may offer more contemporary perspectives, particularly on technological adoption and innovation. This mixture of experience and modern outlook enhances the relevance of the findings for improving vendor management practices.

Table 3.1 Demographic profile

Age		
	Frequency	Per cent: %
18–24	3	1.4
25–34	67	32.1
35–44	93	44.5
45–54	35	16.7
55 and above	11	5.3
Gender		
Male	97	46.4
Female	112	53.6
Years of experience		
Less than 1 year	3	1.4
1–3 years	14	6.7
4–6 years	21	10
7–10 years	45	21.5
More than 10 years	126	60.3
Department		
Procurement/supply chain	4	1.9
Operations	42	20.1
Engineering/technical	87	41.6
Sales	16	7.7
Other	60	28.7

The youngest group, aged 18 to 24, accounted for only 1.4% of respondents, suggesting limited representation of early-career professionals. Although smaller in number, this group may offer insights into emerging expectations in vendor engagement, especially regarding agility and digital integration. Overall, the demographic distribution supports a balanced understanding of vendor performance, with a focus on experienced professionals, while also capturing perspectives from younger employees.

As shown in Table 3.1, 97 of the 209 respondents were male (46.4%) and 112 were female (53.6%). This indicates a slightly higher proportion of female participants, suggesting the study may capture a wider range of perspectives, including those that are often underrepresented in research. The relatively balanced gender distribution strengthens the sample's inclusiveness and improves the applicability of the findings to both male and female respondents.

Gender representation is important because differences in experiences and expectations can influence how individuals assess organisational practices and performance outcomes. A sample dominated by one gender may limit the diversity of views and reduce the overall depth of the findings. In contrast, the gender diversity reflected in this study enriches the data and supports a more equitable understanding of vendor performance and management practices. Therefore, the achieved demographic balance supports a stronger interpretation and underscores the importance of maintaining gender diversity in future research.

The departmental distribution of the 209 respondents provides useful insights into workforce composition and organisational priorities. The largest proportion of participants comes from engineering or technical departments, with 87 respondents representing 41.6% of the sample. This suggests that the organisation places strong emphasis on technical expertise, reflecting the growing importance of engineering capabilities in industries where innovation and technology-driven development are essential. A strong technical workforce often supports improved project execution, operational efficiency and faster development cycles.

The operations department accounts for 42 respondents, representing 20.1% of the total. This indicates the presence of a substantial operational workforce responsible for coordinating processes, optimising resources and ensuring smooth service delivery. Effective operations management is closely linked to increased efficiency, reduced costs and improved customer value, particularly in highly competitive environments.

Sales has the smallest departmental representation, with 16 respondents (7.7%). Although relatively low, the sales function remains critical in driving revenue, building customer relationships and supporting organisational growth. Limited representation may suggest a lean approach to sales operations or indicate an area that requires stronger staffing support to meet business targets.

The ‘other’ category includes 60 respondents, accounting for 28.7%, indicating a wide range of additional roles, including finance, human resources, administration and other support functions. This diversity highlights the importance of supporting departments in strengthening core operations and enabling overall organisational performance.

Overall, the distribution shows a workforce heavily concentrated in technical roles, supported by operational and cross-functional units. This structure underscores the importance of technical and operational capacity for organisational effectiveness, while suggesting that departments such as sales and procurement may require greater attention to ensure balanced performance across all functions.

Employee work experience plays an important role in shaping vendor performance by influencing negotiation, industry and relationship management skills (Sposito *et al.*, 2024). The data show that the majority of respondents, 126 out of 209 (60.3%), have more than 10 years of experience in vendor-related work. This suggests that vendor management in the organisation is largely handled by highly experienced personnel, likely to contribute to stronger vendor relationships and improved performance outcomes (Owusu Kwateng *et al.*, 2022).

In contrast, only 1.4% of respondents have less than 1 year of experience, indicating limited involvement of entry-level employees in vendor management roles.

The results also show that 21.5% of respondents have 7 to 10 years of experience, while 10% have 4 to 6 years. These groups represent employees with moderate experience who can combine practical knowledge with adaptability to changing vendor requirements. However, employees with 1 to 3 years of experience account for 6.7% of the sample and may require additional training and mentorship to strengthen their vendor management competencies.

Overall, the findings highlight the importance of maintaining a skilled, experienced workforce while investing in development programmes for less-experienced employees. Strengthening staff capability at all levels can enhance vendor management effectiveness, support better vendor performance and drive organisational success.

3.3.3 Reliability analysis

Reliability analysis was used to determine how well a set of measurement items consistently captures a particular construct, thereby strengthening confidence in the validity of research findings. A commonly used method for assessing reliability is Cronbach's alpha (α), which measures the internal consistency of questionnaire items. Cronbach's alpha values range from 0 to 1, with higher values indicating stronger reliability. In most research contexts, an alpha value of 0.70 or above is considered acceptable, while values above 0.80 indicate a high level of reliability.

The reliability results given in Table 3.2 show high internal consistency across all constructs. The determinants affecting vendor performance recorded a Cronbach's alpha of 0.888 across five items, indicating that the items measure the construct consistently. Vendor relationship management achieved a Cronbach's alpha of 0.897 across eight items, indicating strong agreement and consistency in measuring communication, trust and collaboration in vendor relationships. The comprehensive vendor rating systems construct produced the highest reliability score, with a Cronbach's alpha of 0.922 across eight items, demonstrating excellent consistency in assessing vendor evaluation dimensions such as quality, cost, delivery and service support. Overall, Cronbach's alpha values ranging from 0.888 to 0.922 confirm that the survey instruments used in this study are reliable and provide consistent measurement across the main variables, supporting the credibility of the research findings.

Table 3.2 Reliability test

Variable	Cronbach's alpha, α	No. of questions
Factors affecting vendor performance	0.888	5
Technological expertise	0.890	8
Vendor relationship management	0.897	8
Comprehensive vendor rating systems	0.922	8

3.3.4 Normality analysis

Assessing normality is important to determine whether the data follow a normal distribution, typically represented by a symmetrical bell-shaped curve in which most values cluster around the mean. In this study, normality was evaluated using two commonly used statistical tests: the Kolmogorov–Smirnov (KS) and Shapiro-Wilk (SW) tests. These tests examine whether the data significantly deviate from a normal distribution; the KS test is generally used for larger sample sizes, while the SW test is more sensitive, especially for smaller samples.

The Kolmogorov–Smirnov test compares the sample distribution to a normal distribution and measures the maximum difference between the two distributions. A significant result ($p < 0.05$) indicates that the data do not follow a normal distribution. The Shapiro–Wilk test evaluates how closely the data fit a normal distribution by producing a statistic closer to 1 when the data are normal. Lower values and significant p-values suggest deviations due to skewness or kurtosis.

Table 3.3 shows that the Kolmogorov–Smirnov (KS) test for the dependent variable factors affecting vendor performance (DVfactors) yields a statistic of 0.177 with $p < 0.001$, indicating that the data significantly deviate from a normal distribution and the null hypothesis of normality is rejected. Similarly, vendor performance (KS = 0.177), technological expertise (KS = 0.162), vendor relationship management (KS = 0.167) and comprehensive vendor rating system (KS = 0.175) all show $p < 0.001$, confirming non-normality across all variables. These consistent findings reinforce that the assumption of normality is not met. Table 3.3 also shows that technological expertise is not normally distributed (KS = 0.162, $df = 209$, $p < 0.001$), indicating significant variation in vendors' technological capabilities. This suggests that while some vendors possess advanced tools, knowledge and capacity for innovation, others lag due to limited investment in technology upgrades, research and development (R&D) or training. Vendors with stronger technological expertise are more likely to manage complex telecommunication projects effectively and deliver higher performance, whereas less capable vendors

Table 3.3 Normality test

Normality tests	Kolmogorov–Smirnov[a]			Shapiro–Wilk		
	Statistic	*df*	Sig.	Statistic	*df*	Sig.
DV_factors affecting vendor performance[b]	0.177	209	<0.001	0.858	209	<0.001
IV1_technological expertise[c]	0.162	209	<0.001	0.870	209	<0.001
IV2_vendor relationship management	0.167	209	<0.001	0.820	209	<0.001
IV3_vendor rating system	0.175	209	<0.001	0.858	209	<0.001

[a]Lilliefors significance correction; [b]DV: dependent variable; [c]IV: independent variable

may face delays and inefficiencies. These findings highlight the importance of accounting for technological gaps in vendor selection and of implementing targeted interventions, such as training programmes, to improve overall project outcomes.

The Kolmogorov–Smirnov test indicates that vendor relationship management deviates significantly from normality (KS = 0.167, *df* = 209, $p < 0.001$), suggesting wide variation in how vendors manage relationships. Effective relationship management supports communication, trust, collaboration and conflict resolution, all of which are essential for timely and successful project delivery. The non-normal distribution suggests that while some vendors perform strongly in this area, others may struggle, leading to miscommunication and delays. Therefore, organisations should prioritise assessing and strengthening vendor relationship practices through initiatives such as training and development to improve overall performance and project outcomes. The results of the Kolmogorov–Smirnov and Shapiro–Wilk tests, which confirm that the data for all variables deviate from normality ($p < 0.001$), are shown in Figure 3.2. These histograms demonstrate the skewness and non-normal patterns across the dependent variable and the three independent factors, reinforcing the decision to utilise non-parametric assessments like Spearman's rank-order correlation.

The findings indicate a non-normal distribution in the comprehensive vendor rating system, reflecting wide variation in how vendors are assessed across multiple performance

Figure 3.2 Histogram for normality test

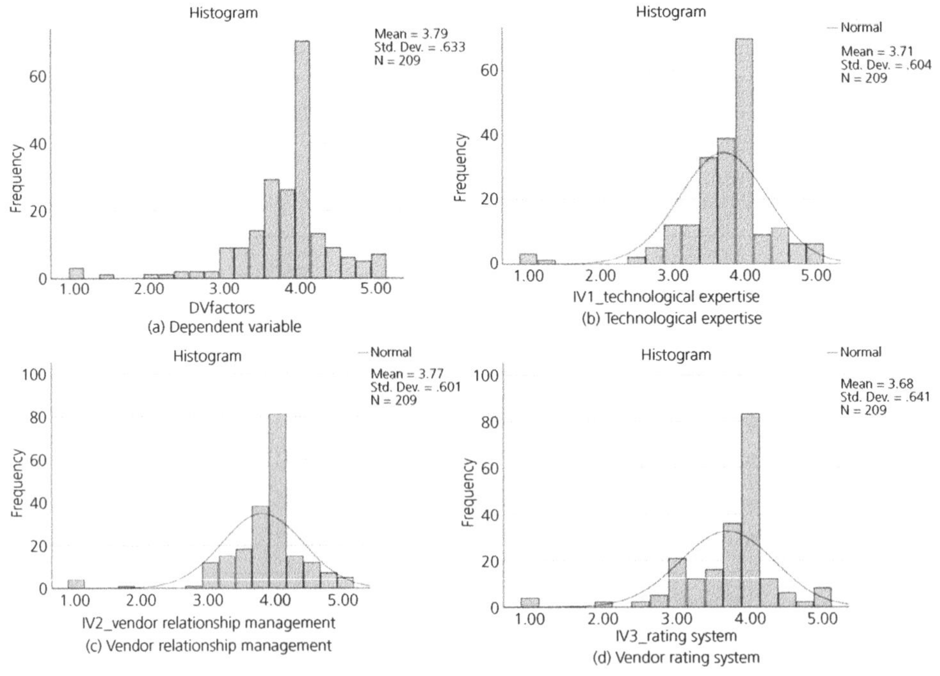

dimensions. This highlights the need for more consistent and standardised rating practices to ensure fair evaluation, improve accountability and support continuous feedback for vendor development. Strengthening the comprehensiveness and consistency of vendor ratings can help organisations identify high-performing vendors, address underperformance and ultimately improve project outcomes.

3.3.5 Descriptive analysis

A descriptive analysis of factors affecting vendor performance is essential for managing vendor relationships and supporting smooth operations. In this study, descriptive statistics, including sample size (n), minimum, maximum, mean and standard deviation (SD), were used to examine key factors, including technological expertise, vendor relationship management and vendor rating systems. These measures help explain respondents' perceptions of the importance of each factor, providing a basis for more informed decision-making in supply chain management.

A comprehensive summary of the descriptive statistics for both the dependent and independent variables, including the sample size, minimum and maximum values, mean and standard deviation, is provided in Table 3.4. The sample size for all variables is 209, which provides a strong basis for reliable analysis. A larger sample improves the accuracy of results by reducing error and increasing the ability to generalise findings to the wider population. The fact that all variables share the same sample size also suggests that the dataset is complete and consistent, enabling a more reliable assessment of the factors affecting vendor performance.

The minimum and maximum values for all variables range from 1 to 5, indicating that respondents expressed a wide variety of opinions. A minimum score of 1 reflects disagreement with the relevance of certain aspects, while a maximum score of 5 reflects strong agreement. This wide range highlights that perceptions of vendor performance factors differ across respondents, reinforcing the need for organisations to remain flexible when managing vendors and responding to varying expectations.

The mean scores indicate that respondents generally agree on the importance of the factors studied. The dependent variable has a mean of 3.79, suggesting that most participants tend to

Table 3.4 Descriptive statistics

Descriptive statistics					
	n	Min	Max	Mean	SD
DV_factors affecting vendor performance	209	1	5	3.79	0.63
IV1_technological expertise	209	1	5	3.71	0.60
IV2_vendor relationship management	209	1	5	3.77	0.60
IV3_vendor rating system	209	1	5	3.68	0.64
Valid *n* (listwise)	209				

agree that these factors influence vendor performance. The independent variables also show mean values close to 4, with technological expertise at 3.71, vendor relationship management at 3.77 and the vendor rating system at 3.68. These results indicate that respondents recognise the importance of technological capability, relationship management and structured evaluation systems in improving vendor performance, which aligns with previous studies.

Standard deviation values are relatively low, showing that responses are fairly consistent across participants. The dependent variable has a standard deviation of 0.63. In contrast, technological expertise and vendor relationship management both have a standard deviation of 0.60, and the vendor rating system has a standard deviation of 0.64. These values suggest that although individual experiences may differ slightly, respondents broadly agree on the importance of these dimensions in vendor performance.

Overall, the descriptive findings show that respondents consistently view technological expertise, vendor relationship management and comprehensive rating systems as important contributors to vendor performance. Mean scores around 3.7 indicate general agreement, and the low-to-moderate standard deviations confirm a stable level of consensus. These results support earlier research and suggest that organisations can enhance vendor performance by investing in technology, strengthening relationships and implementing structured evaluation systems that encourage accountability and continuous improvement.

3.3.6 Spearman's rank correlation

Correlation analysis measures the strength and direction of the linear relationship between variables and was applied in this study to examine how technological expertise, vendor relationship management and the comprehensive vendor rating system are associated with vendor performance. This analysis helps quantify the extent to which improvements in these independent variables are associated with better vendor performance on telecommunication infrastructure projects. Two correlation methods were used: Pearson's correlation coefficient and Spearman's rank-order correlation. Pearson's correlation is appropriate when the data meet the normality assumption, while Spearman's correlation is used when the data are not normally distributed. Both methods identify whether a relationship exists between variables and how strong it is. Pearson's correlation ranges from -1 to $+ 1$, where values closer to $+1$ indicate a strong positive relationship, values closer to -1 indicate a strong negative relationship and a value of 0 indicates no linear relationship.

Since the normality assumption was not met, Spearman's rank-order correlation was used to examine the relationships between the independent variables and vendor performance. Spearman's correlation measures the strength and direction of a monotonic relationship, making it suitable for non-normal or ordinal data. As shown in Table 3.5, all correlations are significant at the 0.01 level (2-tailed), indicating meaningful associations between the study variables.

The strongest relationship with vendor performance is technological expertise ($r = 0.633$, $p < 0.01$), showing a strong positive association. This suggests that vendors with higher technological capability tend to achieve better performance in telecommunication infrastructure projects. The strength of this relationship underscores the importance of technology-driven

Table 3.5 Spearman's correlation coefficient

Variable	DVfactors	Technological expertise	Vendor relationship management	Comprehensive vendor rating system
Vendor Performance (DVfactors)	1	0.633[a]	0.449[a]	0.490[a]
Technological expertise	0.633[a]	1	0.546[a]	0.541[a]
Vendor relationship management	0.449[a]	0.546[a]	1	0.630[a]
Comprehensive vendor rating system	0.490[a]	0.541[a]	0.630[a]	1

[a]Correlation is significant at the 0.01 level (2-tailed).

competence in supporting efficiency, innovation and effective project delivery, consistent with prior literature findings (Chockalingam, 2025; Katuri, 2025).

Vendor relationship management also shows a significant positive relationship with vendor performance ($r = 0.449, p < 0.01$), indicating a moderate association. This implies that vendors who manage client relationships effectively through communication, collaboration and trust tend to perform better. Although the relationship is weaker than technological expertise, it remains important, reinforcing the role of relationship quality in supporting smoother project implementation and improved outcomes.

Similarly, the comprehensive vendor rating system demonstrates a moderate positive relationship with vendor performance ($r = 0.490, p < 0.01$). This suggests that vendors assessed through structured, comprehensive evaluation systems tend to perform better. Such systems provide clear expectations, measurable indicators and regular feedback, which support accountability and continuous improvement. The consistency of this relationship, even under non-normal conditions, supports the view that evaluation mechanisms contribute to improved vendor capability and project success.

3.3.7 Regression analysis

Although the Kolmogorov–Smirnov and Shapiro–Wilk tests confirmed that the dataset does not follow a normal distribution, multiple linear regression was utilised as the primary tool for hypothesis testing. From a strategic leadership perspective, this choice is justified because regression yields higher predictive performance than correlation alone, enabling the quantification of the simultaneous influence of technical, relational and governance factors on overall vendor performance. Furthermore, regression analysis isolates the unique contribution of each independent variable, enabling the identification of the primary driver of project success among competing factors. The model's stability and high explanatory power further validate its application, ensuring that the predictive strength of these leadership variables is accurately captured. Finally, the substantial sample of industry experts provides the necessary statistical power, as linear regression is typically robust to violations of normality in larger datasets, thereby supporting the generation of reliable, evidence-based leadership strategies.

Table 3.6 Regression analysis

R	R^2	Adjusted R^2	Std. error of the estimate	R^2 change	F change	$df1$	$df2$	Sig. F change
0.816	0.666	0.661	0.36859	0.666	136.02	3	205	< 0.001

Note: predictors: (constant), IV3_rating system, IV1_technological expertise, IV2_vendor relationship management.

The R-value of 0.816, as shown in Table 3.6, indicates a strong positive relationship between the three independent variables (technological expertise, vendor relationship management, and comprehensive vendor rating system) and vendor performance, suggesting that these factors collectively exert a strong influence on vendor outcomes in telecommunication infrastructure projects. The R^2 value of 0.666 further indicates that 66.6% of the variance in vendor performance can be explained by the model, demonstrating strong explanatory power and confirming that these variables are important predictors of vendor success.

The adjusted R^2 of 0.661 provides a more accurate estimate by accounting for the number of predictors. Since it remains close to the R^2 value, the model appears stable and does not overestimate its predictive strength, reinforcing the significant contribution of the independent variables to vendor performance.

These findings carry important implications for telecommunication firms. They suggest that selecting vendors with strong technological capability is crucial for delivering complex infrastructure projects efficiently. At the same time, effective relationship management remains essential because even technologically capable vendors may underperform if communication and collaboration with clients are weak. In addition, implementing a comprehensive vendor rating system can support continuous monitoring, accountability and performance improvement. By strengthening these three areas, firms can enhance vendor performance, improve project delivery and achieve better long-term outcomes.

Overall, the regression results confirm that technological expertise, vendor relationship management and comprehensive vendor rating systems significantly influence vendor performance. The strong model fit and high variance support the conclusion that these factors should be prioritised in vendor selection and management, with technological expertise showing the strongest influence, supported by relationship practices and structured performance evaluation systems.

3.3.8 Hypothesis testing

The hypothesis-testing results examine how various factors influence vendor performance. Three independent variables were assessed: technological expertise, vendor relationship management and the rating system. The analysis evaluates the significance of each variable in predicting vendor performance, as presented in Table 3.7, where the p-values indicate the statistical importance of each factor.

Table 3.7 Hypothesis testing

Hypothesis	*B*	Std. error	*β*	t	Sig.	Support
H1: technological expertise has a significant impact on vendor performance	0.585	0.069	0.559	8.473	< 0.001	Supported
H2: vendor relationship management has a significant impact on vendor performance	0.166	0.078	0.157	2.129	0.034	Supported
H3: vendor rating system has a significant impact on vendor performance	0.156	0.07	0.158	2.229	0.026	Supported

The results provide strong support for H1, which proposes that technological expertise significantly affects vendor performance. The findings show a high standardised coefficient ($\beta = 0.559$), a strong *t*-value ($t = 8.473$) and a highly significant *p*-value ($p < 0.001$). This indicates that technological expertise is the most influential predictor of vendor performance in the model. Vendors equipped with advanced technological capabilities are more likely to improve operational efficiency and deliver higher-quality outcomes, which aligns with previous research highlighting the importance of technological competence in vendor selection and performance management (Rani *et al.*, 2024; Stark, 2022). Vendors with strong technical knowledge are also better able to adapt to changing client needs, thereby strengthening their ability to perform consistently in dynamic project environments.

The findings also support H2, which states that vendor relationship management significantly affects vendor performance. The results show a positive and significant effect ($\beta = 0.157$, $t = 2.129$, $p = 0.034$), indicating that vendors who manage relationships effectively tend to achieve better performance. Strong communication, trust and collaboration improve coordination and reduce the likelihood of misunderstandings or delays, leading to better project outcomes. This result is consistent with evidence that effective relationship management reduces project disruptions and increases client satisfaction, while transparency and open communication help vendors maintain consistent service quality (Sepasgozar and Loosemore, 2017; Sharma *et al.*, 2016).

Similarly, H3 is supported, confirming that the vendor rating system has a significant impact on vendor performance. The results show a meaningful positive effect ($\beta = 0.158$, $t = 2.229$, $p = 0.026$), suggesting that structured vendor rating systems contribute to performance improvement. A robust rating system promotes accountability and encourages vendors to meet defined standards, while regular feedback supports continuous improvement. Rating systems that incorporate client input also help vendors identify weaknesses and adjust practices to enhance outcomes. In addition, transparent evaluation mechanisms can motivate vendors to improve performance through healthy competition and clearer expectations.

3.4. Practical insights

The implications of these findings are significant for both researchers and practitioners in the telecommunications industry. For academics, this study strengthens the existing literature on vendor management by providing empirical evidence that supports current theories and encourages further investigation into additional factors that may influence vendor performance. For practitioners, the results highlight the value of prioritising technological expertise, strengthening vendor relationships and applying comprehensive evaluation systems to improve operational efficiency and service quality. The adoption of technology standards and vendor certification processes may also support a more competitive and accountable vendor environment.

This study recommends that telecommunications firms enhance vendor technological capability through structured training and continuous development initiatives. As the industry evolves through developments such as 5G, Internet of Things integration and stronger cybersecurity requirements, vendors must maintain updated technical skills to meet project demands. Firms can support this by offering specialised programmes focused on key technologies and by collaborating with universities and technology providers to deliver practical learning opportunities. Continuous professional development should also be encouraged through conferences, webinars and certification courses to promote adaptability and long-term capability improvement. To ensure these investments lead to measurable outcomes, organisations should implement technology-focused performance indicators and link vendor performance reviews to training results, using metrics such as service quality, response time and customer satisfaction.

In addition, strengthening vendor relationships is essential for improving performance. Telecommunications organisations should establish consistent relationship management practices that promote open communication and collaboration through regular meetings, performance reviews and structured feedback. Digital collaboration tools can further enhance transparency, improve coordination and support faster problem resolution, resulting in better alignment between vendor performance and organisational expectations. Firms are also encouraged to build long-term partnerships by treating vendors as strategic collaborators rather than transactional service providers. Joint initiatives, innovation workshops and shared risk approaches can increase commitment, accountability and innovation across both parties. Recognising and rewarding high-performing vendors through incentives, public recognition or participation in strategic projects can further motivate continuous improvement and enhance service delivery.

Overall, the study reinforces that vendor performance in telecommunications is shaped by the combined influence of technological expertise, effective relationship management and structured evaluation systems. For researchers, the findings provide evidence supporting frameworks such as the resource-based view and social exchange theory, which emphasise the importance of capability and relationship-based advantages. For practitioners, the results offer clear guidance: investing in vendor training, strengthening partnerships and implementing comprehensive performance evaluation systems can improve vendor outcomes and enhance the efficiency of the telecommunications supply chain.

3.5. Conclusion and recommendations

In conclusion, this study shows that vendor performance in the telecommunications industry is shaped by the combined influence of technological expertise, effective relationship management and structured evaluation systems. Strengthening technological capability remains essential for improving service quality and operational efficiency. At the same time, organisations must treat vendors as strategic partners by promoting communication, trust and collaboration through strong relationship management practices.

The study also highlights the importance of comprehensive evaluation mechanisms that assess vendors across key performance dimensions, such as service quality, cost-effectiveness and technological capabilities. These systems support accountability and continuous improvement by providing clear and transparent feedback.

Overall, the findings contribute to vendor management literature by supporting frameworks such as the resource-based view and social exchange theory, while also offering practical guidance for industry practitioners. By prioritising training, relationship development and structured vendor evaluations, telecommunications firms can improve vendor outcomes, strengthen supply chain performance and support long-term success in a competitive, fast-changing industry.

3.5.1 Theoretical contributions

This study strengthens the theoretical understanding of vendor performance in the telecommunications industry by examining how technological expertise, vendor relationship management and comprehensive vendor rating systems interact to influence vendor outcomes. The findings provide empirical support for established theories in management and organisational behaviour, while offering a clearer explanation of how these factors collectively shape vendor performance in a technology-driven environment.

A key contribution is the strong effect of technological expertise on vendor performance, shown by a β value of 0.559 and $p < 0.001$. This result reinforces the resource-based view, which argues that unique resources, such as technological capability, create sustained competitive advantage. The study extends this theory by showing that competitive advantage depends not only on internal organisational resources but also on the technological capability of external vendors. This supports the view that firms must prioritise vendor technological development through training, innovation initiatives and collaboration to remain competitive in a rapidly evolving telecommunications market.

The study also confirms the theoretical importance of vendor relationship management, with a positive and significant effect on vendor performance ($\beta = 0.157$, $p = 0.034$). This aligns with social exchange theory, which explains that relationships develop through mutual benefits, trust and ongoing interaction. The findings suggest that effective relationship management improves collaboration, problem-solving and project execution, supporting the idea that strong partnerships create value for both vendors and organisations.

In addition, comprehensive vendor rating systems significantly influenced vendor performance ($\beta = 0.158$, $p = 0.026$), highlighting the role of structured performance evaluation.

This supports total quality management principles, which emphasise continuous improvement through systematic measurement and feedback. It also aligns with performance management systems theory, which holds that regular monitoring and evaluation help align performance with organisational objectives and encourage accountability. The findings suggest that well-designed rating systems provide clear expectations and feedback that motivate vendors to improve performance over time.

Overall, this study contributes to the vendor performance literature by empirically confirming the significance of technological capability, relationship quality and structured evaluation systems, and by demonstrating how these factors reinforce one another in shaping vendor performance in the telecommunications industry.

3.5.2 Limitations

Although this study provides valuable insights into the factors influencing vendor performance in the telecommunications industry, several limitations should be acknowledged. First, the geographical scope was limited to a specific region, which may reduce the ability to generalise the findings to other contexts. Telecommunications markets differ across regions due to variations in regulation, infrastructure development, technology adoption and customer expectations, so vendor performance drivers may not operate the same way across all markets.

Second, the study used a cross-sectional design, measuring vendor performance at a single point in time. While useful for identifying relationships, this approach does not capture how vendor performance may change over time or in response to industry shifts and organisational interventions. It also limits the ability to establish causal relationships between variables.

Third, the study focused only on three key predictors of vendor performance, which may not fully represent the complexity of vendor management in telecommunications. Other factors, such as pricing strategies, regulatory compliance, market competition and broader economic conditions, may also influence vendor performance.

Finally, the study may be affected by response bias, as the data were collected by way of surveys. Participants may provide socially desirable answers or overestimate vendor capabilities, thereby reducing the accuracy of the results. Future research could strengthen reliability by ensuring anonymity and by incorporating qualitative methods, such as interviews or focus groups, to validate and enrich the quantitative findings.

Overall, addressing these limitations through broader geographic coverage, longitudinal methods, expanded variables and mixed data approaches would enhance the depth and applicability of future research on vendor performance and support the development of more effective vendor management strategies in the telecommunications industry.

3.5.3 Recommendations and future research

The findings and limitations of this study suggest several directions for future research to strengthen the understanding of vendor performance in the telecommunications industry and beyond. One important recommendation is the use of longitudinal research designs that track vendor performance over time. Longitudinal studies would provide a clearer view of how vendor performance changes in response to evolving technology, regulation and market conditions, and would allow researchers to identify patterns and potential cause-and-effect relationships that cross-sectional approaches cannot capture. This approach may also help organisations recognise key stages in vendor partnerships where interventions such as training or resource support are most effective.

Future research should also include cross-industry comparisons to examine whether the factors identified in this study apply beyond telecommunications. Investigating vendor performance in sectors such as healthcare, manufacturing and retail could reveal industry-specific challenges and best practices, as well as shared patterns that strengthen the broader theory of vendor management. Such comparisons could help organisations adopt successful vendor management strategies from other industries and improve their own practices.

In addition, future studies should explore other factors that may influence vendor performance, beyond technological expertise, relationship management and vendor rating systems. Variables such as pricing strategies, market competition, regulatory compliance, supply chain disruption and broader economic conditions may significantly shape vendor behaviour and performance outcomes. Examining how these factors interact with the key variables identified in this study would provide a more complete understanding of vendor performance dynamics.

Another important area for further study is the impact of emerging technologies such as artificial intelligence, blockchain and 5G on vendor performance and vendor–client relationships. These technologies may reshape how vendors deliver services, improve decision-making through predictive analytics and increase transparency and accountability through digital tracking systems. Future research should examine both the opportunities and challenges created by these technologies, as well as their influence on vendor collaboration, monitoring and performance improvement.

Finally, incorporating qualitative methods such as interviews, focus groups and case studies would provide deeper insight into the experiences and perceptions of vendors and telecommunications firms. Qualitative approaches can explain why certain relationships exist, uncover hidden challenges in vendor partnerships and validate quantitative findings through real-world evidence. Combining quantitative and qualitative methods would produce a more complete and practical understanding of vendor management and performance.

Overall, pursuing these research directions will support the development of stronger vendor performance frameworks and improve vendor management strategies across industries. By expanding research methods, exploring new variables and examining the role of emerging technologies, future studies can build on this research's findings and contribute to more effective and sustainable vendor partnerships in a competitive, technology-driven environment.

REFERENCES

Basiru JO, Ejiofor CL, Onukwulu EC and Attah RU (2023) The impact of contract negotiations on supplier relationships: a review of key theories and frameworks for organisational efficiency. *International Journal of Multidisciplinary Research and Growth Evaluation* **4(1)**: 788–802.

Blanutsa VI (2024) Identification of critical telecommunications infrastructure in Russia: a geographical approach. *Geography and Natural Resources* **45(1)**: 1–8.

Chockalingam D (2025) Advanced Vendor Management Systems: Strategic Governance Through Unified Platform Innovation. *Journal of Computer Science and Technology Studies* **7(10)**: 452–466.

Ezeigweneme CA, Umoh AA, Ilojianya VI and Oluwatoyin A (2023) Telecom project management: lessons learned and best practices: a review from Africa to the USA. *World Journal of Advanced Research and Reviews* **20(3)**: 1713–1730.

Hajra A, Ghosh B and Bhowmik S (2025). Cross-sectional studies. In *Translational Cardiology*. Academic Press, London, UK, pp. 221–225.

Huang K, Wang K, Lee PK and Yeung AC (2023) The impact of Industry 4.0 on supply chain capability and supply chain resilience: a dynamic resource-based view. *International Journal of Production Economics* **262**: 108913.

Katuri DRK (2025) ERP-driven vendor management: unlocking efficiency, transparency, and procurement. *International Journal of Scientific Research in Computer Science, Engineering and Information Technology* **11(1)**: 3224–3232.

Larsson C (2025) *5G/5G-Advanced Networks: Planning, Design, and Optimization*. Academic Press, London, UK.

Owusu Kwateng K, Fokuoh B and Tetteh FK (2022) The relationship between vendor managed inventory and operational performance. *Modern Supply Chain Research and Applications* **4(3)**: 202–222.

Rani A, Mishra D and Omerovic A (2024) Multi-criteria decision-making methods: a case of software vendor selection. *TEM Journal: Journal of the Association for Information Communication Technologies, Education and Science* **13(2)**: 1218.

Sarkis J (2020) Supply chain sustainability: learning from the COVID-19 pandemic. *International Journal of Operations and Production Management* **41(1)**: 63–73.

Sepasgozar SME and Loosemore M (2017) The role of customers and vendors in modern construction equipment technology diffusion. *Engineering, Construction and Architectural Management* **24(6)**: 1203–1221.

Sharma RR, Chadee D and Roxas B (2016) Effects of knowledge management on client-vendor relationship quality: the mediating role of global mindset. *Journal of Knowledge Management* **20(6)**: 1268–1281.

Sposito L, Scafuto IC, Serra FR and Ferreira MP (2024) Influence of the project managers' expertise and experience in the success of projects: the moderating effect of emotional intelligence. *International Journal of Managing Projects in Business* **17(1)**: 1–26.

Stark R (2022) The role of digital technology vendors. In *Virtual Product Creation in Industry: The Difficult Transformation from IT Enabler Technology to Core Engineering Competence*. Springer, Berlin, Germany, pp. 465–506.

Tajouri O, El Manzani Y and Lakhal L (2025) The interplay between total quality management (TQM) and organizational learning: similarities or complementarities? *Development and Learning in Organizations: An International Journal* **39(1)**: 39–42.

Vargas P and Tien I (2023) Impacts of 5G on cyber-physical risks for interdependent connected smart critical infrastructure systems. *International Journal of Critical Infrastructure Protection* **42**: 100617.

Advancing Leadership and Collaboration in Project Management: Culture, Knowledge and Team Dynamics

emerald PUBLISHING ice Publishing

Mohammed Reyasudin Basir Khan
ISBN 978-1-83742-803-8
https://doi.org/10.1108/978-1-83742-802-120261006
Emerald Publishing Limited: All rights reserved

Chapter 4
Leadership in virtual project management: driving success in digital logistics operations

Shatana Manoharan and Mohammed Reyasudin Basir Khan

Tun Razak Graduate School, Universiti Tun Abdul Razak (UNIRAZAK), Kuala Lumpur, Malaysia

4.1. Introduction

Modern businesses increasingly rely on virtual project management to handle the growing complexity of logistics operations. Logistics involves coordinating and moving goods and services across locations, making it vital for supply chains and global trade. As companies pursue faster delivery, cost savings and efficiency, managing logistics projects has become more demanding. Virtual project management supports this need by enabling managers to supervise operations across multiple regions without being physically present, using digital tools, communication platforms and project management software to coordinate activities remotely.

In logistics, where activities span countries, time zones and stakeholder networks, virtual project management reduces location barriers and improves flexibility. Technologies such as cloud computing, artificial intelligence (AI) and big data analytics have further transformed logistics, enabling real-time visibility into inventory, shipments and delivery schedules (Cichosz *et al.*, 2020). AI and analytics also support demand forecasting, inventory optimisation, route planning and the automation of routine tasks, enabling managers to focus on strategic decisions (Cichosz *et al.*, 2020; Zhang and Dhaliwal, 2009).

Despite these advantages, virtual logistics projects face challenges, including trust-building, cultural and language differences, time zone constraints and technical disruptions (Alkoud *et al.*, 2023; Smith, 2019). Poor communication and weak collaboration can lead to delays, errors and reduced stakeholder satisfaction. These issues are critical in logistics because project performance is closely tied to on-time delivery and cost control. Many organisations also lack adequate training and tools to support effective virtual teamwork, contributing to inconsistent project outcomes (Alkoud *et al.*, 2023).

The study presented in this chapter addresses these challenges by examining the key factors influencing the success of virtual project management in logistics operations, namely leadership style, virtual communication effectiveness, stakeholder management, team collaboration and project risk management. Using a quantitative approach, the research investigates how these factors affect project outcomes, including timely delivery, stakeholder satisfaction and budget adherence. The findings contribute to academic knowledge by applying leadership and collaboration perspectives to virtual logistics projects. Practically, the study will guide logistics managers in improving leadership practices, strengthening communication, managing stakeholder expectations, enhancing collaboration and reducing project risks in virtual environments. It will also support technology developers in identifying system features that improve virtual project execution, such as integrated dashboards and AI-driven tracking, helping organisations remain competitive in an increasingly digital logistics landscape (Cichosz *et al.*, 2020).

Globalisation and rapid advances in digital technology have reshaped logistics operations, pushing firms to adopt innovative strategies to remain competitive (Zhang and Dhaliwal, 2009). Virtual project management has emerged as a critical approach for logistics companies to plan, coordinate and control projects across dispersed locations using digital tools, communication platforms and project management systems (Batırlık *et al.*, 2022). These technologies support flexibility, accountability and real-time coordination, which are essential for managing global supply chains and meeting rising expectations for speed, accuracy, cost efficiency and customer satisfaction. At the same time, the growth of big data analytics, artificial intelligence and cloud solutions has strengthened decision-making, improved visibility across operations and enabled logistics firms to optimise routes, anticipate demand and automate routine processes (Cichosz *et al.*, 2020). However, the shift to virtual project environments also brings challenges, particularly in leading remote teams, maintaining clear communication and managing diverse stakeholder expectations (Alkoud *et al.*, 2023; Smith, 2019).

Within this context, leadership, communication, collaboration, stakeholder engagement and risk management become essential drivers of performance. Virtual logistics teams depend heavily on leadership that can build trust, sustain motivation and coordinate work across cultural and geographic boundaries. Effective leadership in virtual settings also requires technological competence, ensuring that digital tools support transparency and team alignment.

Communication quality is another determinant of virtual project success. Research shows that communication effectiveness in virtual environments depends on the appropriate use of digital platforms that match task complexity. In logistics projects, where coordination is continuous and errors can trigger disruptions, selecting appropriate communication tools helps reduce misunderstandings and delays (Alkoud *et al.*, 2023).

Stakeholder engagement is equally critical in logistics operations, where projects involve customers, suppliers, regulators and internal teams. Effective virtual stakeholder management requires structured communication, real-time updates and feedback mechanisms that support alignment and timely decision-making. Strong stakeholder coordination improves delivery performance and satisfaction in digital supply chain environments (Cichosz *et al.*, 2020).

Team collaboration is strengthened through trust, mutual support and shared responsibility, which are essential for virtual knowledge exchange (Powell, 2016). In virtual logistics settings, collaboration depends on transparent workflows, regular interaction and tools that support shared task tracking and information exchange. Cultural diversity and time zone differences may hinder cohesion, so structured coordination routines are needed to sustain teamwork and prevent fragmentation (Alkoud *et al.*, 2023).

Project risk management is particularly important in virtual logistics environments due to supply chain disruption, system failures, cybersecurity threats and regulatory changes (Cichosz *et al.*, 2020). Emerging digital systems and AI-driven monitoring enhance early detection and mitigation of risks (Spulbar, 2025). The effective adoption of these technologies depends on organisational readiness and perceived usefulness, which influence performance outcomes in digital supply chains (Zhang and Dhaliwal, 2009).

Project success remains the key outcome, commonly evaluated based on time, cost, quality, stakeholder satisfaction and adaptability. Drawing from existing research on virtual collaboration and digital transformation in logistics (Cichosz *et al.*, 2020), this study positions leadership style, virtual communication effectiveness, stakeholder management, team collaboration and project risk management as core predictors of success. Based on the literature, the study proposes that each independent variable is positively and significantly related to virtual project success in logistics operations.

Figure 4.1 presents the conceptual framework of the study, illustrating the intricate relationships between the independent, mediating and dependent variables involved in virtual project management within the Malaysian logistics sector. The model identifies five core independent variables – leadership style, virtual communication effectiveness, stakeholder management, team collaboration and project risk management – and examines their influence on the dependent variable, project success.

Figure 4.1 Conceptual framework

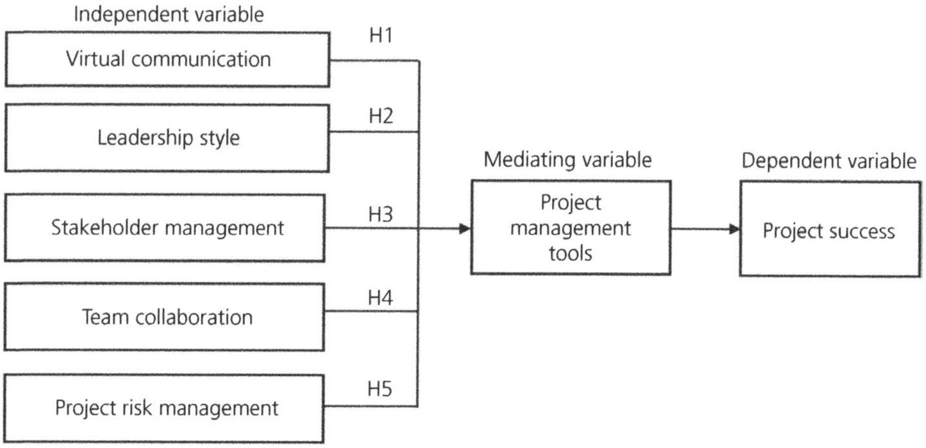

Central to this framework is the role of project management tools (such as SAP Transportation Management and Microsoft Project), which function as a mediating variable. The model proposes that the success of virtual projects does not rely solely on managerial behaviours but is significantly influenced by how these behaviours are channelled through and enhanced by digital platforms. By integrating these tools, organisations can optimise real-time coordination, transparency and decision-making, which are essential for managing complex, geographically dispersed logistics operations.

Ultimately, the framework defines project success through a multi-dimensional lens, including timely completion, cost control, quality performance and stakeholder satisfaction. This integrated approach highlights that achieving effective virtual project management requires a synergy between human leadership and technological mediation.

Based on the study's theoretical and conceptual framework, the formal hypotheses are proposed as follows.

- H1 (leadership style): leadership style has a significant positive relationship with project success, which is fully mediated by project management tools. This indicates that leadership impacts success primarily by facilitating the effective use of digital coordination systems.
- H2 (virtual communication effectiveness): virtual communication effectiveness has a significant positive relationship with project success, which is fully mediated by project management tools. The impact of communication on project outcomes is entirely channelled through the interaction with media-rich technological platforms.
- H3 (stakeholder management): stakeholder management has a significant positive relationship with project success, demonstrating partial mediation through project management tools. It influences success both directly through engagement and indirectly by integrating stakeholder feedback into project tools.
- H4 (team collaboration): team collaboration has a significant positive relationship with project success, demonstrating partial mediation through project management tools. Success is driven both by direct interpersonal cooperation and the enabling role of shared digital resources.
- H5 (project risk management): project risk management has a significant positive relationship with project success, demonstrating partial mediation through project management tools. Effective risk mitigation contributes to success both as a direct managerial practice and through real-time monitoring and early detection tools.

4.2. Methodology

4.2.1 Research design

This section outlines the overall research design and the timeline for conducting data collection, sampling procedures, measurement development and statistical analyses, including reliability and validity testing. A quantitative research approach was adopted to examine relationships and patterns between virtual project success and the identified variables. Given the complexity of virtual work settings, the methodology was designed to capture the diverse technological, cultural and geographic conditions that remote teams face. Furthermore, the rationale for the selected sampling strategy and data collection instruments is presented to ensure the study achieves strong validity and reliability.

4.2.2 Population and sampling procedures

Logistics is a major sector in Malaysia, supporting e-commerce, trade and manufacturing. The country has a large number of logistics firms, with estimates suggesting more than 1500 registered companies operating across freight forwarding, warehousing and transportation services. An increasing proportion of these organisations have adopted virtual project management in response to demands for cost efficiency, operational effectiveness and rapid customer responsiveness. Although the exact number fluctuates, it is reported that approximately 30% to 40% of Malaysian logistics firms have implemented virtual project management using digital communication tools, cloud-based platforms and Internet of Things (IoT)-enabled tracking systems.

To ensure adequate representation across job roles and organisational contexts, this study employs non-probability purposive sampling. This approach strengthens representativeness by ensuring that respondents reflect the diversity of roles and organisational structures within the Malaysian logistics industry. The target sample size for this study is 106 respondents, drawn from logistics organisations in Malaysia that actively practise virtual project management. This sample size is considered appropriate for quantitative analysis, as it supports meaningful statistical testing while remaining practical within the study's timeframe. Participants were selected from a range of roles, including project managers, operations executives, IT professionals and team leaders, as these groups are directly involved in planning, coordinating and executing logistics projects through virtual platforms.

Overall, a sample size of 106 is justified for this study, given its objective of examining relationships between key variables in virtual project management success. This aligns with guidance in the literature suggesting that quantitative studies seeking to establish relationships between variables typically require samples of at least 100 participants to yield reliable and interpretable results.

4.2.3 Data collection method

Data for this study were collected using a structured online questionnaire administered through Google Forms. This platform was selected for its accessibility, ease of use and suitability for reaching respondents across geographically dispersed logistics organisations that practise virtual project management. The online format allows participants to complete the survey at their convenience, regardless of location, aligning with the nature of virtual project environments, where teams often operate across multiple regions.

The questionnaire consisted of closed-ended items designed to measure the study variables, namely leadership style, virtual communication effectiveness, stakeholder management, team collaboration, project risk management, project management tools and virtual project success. Responses were captured using a five-point Likert scale ranging from 1 (strongly disagree) to 5 (strongly agree), enabling the quantification of perceptions and facilitating statistical analysis. Structured surveys are appropriate for this purpose because they provide uniform data collection across participants and allow for consistent comparison and analysis.

Google Forms also supports efficient data handling through automated response collation and export to statistical software such as the statistical package for the social sciences (SPSS).

Before full administration, a pilot test involving ten participants was conducted to assess the clarity, relevance and structure of the survey items. Pilot testing is essential in survey research as it helps identify ambiguities and improves instrument quality, thereby increasing the reliability of the final data. Feedback from the pilot was used to refine the questionnaire before distribution to the full sample.

After finalisation, survey invitations were distributed by way of email and through relevant communication channels, such as WhatsApp and Instagram. Reminder messages were issued during the data collection period to improve participation and response rates. The survey was open for four weeks to ensure sufficient time to complete it.

To strengthen validity, questionnaire items were developed in line with established theoretical and empirical literature, ensuring alignment with the research objectives and constructs. Reliability was assessed using Cronbach's alpha (α); values of 0.70 or higher indicate acceptable internal consistency. Overall, this data collection procedure was designed to generate reliable and valid evidence on the determinants of virtual project success within the Malaysian logistics industry.

4.2.4 Operationalisation and measurement

This study operationalises key variables to examine the determinants of virtual project performance within the Malaysian logistics industry. A quantitative approach was employed to measure the effects of five independent variables, namely team collaboration, leadership style, virtual communication effectiveness, stakeholder management and project risk management, on the dependent variable of project success. In addition, the mediating role of project management tools was assessed to determine how digital platforms strengthen the relationship between these predictors and project outcomes. All constructs were measured using a five-point Likert scale ranging from strongly disagree to strongly agree, allowing respondents' perceptions to be quantified for statistical analysis.

The study population comprises employees from major logistics organisations in Malaysia, including DHL and City-Link, which have incorporated virtual project management into their daily operations. These organisations provide an appropriate context for investigating virtual project success, given their extensive reliance on remote coordination and digital systems. The target sample size is 106 respondents, comprising project managers, team leaders, IT personnel and operations executives directly involved in planning and executing virtual projects. This sample size is supported by prior empirical studies indicating that participant ranges between 50 and 106 provide sufficient statistical power and credible evidence for analysing virtual team performance and project success in logistics settings.

Team collaboration refers to the extent to which team members cooperate effectively in virtual environments to share knowledge, solve problems and complete tasks. This construct was measured through items assessing coordination frequency, willingness to support colleagues and the effectiveness of collaboration tools in facilitating teamwork. Leadership style focuses on the extent to which leaders provide guidance, motivate team members and maintain accountability in virtual settings. Measurement items examine clarity of direction, adaptability to remote challenges and the leader's role in promoting trust and engagement. Virtual communication effectiveness assesses the clarity, timeliness and reliability of information exchange through digital channels. Items capture the suitability of communication tools, the extent to which misunderstandings are reduced and the effectiveness of virtual meetings. Stakeholder management examines how well organisations engage stakeholders, manage expectations and address concerns using digital platforms. This construct was measured through items related to stakeholder updates, feedback integration and decision-making involvement. Project risk management refers to an organisation's ability to identify, monitor and mitigate risks in virtual projects. Measurement items focus on proactive risk identification, contingency planning, and regular review of risk strategies.

Project management tools act as the mediating variable, representing the use of platforms such as SAP Transportation Management and Microsoft Project to support task tracking, resource allocation, reporting and coordination. This construct assesses tool usability, integration with communication platforms and the extent to which the tools enhance efficiency and transparency in virtual project execution.

Project success is operationalised as the degree to which virtual projects achieve their objectives within defined time, cost and quality constraints while meeting stakeholder expectations. It was measured using items that assessed the achievement of project goals, stakeholder satisfaction, adherence to budgets and schedules, and delivery quality. Collectively, the operationalisation of these constructs enables a systematic evaluation of how leadership, communication, collaboration, stakeholder engagement and risk management contribute to the success of virtual logistics projects and how project management tools strengthen these relationships.

4.2.5 Pre-analysis data screening

Before conducting the primary statistical analyses, a rigorous pre-analysis data screening was performed on the 106 valid responses to ensure the integrity and quality of the dataset. First, the data were thoroughly checked for missing values; the assessment confirmed that all 106 records were complete, as evidenced by the 100% response rate, providing a comprehensive dataset for examining the relationships between variables. Second, an outlier analysis was conducted to identify any extreme responses that could distort regression coefficients or inflate error terms in the mediation analysis. Finally, normality testing was performed to evaluate the distribution of the independent, mediating and dependent variables. The PROCESS Macro (Model 4), utilised for hypothesis testing, employs bootstrapping. This robust nonparametric technique does not strictly require the assumption of normality; this screening confirmed that the data remained well-suited for the subsequent multivariate procedures. This preliminary phase ensures that the findings from the factor analysis and reliability testing are derived from a clean, stable and statistically sound foundation.

4.3. Results and analysis

4.3.1 Response rate

A total of 106 respondents completed the questionnaire, resulting in a 100% response rate. This indicates that all individuals who were invited and agreed to participate provided usable responses within the data collection period. Achieving a full response rate strengthens the study by ensuring the intended sample size is met, minimising non-response bias and improving the reliability of the statistical analysis. Consequently, the dataset can be considered comprehensive for the selected sample and suitable for examining the relationships between the study variables.

4.3.2 Demographic profile

The demographic profile of the 106 respondents provides a critical lens for understanding how project management tools mediate the success of logistics operations. Table 4.1 illustrates the demographic profile of the respondents. The sample is characterised by a predominantly young and early-career workforce, with 55.7% of participants in the 25–34 age bracket and nearly 10% in the 18–24 age bracket. This demographic profile suggests a high level of digital

Table 4.1 Demographic profile of respondents

Variable	Category	Frequency	Percentage: %
Age group	18–24 years	10	9.4%
	25–34 years	59	55.7%
	35–44 years	27	25.5%
	45–54 years	8	7.5%
	55 years and above	2	1.9%
Gender	Male	46	43.4%
	Female	60	56.6%
Education level	High school/certificate	25	23.6%
	Diploma	31	29.2%
	Bachelor's degree	40	37.7%
	Master's degree	8	7.5%
	Doctorate	2	1.9%
Organisational role	Project manager	9	8.5%
	Team leader	26	24.5%
	Logistics coordinator	14	13.2%
	Operation staff	28	26.4%
	Other	29	27.4%
Virtual PM adoption	Yes	77	72.6%
	No	29	27.4%
Primary virtual tools	Microsoft Teams	58	54.7%
	Zoom	29	27.4%
	Skype	2	1.9%
	Other	17	16.0%

literacy, which is essential, given that leadership style and communication effectiveness rely entirely on digital tools to achieve project success.

Educational and professional backgrounds further reinforce the validity of the findings. Most respondents are well-educated, with 37.7% holding Bachelor's degrees and 29.2% holding diplomas, ensuring they possess the technical proficiency needed to utilise platforms such as SAP TM and Microsoft Project. While the sample includes project managers (8.5%) and team leaders (24.5%), the largest segments are operation staff and other support roles (combined 53.8%), providing a grounded, operational perspective on how team collaboration and risk management directly and indirectly influence project outcomes. Furthermore, with 72.6% of organisations already implementing virtual project management, the reliance on tools such as Microsoft Teams (54.7%) and Zoom (27.4%) underscores the industry's shift toward the integrated technological framework proposed in the study (Tajudeen *et al.*, 2025).

4.3.3 Factor analysis

Factor analysis was conducted to evaluate the underlying structure of the research constructs and confirm that the measured items accurately align with the study's theoretical framework. This statistical technique reduced the dataset to meaningful dimensions, clarifying the relationships between the five independent variables (IV) (leadership style, virtual communication, stakeholder management, team collaboration and project risk management), the mediating variable (MV) (project management tools (PMT)) and the dependent variable (DV) (project success). Table 4.2 shows the summary of the factor analysis.

The analysis confirmed excellent sampling adequacy across all constructs, with Kaiser-Meyer-Olkin (KMO) values ranging from 0.864 to 0.901, well above the acceptable threshold. Furthermore, Bartlett's test of sphericity was highly significant ($p < 0.001$) for every variable, validating that the correlation matrices were suitable for factor analysis.

For each construct, a single component was extracted, with all items demonstrating strong factor loadings (ranging from 0.793 to 0.920). These high loadings indicate that the items

Table 4.2 Factor analysis summary

Construct	KMO measure	Bartlett's sig., *p*	Loading range	No. of items
Team collaboration (IV1)	0.890	<0.001	0.822–0.892	5
Leadership style (IV2)	0.864	<0.001	0.844–0.899	5
Virtual communication (IV3)	0.896	<0.001	0.793–0.879	5
Stakeholder management (IV4)	0.875	<0.001	0.872–0.920	5
Project risk management (IV5)	0.901	<0.001	0.863–0.910	5
Project management tools (MV)	0.883	<0.001	0.800–0.903	5
Project success (DV)	0.873	<0.001	0.859–0.889	5

consistently represent their respective constructs. Specifically, the results reinforce the importance of these factors in logistics: effective communication and collaboration tools reduce misunderstandings and support real-time coordination. At the same time, robust risk and stakeholder management frameworks are essential for addressing the uncertainties of geographically dispersed virtual projects.

4.3.4 Reliability analysis

Reliability analysis was conducted to assess the internal consistency of the measurement instruments, ensuring that the items within each construct yield stable, dependable results. In this study, Cronbach's alpha (α) served as the primary metric for assessing reliability, with a value of 0.70 or higher considered acceptable for internal consistency (Izah *et al.*, 2023).

Table 4.3 presents the summary of the reliability analysis for this study. The analysis revealed excellent reliability across all research constructs, with Cronbach's alphas ranging from 0.907 to 0.940, well above the required benchmarks. This high level of consistency is critical given the study's focus on how project management tools mediate the relationship between managerial factors and project outcomes. Specifically, the independent variables (IVs) for team collaboration, leadership style, virtual communication, stakeholder management and project risk management all demonstrated very strong internal consistency.

Furthermore, the mediating variable (project management tools (PMT)) and the dependent variable (project success) yielded alpha values of 0.915 and 0.923, respectively, confirming that the instruments are robust enough to capture the complex full and partial mediation effects proposed in the revised hypotheses. Across all scales, corrected item-total correlations were consistently high, and 'alpha if item deleted' tests indicated that removing any single item would not significantly improve the overall scale's reliability. Collectively, these findings provide a highly stable foundation for the subsequent hypothesis testing and validation of the conceptual framework.

Table 4.3 Reliability analysis summary

Variable	Construct	No. of items	Cronbach's alpha, α	Internal consistency
Independent (IV1)	Team collaboration	5	0.917	Excellent
Independent (IV2)	Leadership style	5	0.924	Excellent
Independent (IV3)	Virtual communication Effectiveness	5	0.907	Excellent
Independent (IV4)	Stakeholder management	5	0.940	Excellent
Independent (IV5)	Project risk management	5	0.934	Excellent
Mediating (MV)	Project management tools	5	0.915	Excellent
Dependent (DV)	Project success	5	0.923	Excellent

4.3.5 Hypothesis testing

The hypothesis testing for this study was conducted to evaluate how five key managerial factors, namely leadership style, virtual communication, stakeholder management, team collaboration and project risk management, influence the success of virtual project management in Malaysian logistics operations. To provide a more robust analysis, the study utilised the PROCESS Macro (Model 4) to examine the mediating role of project management tools (such as SAP TM and Microsoft Project).

Tables 4.4 and 4.5 and the following discussion present the statistical results and the nature of the relationships found between these variables. Table 4.4 summarises the total, direct and indirect effects and summarises the regression coefficients (β) for each pathway. Mediation is confirmed when the bootstrapped confidence interval (CI) for the indirect effect excludes zero.

Table 4.4 Summary of total, direct and indirect effects

Hypothesis	Independent variable (IV)	Total effect (β)	Direct effect (β)	Indirect effect (β)	Bootstrapped 95% CI (indirect)
H1	Leadership style	0.5898***	0.0490 (ns)	0.5408	[0.3489, 0.6786]
H2	Virtual communication	0.7559***	0.1106 (ns)	0.6453	[0.4915, 0.8097]
H3	Stakeholder management	0.7593***	0.1754*	0.5839	[0.3876, 0.7765]
H4	Team collaboration	0.8463***	0.2886**	0.5577	[0.3505, 0.7702]
H5	Project risk management	0.8460***	0.3074***	0.5385	[0.3248, 0.7611]

Note: [ ] = bootstrapped 95% confidence interval; *** $p < 0.001$; ** $p < 0.01$; * $p < 0.05$; (ns) = not significant.

Table 4.5 Final results of hypothesis testing and mediation type

Hypothesis	Relationship proposed	Outcome	Mediation type
H1	Leadership style project success	Supported	Full mediation
H2	Virtual communication project success	Supported	Full mediation
H3	Stakeholder management project success	Supported	Partial mediation
H4	Team collaboration project success	Supported	Partial mediation
H5	Project risk management project success	Supported	Partial mediation

Based on the statistical findings, all five hypotheses were supported, although they exhibited different patterns of mediation through project management tools, as shown in Table 4.5.

The results of this research reveal a critical insight into virtual logistics by demonstrating the dominance of full mediation in the relationship between leadership style (H1) and virtual communication effectiveness (H2). Specifically, the statistical analysis showed that the direct effect of leadership on project success was non-significant ($\beta = 0.0490, p = 0.3702$), suggesting that in a remote environment, a leader's ability to motivate or align a team only translates into success when it is channelled through the effective use of project management tools. Similarly, once the mediator is considered, virtual communication effectiveness does not independently influence success, confirming that the 'human' element of information exchange requires a robust 'technological bridge' to be effective in geographically dispersed settings.

In contrast, the study identified partial mediation by stakeholder management (H3), team collaboration (H4) and project risk management (H5). This indicates that, while digital platforms significantly enhance these three factors, they also retain a direct, independent impact on project success. For example, team collaboration drives success through both direct interpersonal coordination and the enabling function of shared digital resources. Notably, project risk management was identified as an exceptionally strong driver ($\beta_{total} = 0.8460$), underscoring the need for proactive mitigation as both a standalone managerial practice and a process supported by real-time monitoring and predictive tools.

These findings carry profound strategic implications for the Malaysian logistics industry, illustrating that project success is an integrated outcome of managerial behaviours and technological proficiency. Because every variable tested demonstrated significant total effects, organisations such as DHL Malaysia and City-Link cannot afford to focus solely on 'soft skills' like leadership and collaboration in isolation. Instead, they must ensure these skills are deeply integrated into advanced platforms such as SAP Transportation Management and Microsoft Project to manage the complexities of dispersed supply chains. This holistic approach, combining human expertise with digital mediation, is the primary driver of timely completion, cost control, quality performance and stakeholder satisfaction in a rapidly digitalising industry.

4.4. Practical insights

The research findings offer a detailed roadmap for the Malaysian logistics sector, emphasising that success in a virtual environment depends on a combination of 'soft' managerial skills and the 'hard' implementation of digital platforms. The study shows that success is not just the result of traditional oversight but also a sociotechnical synergy in which management behaviours are shaped by and bolstered through sophisticated project management tools.

One of the primary insights for practitioners is the need for a transition toward technology-mediated leadership. Because project management tools fully mediated leadership style (H1), managers must move beyond traditional command-and-control styles to become technologically competent facilitators. In a remote environment, a leader's ability to inspire or align a team translates into measurable success only when it is channelled through effective digital coordination systems, such as SAP Transportation Management or Microsoft Project.

Leadership development in logistics firms should therefore prioritise training in virtual team coordination and the use of technology to maintain transparency and accountability.

Similarly, the full mediation of virtual communication effectiveness (H2) suggests that the human element of information exchange requires a robust 'technological bridge' to be effective in geographically dispersed settings. Applying media richness theory, logistics organisations should ensure that communication practices are matched to task complexity. For routine updates, lean digital media may suffice, but for complex decision-making and resolving ambiguities, managers must leverage high-richness tools like video conferencing to minimise misunderstandings and delays. Success depends on ensuring these communication methods are deeply integrated with task-tracking platforms so that information flows seamlessly across time zones.

For factors such as stakeholder management (H3), team collaboration (H4) and project risk management (H5), the study identified a partial mediation pattern, indicating that these drivers maintain a significant direct and independent impact on project success. Organisations like DHL Malaysia and City-Link should prioritise integrated stakeholder engagement strategies that utilise shared digital dashboards and automated reporting to foster transparency and trust. Fostering a collaborative culture likewise requires a dual commitment to building interpersonal trust and the strategic use of shared digital resources that enable real-time file sharing and collective problem-solving among team members.

Finally, as project risk management emerged as the strongest predictor of success ($\beta = 0.8460$), logistics firms must evolve from a reactive to a proactive stance. The practical implication is a move toward predictive analytics and real-time monitoring systems that enable early detection of supply chain disruptions or system failures. By integrating risk mitigation frameworks with adaptive leadership and clear communication, virtual project teams can navigate the uncertainties inherent in global supply chains. Ultimately, by ensuring every managerial action is funnelled through sophisticated digital tools, Malaysian logistics providers can achieve superior results in timely delivery, cost control and overall stakeholder satisfaction

4.5. Conclusion and recommendations

The demographic characteristics of the 106 respondents provide a robust foundation for interpreting these findings, as the sample was drawn directly from professionals involved in virtual project management within the Malaysian logistics sector. The dominance of a young, well-educated workforce with 55.7% aged between 25 and 34 and over 37% holding Bachelor's degrees suggests a high level of digital literacy essential for navigating the technological mediation required for project success. The high response rate and the representation of various roles, from operations staff to project managers, ensure that the results reflect a grounded, credible operational perspective on virtual project performance.

Factor analysis successfully validated the study's core constructs by reducing the dataset into meaningful dimensions and confirming strong correlations between the observed variables and their respective factors. The identified constructs – leadership style, virtual communication effectiveness, stakeholder management, team collaboration and project risk management – demonstrated high KMO values (0.864 to 0.901) and significant Bartlett's test results, providing a solid statistical basis for the theoretical framework (Zhang *et al.*, 2024). Furthermore, reliability testing confirmed the measurement instruments' internal consistency, with Cronbach's alphas for all constructs ranging from 0.907 to 0.940, well above the 0.70 threshold. This reinforces the data's reliability and the robustness of the conclusions regarding the drivers of virtual project success.

The results of the hypothesis testing highlight a sophisticated interplay between managerial factors and technological mediation. The study found that leadership style (H1) and virtual communication effectiveness (H2) do not have significant direct effects on project success but are fully mediated by project management tools. This indicates that, in a virtual environment, leadership and communication translate into success only when they facilitate the effective utilisation of digital coordination systems. Conversely, stakeholder management (H3), team collaboration (H4) and project risk management (H5) exhibited partial mediation, meaning they influence project outcomes both directly through managerial excellence and indirectly through enhanced performance enabled by integrated digital platforms. Project risk management emerged as a particularly critical driver, demonstrating the strongest total effect on success.

Based on these findings, several practical recommendations are proposed for logistics organisations. First, firms should prioritise adaptive leadership training that teaches managers to lead virtual teams specifically by leveraging technology to enhance transparency and accountability. Second, investment in communication tools should be paired with strategies that match media richness to task complexity to minimise misunderstandings. Third, stakeholder engagement should be modernised by using integrated digital dashboards and real-time feedback mechanisms to maintain alignment across dispersed networks. Fourth, team collaboration must be supported by deliberate virtual trust-building initiatives and shared digital resources that facilitate seamless coordination. Finally, risk management frameworks should be updated to include predictive analytics and real-time monitoring tools to identify and mitigate threats early in the project lifecycle.

Overall, the study demonstrates that virtual project success in the Malaysian logistics industry is an integrated outcome of managerial behaviours and technological proficiency. By recognising the specific full and partial mediation effects of project management tools, organisations can better optimise their resources. These insights offer a practical framework for logistics firms to improve their virtual project performance and remain competitive in an increasingly digitalised global operating environment.

4.5.1 Implications and strategic considerations

The findings of this study have significant implications for the Malaysian logistics sector, which is rapidly adopting virtual project management to navigate the pressures of globalisation and technological advancement. A primary strategic consideration is that adaptive virtual leadership must be treated as a core organisational competency. For firms such as DHL Malaysia and City-Link, success in dynamic environments requires leaders with high emotional intelligence, proficiency in coordinating virtual teams and the ability to integrate advanced digital systems into daily operations seamlessly.

Furthermore, the study highlights a critical need for sustained investment in sophisticated communication platforms that facilitate real-time coordination across diverse geographic locations and time zones. Organisations should prioritise tools such as Microsoft Teams, Zoom or industry-specific solutions that unify communication and project management functions to ensure timely information sharing and efficient decision-making.

Stakeholder management must also be modernised to reflect the complexity of logistics networks involving suppliers, clients and regulators. The findings suggest that implementing digitally populated dashboards, automated reporting and structured feedback channels is essential for maintaining alignment and stakeholder trust throughout the project lifecycle. Additionally, fostering a collaborative culture through structured virtual team-building and the strategic use of cloud-based tools such as SAP Transportation Management and Microsoft Project will enhance operational resilience. Finally, firms are encouraged to adopt predictive analytics and real-time monitoring systems to identify and mitigate risks early, thereby maintaining a competitive edge in the evolving digital landscape.

4.5.2 Limitations of the study

Despite providing valuable insights, this research is subject to several limitations. First, the study relies on self-reported survey data, which may be susceptible to social desirability bias or recall inaccuracies. Respondents might have provided answers based on perceived organisational expectations rather than their actual day-to-day experiences, potentially skewing the results.

Second, the sample size and scope restrict the broader generalisability of the findings. While 106 respondents provided sufficient statistical power for the current analysis, the focus was primarily on large, well-established organisations like DHL and City-Link. Consequently, the results may not accurately reflect the challenges faced by small and medium-sized enterprises (SMEs) or regional logistics operators with varying resource constraints.

Third, the research is primarily quantitative, which limits the ability to explore deeper contextual and behavioural factors. The absence of qualitative data means that the nuanced 'lived experiences' of virtual team members regarding leadership and collaboration dynamics were not fully captured. Finally, the study did not examine external variables such as shifting economic conditions, regulatory frameworks or disparities in regional technological infrastructure, all of which can significantly influence virtual project outcomes.

4.5.3 Recommendations for future research

To address the identified limitations, future research should consider the following avenues.

- *Mixed-methods approach*: combining quantitative surveys with qualitative techniques such as semi-structured interviews, case studies or ethnographic observations would provide a richer, more contextualised understanding of the practical challenges faced by virtual project managers.
- *Expanded sample and sector scope*: future studies should broaden their scope to include a more diverse range of organisations, particularly SMEs, to improve external

validity. Additionally, replicating this research in other sectors, such as manufacturing, healthcare or education, would enable valuable cross-sectoral comparisons of virtual project success drivers.

- *Longitudinal research designs*: implementing studies that track virtual project practices over multiple project cycles would help clarify how leadership styles, team dynamics and technological proficiency evolve as organisations gain experience with digital tools.
- *Contextual and external factors*: investigating the impact of national digital transformation policies, regulatory frameworks and varied economic conditions would provide a more holistic view of virtual project performance.
- *Emerging technologies*: further investigation is needed into the role of artificial intelligence (AI), machine learning and blockchain. Future research could specifically assess how these innovations enhance real-time decision-making, optimise resource allocation and strengthen risk management within virtual logistics environments.

REFERENCES

Alkoud S, Zainudin D and Sarif SM (2023) Challenges, barriers, and obstacles facing virtual teams: a conceptual study. *Sciences* **13(4)**: 1473–1487.

Batırlık SN, Gencer YG and Akkucuk U (2022) Development of the Global Virtual Team Leadership Scale (GVTLS) in multinational companies. *Sustainability* **14(2)**: 1038.

Cichosz M, Wallenburg CM and Knemeyer AM (2020) Digital transformation at logistics service providers: barriers, success factors and leading practices. *The International Journal of Logistics Management* **31(2)**: 209–238.

Izah SC, Sylva L and Hait M (2023) Cronbach's alpha: a cornerstone in ensuring reliability and validity in environmental health assessment. *ES Energy and Environment* **23**: 1057.

Powell DJ (2016) *Social Knowledge Creation and Emergent Digital Research Infrastructure for Early Modern Studies*. PhD thesis. University of Victoria, Victoria, BC, Canada.

Smith VK (2019) *The Challenges of Virtual Leadership Affecting Project Delivery in the Public Sector*. Walden University, Minneapolis, MN, USA.

Spulbar LF (2025) Legal frameworks for AI-driven markets and their challenges and opportunities in the digital economy. *Revista de Ştiinţe Politice. Revue des Sciences Politiques* **86**: 288–304.

Tajudeen FP, Moghavvemi S, Thirumoorthi T, Phoong SW and Bahri ENBA (2025) *Digital Transformation of Malaysian Small and Medium Enterprises*. Emerald, Bingley, UK.

Zhang C and Dhaliwal J (2009) An investigation of resource-based and institutional theoretic factors in technology adoption for operations and supply chain management. *International Journal of Production Economics* **120**: 252–269.

Zhang Z, Sangsawang T, Vipahasna K and Pigultong M (2024) A mixed-methods data approach integrating importance-performance analysis (IPA) and Kaiser-Meyer-Olkin (KMO) in applied talent cultivation. *Journal of Applied Data Sciences* **5(1)**: 256–267.

Mohammed Reyasudin Basir Khan
ISBN 978-1-83742-803-8
https://doi.org/10.1108/978-1-83742-802-120261008
Emerald Publishing Limited: All rights reserved

Chapter 5
Leading digital transformation: the strategic role of managerial capabilities in 5G and cloud adoption

Kuah Shen Hsing and Mohammed Reyasudin Basir Khan

Tun Razak Graduate School, Universiti Tun Abdul Razak (UNIRAZAK), Kuala Lumpur, Malaysia

5.1. Introduction

Malaysia is undergoing a rapid digital transformation, with cloud computing and 5G technology serving as the fundamental pillars for modernising business models and service delivery. A landmark event in this transition is the planned establishment of a local Amazon Web Services (AWS) cloud region, which signifies Malaysia's growing infrastructure readiness and a sustained national demand for scalable digital services (Mohamed *et al.*, 2025). These technologies offer substantial organisational incentives, including cost optimisation, operational agility and reduced reliance on physical hardware. While high-regulation sectors like finance utilise these capabilities for service continuity, the adoption of such solutions is also reshaping healthcare, retail and manufacturing. This shift is further propelled by national policies such as MyDigital and the Malaysia Digital Economy Blueprint, which aim to strengthen digital governance and cybersecurity.

Despite these benefits, Malaysian businesses face significant hurdles in translating technological promise into operational reality. Many organisations focus heavily on technical procurement while underestimating the managerial and human factors necessary for success. This is particularly evident in the 'digital divide' affecting small and medium-sized enterprises (SMEs), which often lack the resources and expertise found in large corporations like PETRONAS (Asiaei and Ab. Rahim, 2019). Consequently, the application of structured project management practices has become a decisive factor in determining whether technology adoption yields measurable benefits. To ensure long-term viability, it is increasingly important to integrate sustainability into these management frameworks (Silvius and Schipper, 2014).

The primary objective of the study presented in this chapter is to evaluate how specific project management practices contribute to the success of adoption. This research focuses on four critical dimensions.

- *Project planning*: acting as the link between strategy and execution, effective planning ensures alignment with strategic objectives and proper resource allocation (Dvir *et al.*, 2003).
- *Risk management*: essential for navigating uncertainties like cybersecurity and vendor dependency, risk management must be dynamic across the project lifecycle (Cervone, 2006). This involves identifying systemic risk and developing response strategies (Dvir *et al.*, 2003).
- *Stakeholder engagement*: grounded in stakeholder theory, this involves actively addressing the needs and influence of all affected parties to reduce resistance and build trust (Freeman, 2010).
- *Change management*: this focuses on communication, training and employee readiness to ensure smooth transitions. The awareness, desire, knowledge, ability and reinforcement (ADKAR) model emphasises that leadership and reinforcement are essential for these change processes (Hiatt, 2006).

Theoretically, this research is underpinned by Rogers' diffusion of innovations theory (Rogers *et al.*, 2014), which explains how innovations spread based on their relative advantage and compatibility. It also utilises the task-technology fit model (Goodhue and Thompson, 1995) to explain how value is created when system capabilities align with operational requirements. Additionally, the study incorporates the technology acceptance model (TAM), in which perceived usefulness and ease of use (Davis, 1989) and external influences (Venkatesh and Davis, 2000) determine organisational acceptance. By employing tailored survey principles (Dillman, 2011) to collect data from 200 respondents, the study ensures high response quality across diverse organisational strata. It provides practical, evidence-based recommendations for Malaysia's digital future.

The reviewed literature supports a conceptual framework in which project planning, risk management, stakeholder engagement and change management are expected to influence the adoption of 5G and cloud technologies positively. The conceptual framework for this study is shown in Figure 5.1. The hypothesised relationships reflect the logic that structured management practices address both the technical execution and the human factors essential for sustained technology use. This framework is evaluated through four primary hypotheses.

- H1: perceived risk management has a positive effect on technology adoption. This hypothesis posits that organisations that systematically identify, communicate and respond to uncertainties, such as cybersecurity threats and vendor dependencies, are more likely to achieve successful implementation. Risk management serves as a value-creating capability by enabling secure scaling and stable performance amid shifting project lifecycles.
- H2: perceived project planning has a positive effect on technology adoption. The logic behind this hypothesis is that effective planning connects organisational strategy with execution by clearly defining scope, allocating resources and managing

Figure 5.1 Conceptual framework

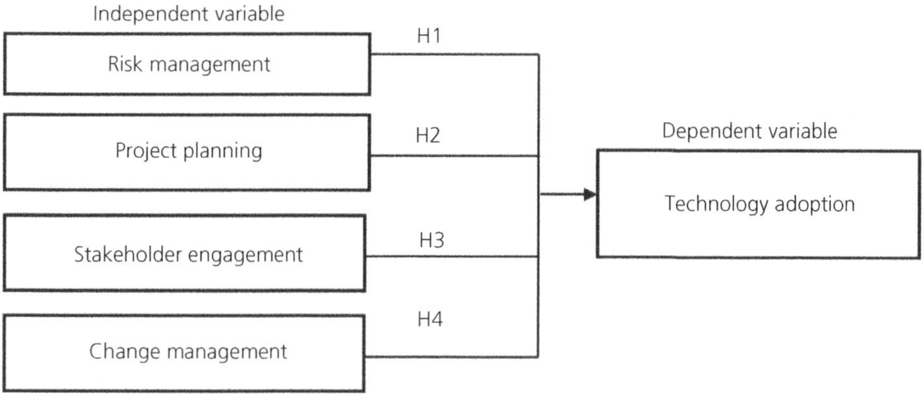

timelines. High-quality planning is essential for converting technological potential into stable organisational performance and ensuring alignment with strategic objectives.

- H3: perceived stakeholder engagement has a positive effect on technology adoption. Grounded in stakeholder theory, this hypothesis suggests that actively addressing the needs and influence of all affected parties reduces resistance and builds trust. Effective engagement facilitates implementation and acceptance by ensuring that communication is inclusive and responsive to the different interests of regulators, providers and internal teams.
- H4: perceived change management has a positive effect on technology adoption. This hypothesis emphasises that managing transitions through leadership, communication and training is vital for ensuring employee readiness and acceptance. Change management is identified as having the strongest relationship with adoption success, as it directly shapes how individuals transition into new systems and ensures that the technology is used effectively over the long term.

Overall, these hypotheses are grounded in established theoretical perspectives, such as Rogers' diffusion of innovations theory and the technology acceptance model, which emphasise that adoption is a sociotechnical process rather than a purely technical decision. The study ultimately confirms that these four practices are critical, interrelated determinants that drive effective technology integration within Malaysian businesses.

Recent empirical studies provide evidence of both opportunities and barriers in Malaysia (Thakur *et al.*, 2024). Research involving SME employees suggests awareness of 5G benefits but also concerns about cost, security risks, the learning curve and uncertainty about adoption outcomes, which can delay adoption until others demonstrate success (Odukoya, 2024). Regional research on cloud adoption in Malaysian enterprises highlights relatively low adoption rates and persistent concerns about security and reliability, while also emphasising that top management support is a critical enabler (Sheikh Mokhtar and Mahomed, 2025).

Studies in sectors such as construction indicate that cloud computing can positively influence time, cost and coordination outcomes, and research on cloud-based IT projects identifies technical team capacity and senior management support as critical success factors, along with soft skills and limitations in contract management (Jaafar, 2024; Waqar *et al.*, 2023).

The primary objective of this study is to examine the current state of 5G and cloud technology adoption across Malaysian business sectors and to evaluate how specific project management practices, such as planning, risk management, stakeholder engagement and change management, contribute to implementation success. Specifically, the research analyses how these practices mitigate resistance and align technology with operational needs within Malaysia's evolving digital economy. Furthermore, the study considers the impact of firm size on adoption capacity, highlighting the unique constraints faced by SMEs compared to large organisations. Ultimately, this research seeks to provide practical, evidence-based recommendations for businesses, policy makers and industry associations to foster an effective, inclusive and sustainable digital transformation in Malaysia.

5.2. Methodology

5.2.1 Research design

The survey was carefully developed, with concise, relevant questions that focus on project management, change management and technology adoption. It was administered electronically to reach respondents across the targeted business sectors through professional networking platforms and direct email invitations to participating organisations. Dillman's tailored design method (Dillman, 2011) was applied to improve response rates and enhance response quality. The online format also supports confidentiality and convenience, allowing participants to respond at a time that suits them and encouraging more honest feedback, particularly on sensitive issues related to organisational change and technology adoption. Overall, this approach aims to provide a clear picture of 5G and cloud adoption among Malaysian businesses and to support rigorous analysis in the later stages of the study.

5.2.2 Population and sampling procedures

This study used stratified random sampling to include both SMEs and large corporations, consistent with mixed methods sampling guidance that supports stratification to ensure adequate representation across population subgroups. The sample was stratified by business size and digitisation stage, including organisations that are contemplating, implementing or have completed digitisation initiatives, as this approach can improve representativeness when examining diverse groups. Within each stratum, organisations were selected by random sampling to minimise bias and enhance the generalisability of the findings. Data were collected through an online survey distributed to the selected organisations, following tailored survey principles that emphasise adapting procedures to population segments to improve response quality and effectiveness (Dillman, 2011).

5.2.3 Data collection method

The data collection method was a structured survey tailored to gather detailed information from Malaysian businesses on the adoption of 5G and cloud technologies. This method is grounded in quantitative research methodology, which enables the collection of measurable, statistically analysable data.

5.2.4 Operationalisation and measurement

Risk management was measured by evaluating the organisation's capability to identify and mitigate risks associated with technology adoption, guided by the risk management framework of Dvir *et al.* (2003), which emphasises risk identification, risk communication and response strategies. Project planning was assessed based on alignment with strategic objectives, adequacy of resource allocation and effectiveness of timeline management, drawing on Dvir *et al.* (2003), who highlight planning alignment as a key factor in project success. Stakeholder engagement was measured by examining the extent and effectiveness of stakeholder involvement throughout the technology adoption process, informed by Freeman's stakeholder theory (Freeman, 2010), which stresses the importance of identifying and actively engaging affected parties. Change management was evaluated through the organisation's management of technology transition, including communication, training and resistance management, based on Lewin's model of unfreezing, changing and refreezing and supported by Kotter's eight-step approach to organisational change (Santos *et al.*, 2023; Engmann *et al.*, 2024).

Reliability was evaluated using Cronbach's alpha (α) to assess internal consistency; values of at least 0.70 indicate acceptable reliability, and key items may also be examined through test–retest procedures. Validity was strengthened through content validation based on an extensive literature review and expert review. Construct validity was assessed using factor analysis to confirm that items load appropriately within their intended constructs. Criterion validity was examined by correlating survey measures with relevant external benchmarks. A pilot study with a small portion of the target population was conducted before the main survey to refine the instrument, identify ambiguities and improve measurement accuracy and consistency.

5.3. Results and analysis

5.3.1 Demographic profile

The demographic profile summary of 200 respondents involved in this study is shown in Table 5.1. The summary provides a comprehensive look at the workforce currently navigating 5G and cloud technology adoption in Malaysia.

Table 5.1 Demographic profile summary (continued on next page)

Category	Sub-category	Frequency (n=200)	Percentage: %
Age	25 and under	20	10.0
	26–35	100	50.0
	36–45	47	23.5
	46–55	18	9.0
	56 and over	15	7.5
Cloud usage	Users	170	85.0
	Non-users	30	15.0
Position	Team leader and above	~120	~60.0
	Executive and below	~80	~40.0

Table 5.1 Continued

Category	Sub-category	Frequency (n=200)	Percentage: %
Organisation size	Large	92	46.0
	Small	74	37.0
	Other (medium/not specified)	34	17.0
Top industries	Manufacturing, finance, healthcare, education, construction	>20 per sector	

The sample is heavily concentrated in the younger-to-middle-aged age bracket, with 73.5% of respondents aged 26–45. This reflects the typical composition of Malaysia's active workforce and suggests that the data come from individuals who are likely at the forefront of digital tool usage and implementation.

An overwhelming 85% of respondents already use cloud technology. This high adoption rate suggests that participants are familiar with the benefits of digital transformation, including operational efficiency, cost reduction and scalability. This familiarity adds weight to their perspectives on the challenges and successes of implementing 5G and cloud services.

Approximately 60% of respondents hold team leader or higher roles. This is significant because the study focuses on project management practices; the majority of respondents with decision-making responsibilities ensure that the feedback reflects the views of those who oversee implementation strategies and manage organisational change.

The study successfully captured data from both large organisations (46%) and small organisations (37%). This balance is crucial for understanding the 'digital divide' in Malaysia, where SMEs often face more severe resource constraints and skill shortages than large corporations like PETRONAS, which can afford private 5G networks.

The high response rates from the manufacturing, finance, healthcare, education and construction sectors (each over 10%) align with areas where 5G and cloud technology are most transformative. For instance, the finance sector utilises cloud-native capabilities for agility, while manufacturing leverages these technologies for automation and real-time analytics.

5.3.2 Reliability analysis

Establishing reliability is critical to ensure that survey instruments measure constructs consistently. Reliability coefficients indicate the extent to which observed score variance reflects true score variance, and can be assessed through approaches such as test–retest, interrater, parallel forms and internal consistency. Cronbach's alpha (α) is widely used to assess internal consistency among Likert-scale items. Cronbach's alpha is calculated based on the number of items, the average covariance between item pairs and the average item variance. As a general guideline, alpha values of 0.70 or higher are considered acceptable, although higher thresholds of 0.90 to 0.95 may be preferred for greater confidence. As indicated in Table 5.2, for example, alpha values of 0.73 for risk management, 0.89 for project planning and 0.91 for both stakeholder engagement and change management indicate acceptable to strong reliability, suggesting that the constructs are measured consistently.

5.3.3 Normality analysis

Skewness describes the degree of asymmetry in a distribution around its mean. Values near zero indicate approximate symmetry, positive skewness reflects a longer right tail and negative skewness reflects a longer left tail. Kurtosis indicates the relative heaviness of a distribution's tails compared to a normal distribution. Values near zero suggest similar tail behaviour, positive kurtosis indicates heavier tails and negative kurtosis indicates lighter tails. As shown in Table 5.3, these statistics describe the distribution's shape but, on their own, do not confirm normality.

Table 5.2 Reliability assessment

Type of variable	Variable	α	No. of items
Dependent	Technology adoption	0.93	5
Independent	Risk management	0.73	5
	Project planning	0.89	5
	Stakeholder engagement	0.91	5
	Change management	0.91	5

Table 5.3 Normality analysis

	Descriptive statistics								
	n	Min	Max	Mean	SD	Skewness		Kurtosis	
	Stat	Stat	Stat	Stat	Stat	Statistic	Std. error	Statistic	Std. error
Risk management	200	1.00	5.00	3.4160	0.82361	–0.449	0.172	0.195	0.342
Project planning	200	1.00	5.00	3.5440	0.90740	–0.571	0.172	–0.005	0.342
Stakeholder engagement	200	1.00	5.00	3.6130	0.91401	–0.462	0.172	–0.136	0.342
Change management	200	1.00	5.00	3.5400	0.90559	–0.444	0.172	–0.058	0.342
Adoption of 5G and cloud technology	200	1.00	5.00	3.4370	1.01091	–0.577	0.172	–0.064	0.342
Valid *n* (listwise)	200								

Figure 5.2 presents the normal Q-Q plots for risk management, project planning, stakeholder engagement, change management and technology adoption to assess whether the data for each construct follows a normal distribution. In all plots, the observed data points align closely with the diagonal reference line, indicating that the variables are approximately normally distributed. Although there are slight deviations at the lower and upper tails in some plots, these deviations are minimal and do not suggest serious violations of normality. Overall, the graphical results confirm that the assumption of normality is reasonably satisfied for these variables, supporting the suitability of parametric analyses, such as correlation and regression, for the study.

Figure 5.2 Q-Q plots

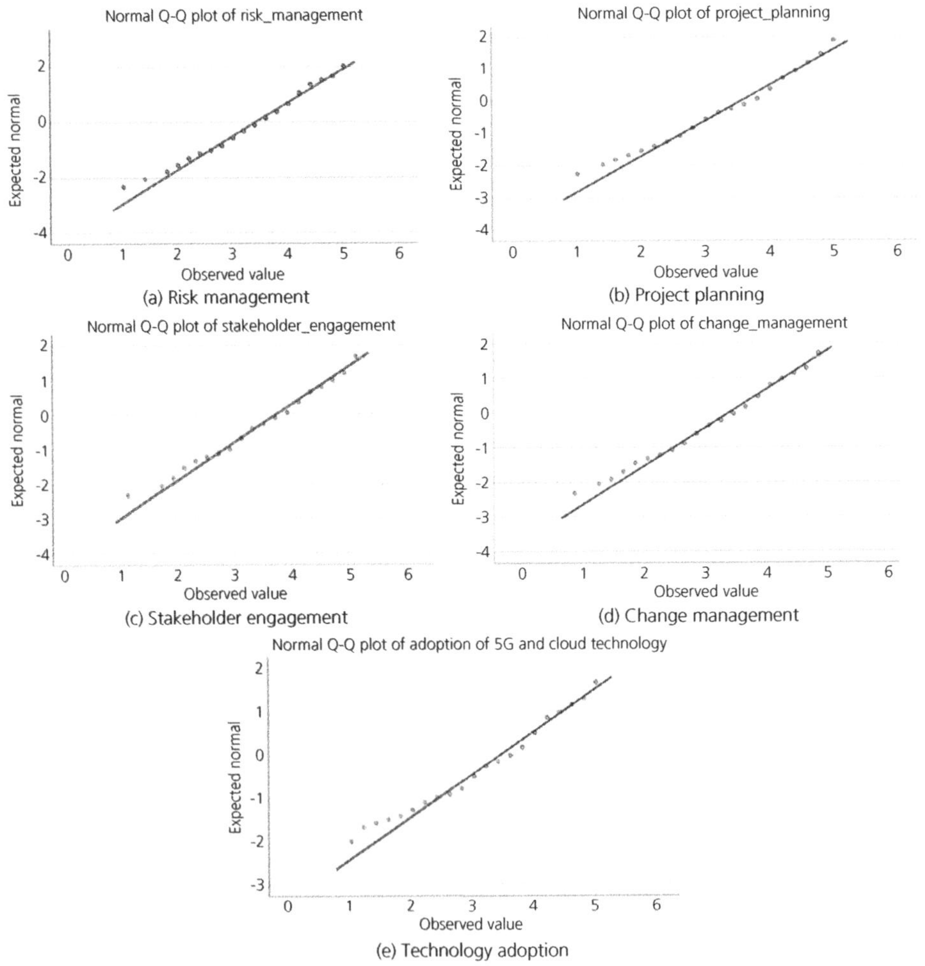

5.3.4 Descriptive analysis

Table 5.4 presents the descriptive statistics for the key study variables based on 200 respondents. The mean scores for all constructs, which are technology adoption (3.4370), risk management (3.4160), project planning (3.5440), stakeholder engagement (3.6130) and change management (3.5400), indicate moderate to relatively high levels of agreement among respondents. Stakeholder engagement recorded the highest mean, suggesting it is perceived as the most strongly practised factor, while risk management shows the lowest mean, although still within a moderate range. The standard deviation (SD) values, ranging from 0.82361 to 1.01091, indicate acceptable variability in responses, with technology adoption exhibiting slightly greater dispersion than the other variables. Overall, the results suggest consistent perceptions across respondents and provide a sound basis for further inferential statistical analysis.

5.3.5 Pearson correlation analysis

Correlation analysis is a statistical method used to assess the strength and direction of association between two continuous variables. It identifies statistical relationships rather than causal effects, as unmeasured factors may influence observed associations. A correlation indicates that variation in one variable is systematically related to variation in another, with the direction of the relationship determined by the sign of the correlation coefficient.

The Pearson correlation analysis, conducted in the statistical package for the social sciences (SPSS), examines relationships between factors using a dataset derived from an online survey. This analysis categorises the linear relationships into three groups: weak, moderate and strong. The categorisation is based on the value of the correlation coefficient: values ranging from 0.3 to 0.5 are classified as weak, from 0.5 to 0.7 as moderate, and from 0.7 to 1.0 as strong. A value below 0.3 indicates no significant relationship.

The analysis in Table 5.5 and summary in Table 5.6 indicate that the factors risk management, project planning, stakeholder engagement, change management and technology adoption are significantly interrelated. Strong correlations in several pairings suggest that effectiveness in one aspect is often mirrored by effectiveness in another, highlighting the interdependence of these factors within the dataset.

Table 5.4 Descriptive statistics summary

	Descriptive statistics		
	Mean	SD	*n*
Technology adoption	3.4370	1.01091	200
Risk management	3.4160	0.82361	200
Project planning	3.5440	0.90740	200
Stakeholder engagement	3.6130	0.91401	200
Change management	3.5400	0.90559	200

Table 5.5 Pearson correlation analysis

		Correlation				
		Risk management	Project planning	Stakeholder engagement	Change management	Technology adoption
Risk management	Pearson correlation	1	0.772[a]	0.679[a]	0.643[a]	0.648[a]
	Sig. (1-tailed)		<0.001	<0.001	<0.001	<0.001
	n	200	200	200	200	200
Project planning	Pearson correlation	0.772[a]	1	0.817[a]	0.754[a]	0.698[a]
	Sig. (1-tailed)	<0.001		<0.001	<0.001	<0.001
	n	200	200	200	200	200
Stakeholder engagement	Pearson correlation	0.679[a]	0.817[a]	1	0.827[a]	0.716[a]
	Sig. (1-tailed)	<0.001	<0.001		<0.001	<0.001
	n	200	200	200	200	200
Change management	Pearson correlation	0.643[a]	0.754[a]	0.827[a]	1	0.732[a]
	Sig. (1-tailed)	<0.001	<0.001	<0.001		<0.001
	n	200	200	200	200	200
Technology adoption	Pearson correlation	0.648[a]	0.698[a]	0.716[a]	0.732[a]	1
	Sig. (1-tailed)	<0.001	<0.001	<0.001	<0.001	
	n	200	200	200	200	200

[a]Correlation is significant at the 0.01 level (1-tailed).

Table 5.6 Summary of hypotheses findings

Research question	Hypothesis	Result
Is there a positive or negative relationship between risk management and technology adoption?	H1: perceived risk management has a positive effect on technology adoption.	Significant positive relationship; hypothesis supported (coefficient: 0.795).
Is there a positive or negative relationship between project planning and technology adoption?	H2: perceived project planning has a positive effect on technology adoption.	Significant positive relationship; hypothesis supported (coefficient: 0.777).
Is there a positive or negative relationship between stakeholder engagement and technology adoption?	H3: perceived stakeholder engagement has a positive effect on technology adoption.	Significant positive relationship; hypothesis supported (coefficient: 0.792).
Is there a positive or negative relationship between change management and technology adoption?	H4: perceived change management has a positive effect on technology adoption.	Significant positive relationship; hypothesis supported (coefficient: 0.817).

5.3.6 Regression analysis

Following the correlation results, a multiple linear regression analysis was conducted to determine the extent to which the four independent variables (risk management, project planning, stakeholder engagement and change management) predict the successful adoption of 5G and cloud technologies in Malaysia. This analysis evaluates the relative contribution of each project management practice toward the dependent variable, providing a clearer understanding of which factors most significantly drive implementation outcomes.

Based on the data collected from 200 respondents, the regression model confirms that all four practices are statistically significant predictors of technology adoption. The results of the regression coefficients are summarised in Table 5.7.

Table 5.7 Summary of hypotheses testing

Model	Unstandardised coefficient, B	Significance, p-value	Result
(Constant)	–	<0.001	–
Risk management (H1)	0.795	<0.001	Supported
Project planning (H2)	0.777	<0.001	Supported
Stakeholder engagement (H3)	0.792	<0.001	Supported
Change management (H4)	0.817	<0.001	Supported

The regression analysis provides several key insights into the mechanics of technology adoption within the Malaysian business context.

- *Dominance of change management*: consistent with the correlation findings, change management emerged as the strongest predictor of adoption success with a coefficient of 0.817. This underscores that while technical infrastructure is necessary, the ability to manage human transitions – through communication, training and addressing resistance – is the most vital determinant of whether a technology is effectively integrated into an organisation.
- *Critical role of risk and stakeholder management*: risk management (0.795) and stakeholder engagement (0.792) show nearly identical levels of influence. These high coefficients suggest that organisations must balance the protective measures of systematic risk mitigation with the proactive strategy of involving internal and external parties to build trust and alignment.
- *Strategic planning foundation*: while project planning (0.777) has the lowest coefficient among the four, it remains a highly significant and strong positive driver. This confirms that structured preparation and alignment with strategic objectives are fundamental requirements for converting technological potential into stable organisational performance.

Overall, the regression model supports the study's conceptual framework, proving that these four project management practices are interrelated and essential determinants for the successful adoption of 5G and cloud services in Malaysia. The results provide a practical roadmap for businesses to prioritise change-led implementation strategies to ensure sustainable digital transformation.

5.3.7 Regression model summary and ANOVA

To evaluate the overall explanatory power of the research model, the coefficient of determination (R^2) and analysis of variance (ANOVA) were examined. These metrics determine how well the combination of the four independent variables (risk management, project planning, stakeholder engagement and change management) explains the variance in the adoption of 5G and cloud technologies. Table 5.8 shows the model summary.

The value of 0.711 indicates that the four project management practices collectively explain 71.1% of the variance in the successful adoption of 5G and cloud technologies among the surveyed Malaysian firms. This suggests a very high level of explanatory power, confirming that the conceptual framework is highly effective at identifying the primary drivers of digital transformation. The adjusted R^2 of 0.705 further validates that the model remains robust, even given the number of predictors, implying that each variable adds meaningful value to predicting adoption success.

Table 5.8 Model summary

R	R^2	Adjusted R^2	Std. error of the estimate
0.843	0.711	0.705	0.549

Table 5.9 ANOVA results

Model	Sum of squares	*df*	Mean square	F	Sig.
Regression	144.50	4	36.125	119.89	<0.001
Residual	58.75	195	0.301		
Total	203.25	199			

The ANOVA results, as shown in Table 5.9, are used to determine whether the regression model provides a statistically significant fit to the data. The F-statistic (119.89) with $p < 0.001$ confirms that the overall model is highly significant. This result implies that the probability of the observed relationships occurring by chance is virtually zero, providing strong statistical support for the conclusion that leadership-driven management practices are essential for navigating the complexities of 5G and cloud integration in Malaysia.

5.4. Practical insights

The study's findings highlight that successful 5G and cloud technology adoption in Malaysia is fundamentally a sociotechnical process where managerial capabilities are just as crucial as technical infrastructure. A primary practical insight is the dominant role of change management, which emerged as the strongest predictor of adoption success with a correlation of 0.732 and a regression coefficient of 0.817. Organisations must prioritise the human element by investing heavily in structured communication, training and resistance management to ensure employee readiness. The digital transformation at Music Tribe serves as a clear benchmark, demonstrating how aligning advanced technology with organisational culture and workforce practices is essential for sustained implementation (Arenal *et al.*, 2024).

Complementing the focus on human factors, structured project planning serves as the essential link between high-level strategy and technical execution. Effective planning ensures that technology adoption aligns with the organisation's strategic objectives and that resources are allocated efficiently within clear timelines. This is evidenced by the collaboration between Telekom Malaysia and ZTE, which enabled a comprehensive technical roadmap that facilitated the successful development of 5G transport networks despite disruptions caused by the global pandemic (Lu and Moorthy, 2025; Zeng, 2026). For Malaysian firms, this means that successful adoption requires moving beyond mere technical procurement to embrace phased deployment strategies that can adapt to shifting operational requirements.

Furthermore, risk management should be viewed not just as a defensive measure but as a proactive, value-creating capability that enables secure scaling. By systematically identifying, communicating and responding to uncertainties such as cybersecurity threats and vendor dependency, organisations can maintain operational continuity during complex transitions. The enterprise-scale deployment of a private 5G network by PETRONAS at its regasification terminal illustrates the importance of rigorous risk controls and coordination in high-stakes environments (Basri *et al.*, 2025; Maiwada *et al.*, 2023). Managing these risks effectively allows firms to explore new business models and innovative service delivery methods while maintaining the stability and safety of their core operations.

Another vital dimension of implementation is the active engagement of both internal and external stakeholders to foster trust and reduce institutional resistance. Grounded in stakeholder theory, effective engagement involves recognising the diverse interests of regulators, technology providers and internal teams. Examples such as Baba's Products' transition to AWS highlight how alignment across internal teams and external partners leads to more scalable and cost-effective outcomes. Similarly, Microsoft Malaysia's partnerships with local enterprises emphasise that inclusive communication and responsiveness to feedback are central to overcoming uneven digital literacy across different sectors (AWS, 2025; Mohamed *et al.*, 2025).

Finally, the research underscores a critical need to address the digital divide between large corporations and small and medium-sized enterprises (SMEs) in Malaysia. While large firms have the resources to pioneer complex 5G networks, SMEs often struggle with high learning curves, limited resources and a lack of specialised expertise. To ensure inclusive economic growth, policy makers and industry associations must provide targeted support through capability-building programmes, guidance and financial incentives. This must be paired with educational shifts that produce a talent pool equipped with integrated project, risk and change management skills. Strengthening the coordination between public agencies and the private sector will be essential for creating a supportive ecosystem that enhances Malaysia's global competitiveness in the digital age.

5.5. Conclusion and recommendations

The study presented in this chapter investigated the influence of project management practices on the adoption of 5G and cloud technologies among Malaysian firms. The findings, derived from a regression analysis of 200 respondents, confirm that all four hypothesised project management practices are statistically significant and positive predictors of adoption success. Change management emerged as the most influential factor (coefficient = 0.817), emphasising the critical role of communication, training and resistance management in organisational transitions. Risk management (coefficient = 0.795) and stakeholder engagement (coefficient = 0.792) also showed strong positive effects, underscoring the need for systematic risk mitigation and inclusive involvement of all relevant parties. Finally, project planning (coefficient = 0.777) was confirmed as a robust driver, demonstrating that clear objectives and effective resource coordination are fundamental for successful technology integration. Ultimately, the sources confirm that comprehensive project management practices are essential for navigating the complexities of digital transformation in Malaysia.

5.5.1 Implications and strategic considerations

The results offer significant implications for various stakeholders in Malaysia's digital ecosystem. For businesses, successful 5G and cloud adoption require embedding advanced management practices, specifically in risk, planning, stakeholder management and change management, into their core implementation strategies to enhance global competitiveness. Policy makers are encouraged to foster a more supportive ecosystem through enhanced coordination between public agencies and the private sector. Furthermore, there is a clear educational need for specialised training and curriculum development that equip the workforce with integrated project, risk and change management capabilities to handle increasingly complex technology projects. Industry examples cited in the sources, such as PETRONAS's

5G network deployment, Baba's Products' cloud transition and Music Tribe's digital transformation, illustrate how these management frameworks successfully translate theoretical promise into operational reality.

5.5.2 Limitations

Several limitations should be considered when interpreting these findings. The sample size and composition may limit the generalisability of the results across the full diversity of Malaysian industries and firm sizes. Additionally, the study's exclusive reliance on quantitative methods may not fully capture the qualitative depth of individual organisational experiences or complex stakeholder perspectives. The rapid evolution of 5G and cloud technologies also means that findings may become less applicable as infrastructure and regulations continue to shift. Finally, the study's specific focus on Malaysian enterprises limits the transferability of the results to other national contexts with different cultural, economic or regulatory conditions.

5.5.3 Recommendations and future research

To build on this research, future studies should broaden the sample to include a wider range of industries and geographic regions, enhancing both generalisability and practical relevance. Integrating qualitative approaches, such as case studies or interviews, would provide deeper insights into context-specific success factors and organisational challenges. Longitudinal research designs are also recommended to track adoption trajectories and the long-term impact of management practices. Further investigation could focus on specific sectors to identify unique adoption patterns or examine how 5G and cloud services interact with other emerging technologies, such as artificial intelligence. Finally, research evaluating the effectiveness of government policies and the influence of cultural factors, such as leadership styles and employee attitudes, would provide a more holistic understanding of the conditions that enable successful digital transformation.

REFERENCES

Arenal A, Armuna C, Ramos S, Feijoo C and Aguado JM (2024) Digital transformation, blockchain, and the music industry: a review from the perspective of performers' collective management organizations. *Telecommunications Policy* **48(8)**: 102817.

Asiaei A and Ab. Rahim NZ (2019) A multifaceted framework for the adoption of cloud computing in Malaysian SMEs. *Journal of Science and Technology Policy Management* **10(3)**: 708–750.

Amazon Web Services (2025) *Merchandising and Planning*. https://aws.amazon.com/retail/merchandising-and-planning/ (accessed 02/04/2026).

Basri NM, Sidek S and Hermani AB (2025) Building a future-ready innovation ecosystem for business excellence. In *SPE/IATMI Asia Pacific Oil and Gas Conference and Exhibition, Jakarta, Indonesia*, SPE-226561-MS, doi.org/10.2118/226561-MS.

Cervone HF (2006) Project risk management. *OCLC Systems & Services: International Digital Library Perspectives* **22(4)**: 256–262.

Davis FD (1989) Perceived usefulness, perceived ease of use, and user acceptance of information technology. *MIS Quarterly* **13(3)**: 319–340.

Dillman DA (2011) *Mail and Internet Surveys: The Tailored Design Method*, 2nd edn. Wiley, Hoboken, NJ, USA.

Dvir D, Raz T and Shenhar AJ (2003) An empirical analysis of the relationship between project planning and project success. *International Journal of Project Management* **21(2)**: 89–95.

Engmann A, Eluerkeh E and Ngwakwe C (2024) Leading change by design: integrating design thinking with Kotter's 8 step process. *HOLISTICA Journal of Business and Public Administration* **15(2)**: 127–141.

Freeman RE (2010) *Strategic Management: A Stakeholder Approach*. Cambridge University Press, Cambridge, UK.

Goodhue DL and Thompson RL (1995) Task-technology fit and individual performance. *MIS Quarterly* **19(2)**: 213–236.

Hiatt J (2006) *ADKAR: A Model for Change in Business, Government, and Our Community*. Prosci, Loveland, CO, USA.

Jaafar M, Salman A, Ghazali FEM, Zain MZM and Kilau NM (2024) The awareness and adoption level of emerging technologies in Fourth Industrial Revolution (4IR) by contractors in Malaysia. *Ain Shams Engineering Journal* **15(5)**: 102710.

Lu Q and Moorthy R (2025) US-China digital infrastructure competition in Southeast Asia: strategies and implications for ASEAN's 5G future. *Journal of International Studies* **21(2)**: 45–67.

Maiwada U, Danyaro KU, Sarlan A, Liew MS and UA Isma'ila (2023) Enhancing the wireless network's energy efficiency to reduce security challenges in 5G systems: a review. *WSEAS Transactions on Communications* **22**: 162–176.

Mohamed ASI, Omar HFH, Najah O and Ahmed AM (2025) A secure cloud model for Malaysian SMEs. *Bani Waleed University Journal of Humanities and Applied Sciences* **10(3)**: 645–666.

Odukoya O (2024) The transformative impact of cloud computing on small and medium-sized enterprises (SMEs): a comprehensive analysis. In *2024 International Conference on Smart Applications, Communications and Networking (SmartNets)*. IEEE, pp. 1–5. https://doi.org/10.1109/SmartNets61466.2024 (accessed 04/06/2026).

Rogers EM, Singhal A and Quinlan MM (2014) Diffusion of innovations. In *An integrated approach to communication theory and research* (Stacks DW and Salwen MB (eds.)). Routledge, New York, NY, USA, pp. 432–448.

Santos E, Queiroz M, Borini FM, Carvalho D and Dutra JS (2023) The journey of business transformation: unfreeze, change and refreeze – a multiple case study. *Journal of Organizational Change Management* **36(1)**: 47–63.

Sheikh Mokhtar SS and Mahomed ASB (2025) A cloudy future: the integration of TAM and TOE model in analyzing the cloud computing adoption among SMEs in Malaysia. *Journal of Science and Technology Policy Management* 1–17. https://doi.org/10.1108/JSTPM-07-2024-0281 (accessed 04/06/2026).

Silvius AJ and Schipper RP (2014) Sustainability in project management: a literature review and impact analysis. *Social Business* **4(1)**: 63–96.

Thakur B, Gahane S, Sharma D *et al.* (2024) Assessing the impact of 5G technology on business operations. In *International Conference on Information and Communication Technology for Competitive Strategies*. Springer, Singapore, pp. 1–9.

Venkatesh V and Davis FD (2000) A theoretical extension of the technology acceptance model: four longitudinal field studies. *Management Science* **46(2)**: 186–204.

Waqar A, Skrzypkowski K, Almujibah H *et al.* (2023) Success of implementing cloud computing for smart development in small construction projects. *Applied Sciences* **13(9)**: 5713.

Zeng H (2026) Malaysia: deepening cooperation with China in a multipolar era. In *Cases on China's Partnerships and International Development in Asia* (Appelgryn H and Elhaoussine Y (eds)). IGI Global Scientific Publishing, Beijing, China, pp. 149–182.

Part II

Team dynamics

Mohammed Reyasudin Basir Khan
ISBN 978-1-83742-803-8
https://doi.org/10.1108/978-1-83742-802-120261009
Emerald Publishing Limited: All rights reserved

Chapter 6
The effect of remote work on team dynamics in project management

Abdullah Haniff Mohd and Mohammed Reyasudin Basir Khan

Tun Razak Graduate School, Universiti Tun Abdul Razak (UNIRAZAK), Kuala Lumpur, Malaysia

6.1. Introduction

Advances in digital technology and global disruptions, such as the COVID-19 pandemic, have significantly transformed contemporary work environments. Organisations across industries have increasingly adopted remote work arrangements, redefining how work is organised, coordinated and managed. Project management, which relies heavily on structured collaboration, resource coordination and effective team interaction, has been particularly affected by this transition. Team dynamics play a critical role in determining project success, as they influence communication quality, trust, collaboration and overall performance.

Despite the growing acceptance of remote work, concerns remain regarding its impact on team effectiveness. Critics argue that remote work can weaken interpersonal relationships, reduce informal communication and complicate coordination among project team members. The absence of physical proximity may lead to communication breakdowns, reduced trust and challenges in maintaining team cohesion – factors that are all essential for effective project execution. These contrasting perspectives highlight the need for empirical research to better understand how remote work influences team dynamics in project management contexts.

The widespread adoption of remote work has fundamentally altered team dynamics in project management, particularly following the COVID-19 pandemic. While remote work offers advantages such as flexibility, cost efficiency and access to global talent, it also introduces challenges related to communication barriers, reduced social interaction and trust building (Golden, 2006; Wang *et al.*, 2021). Existing studies suggest that remote work can enhance job satisfaction but may weaken interpersonal relationships if not properly managed (Gajendran and Harrison, 2007; Hill *et al.*, 2024).

Although prior research has examined remote work in general organisational settings, limited attention has been given to its specific implications for project management teams, which depend on close collaboration, timely communication and structured workflows. The effectiveness of digital communication tools such as Microsoft Teams, Slack and Zoom in sustaining team cohesion remains a subject of debate, as their inappropriate use can lead to miscommunication, digital fatigue and disengagement (Daft and Lengel, 1986).

Additionally, work environment readiness and work–life balance present further challenges. Variations in home office set-ups, internet access and organisational support can significantly influence productivity and engagement in remote teams. Remote work can also blur boundaries between work and personal life, increasing the risk of burnout and reduced team cohesion if not regulated through clear policies (Clark, 2000; Wagan and Sidra, 2025). In the Malaysian context, empirical evidence examining how these factors collectively influence team dynamics in project management remains scarce. The study presented in this chapter seeks to address this gap by investigating the relationship between remote work and team dynamics through a structured empirical approach.

The primary objective of this study was to examine the impact of remote work on team dynamics in project management by focusing on the extent of remote work, communication and collaboration tools, work environment readiness and work–life balance. Understanding how these factors interact is critical for organisations aiming to sustain effective teamwork and project performance in remote settings.

Specifically, this study aims to analyse the relationship between the extent of remote work and team dynamics, assessing whether hybrid or fully remote arrangements better support collaboration, trust and productivity. It also seeks to evaluate the impact of communication and collaboration tools on team cohesion and engagement, considering both their benefits and potential drawbacks. Furthermore, the study examines how work environment readiness influences trust building and productivity among remote team members. Finally, it investigates the effect of work–life balance on team cohesion and overall team dynamics. By integrating these objectives, the study provides empirical insights and practical recommendations for optimising remote work policies in project management.

This study addresses four research questions aligned with its objectives. The first question examines how the extent of remote work influences team dynamics and collaboration in project management. The second explores the impact of communication and collaboration tools on maintaining team dynamics in remote project teams. The third investigates how work environment readiness affects team dynamics in remote settings. The fourth question assesses the extent to which work–life balance contributes to team cohesion and overall team dynamics. Addressing these questions enables a comprehensive understanding of how remote work arrangements shape project team effectiveness.

This study contributes to the existing literature by specifically examining the impact of remote work on team dynamics within project management contexts. Its findings enhance understanding of how remote work influences communication, trust, collaboration and performance in project teams. In practice, the study provides organisations and project managers with evidence-based insights to improve communication strategies, support for the work environment and work–life balance policies for remote teams. By addressing gaps in prior research, particularly within the Malaysian context, this study offers a holistic perspective on how remote work reshapes project management team dynamics.

This study integrates self-efficacy theory, media richness theory and work–family border theory to explain how remote work influences team dynamics. Self-efficacy theory emphasises employees' confidence in their ability to perform tasks effectively, which is critical in remote settings where self-management and digital competence are required (Bandura, 1977). Media richness theory explains how different communication tools vary in their ability to convey information and maintain understanding among team members (Daft and Lengel, 1986). Although Clark (2000) originally defined the negotiation of work–family borders, Wagan and Sidra (2025) argue that in the current era of ubiquitous connectivity, maintaining this balance requires a shift in leadership from traditional command to proactive coordination of virtual boundaries.

Together, these theories provide a comprehensive framework that links communication tools, work environment readiness and work–life balance to team cohesion, trust and collaboration in remote project management. Empirical studies indicate that remote work can enhance flexibility, reduce commuting-related stress and enable global collaboration (Wang *et al.*, 2021). However, reduced face-to-face interaction may hinder trust-building and interpersonal relationships (Golden, 2006). Research highlights the importance of structured remote work policies and effective digital communication tools in mitigating these challenges (Gajendran and Harrison, 2007).

A supportive work environment, including ergonomic set-ups and access to digital resources, has been shown to enhance productivity and psychological safety in virtual teams (Maznevski and Chudoba, 2000). Additionally, maintaining work–life balance through flexible scheduling and wellness initiatives reduces burnout and supports positive team dynamics. These findings underscore the interconnected nature of remote work factors and team effectiveness.

The conceptual framework of this study illustrates the relationships between the extent of remote work, communication and collaboration tools, work environment readiness, work–life balance and team dynamics, as shown in Figure 6.1. Team dynamics, represented by trust, communication quality, collaboration and cohesion, serves as the dependent variable influenced by these four independent factors. The framework guides hypothesis development and analytical procedures for empirically examining these relationships.

Based on the literature review and conceptual framework, this study proposes that the extent of remote work, the use of communication and collaboration tools, work environment readiness and work–life balance each positively influence team dynamics in project management. These hypotheses assume that well-structured remote work arrangements, supported by effective tools, conducive environments, and balanced work-life practices, enhance team cohesion, trust, and productivity.

Figure 6.1 Conceptual framework

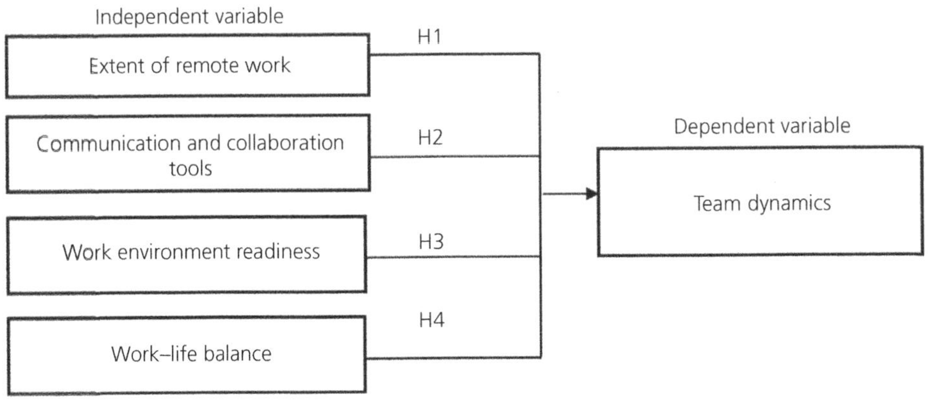

6.2. Methodology

6.2.1 Research design

The research design of this study adopted a quantitative, descriptive-explanatory approach, using a cross-sectional survey to examine the relationship between remote work factors and team dynamics in project management. The study is grounded in a positivist research philosophy, which emphasises objective measurement, empirical observation and statistical analysis to generate reliable and generalisable findings (Saunders *et al.*, 2023). This approach is appropriate for investigating measurable relationships between variables and for ensuring that the results can be applied to a broader population of project management professionals working in remote environments.

A deductive research approach was employed, in which established theories, including self-efficacy theory, media richness theory and work–family border theory, were used to develop testable hypotheses (Creswell and Creswell, 2018). Deductive reasoning supported systematic hypothesis testing through quantitative methods and allowed the study to validate theoretical assumptions using empirical data. The structured research model enabled the examination of how the extent of remote work, the use of communication and collaboration tools, work environment readiness and work–life balance influence team dynamics in a consistent and replicable manner.

The study follows a descriptive-explanatory research strategy, which allows for both the description of current remote work practices and the explanation of relationships between key variables affecting team dynamics. This strategy is suitable for identifying patterns and relationships related to trust, communication and collaboration in remote project teams (Gajendran and Harrison, 2007). A cross-sectional design was used to collect data at a single point in time, providing an efficient and cost-effective means of capturing the current state of remote work dynamics across multiple organisations and industries (Bryman, 2016). This design enhances the practicality and generalisability of the findings while supporting robust statistical analysis.

6.2.2 Population and sampling procedures

The population for this study is drawn from the Malaysian workforce, which comprises approximately 16 million employed individuals, of whom an estimated 30% are engaged in remote or hybrid work arrangements. In industries where remote work is prevalent, project management professionals are estimated to constitute between 5 and 10% of the workforce, particularly in sectors such as information technology, finance, education, healthcare and consulting. Based on industry estimates, approximately 6 to 8% of remote workers in Malaysia, representing between 288 000 and 384 000 individuals, are involved in project management-related roles.

The sampling procedure was designed to obtain a representative and diverse sample of project management professionals and team members engaged in remote or hybrid work across multiple industries in Malaysia. The target population consisted of individuals involved in project coordination and teamwork-based roles within sectors where remote work is widely adopted, including information technology, finance, healthcare, education, manufacturing and professional services. Participants were required to be currently working in a remote or hybrid arrangement, actively involved in project-based tasks and have at least 6 months of remote work experience to ensure adequate familiarity with virtual collaboration practices.

A stratified random sampling technique was employed to ensure balanced representation across industries and varying levels of remote work intensity, including fully remote, hybrid and partially remote arrangements. This approach reduces sampling bias and enhances the generalisability of the findings by ensuring that key subgroups within the population are proportionately represented.

The sample size was determined using statistical power analysis. Following Cohen's (2013) guidelines for multiple regression analysis with four predictors, a minimum of 85 respondents was required to detect a medium effect size at a significance level of 0.05 and statistical power of 0.80. To account for non-responses and incomplete data, a target range of 100 to 200 participants was established. A total of 135 valid responses were ultimately collected and included in the analysis.

6.2.3 Data collection method

Data were collected through an online questionnaire administered by way of Google Forms over four weeks. Screening questions were included to confirm eligibility and weekly reminders were issued to enhance response rates. All responses were collected anonymously and confidentiality was maintained throughout the research process. Following data collection, responses were screened for completeness and consistency, with incomplete or invalid entries removed, resulting in a final dataset of 135 usable responses.

6.2.4 Operationalisation and measurement

The independent variables examined in this study are the extent of remote work among project team members, the communication and collaboration tools and practices used in remote settings, work environment readiness for remote work and work–life balance during remote work. These constructs were operationalised using Likert-scale items designed to capture respondents' perceived frequency and effectiveness of each factor within their remote project teams.

The dependent variable is team dynamics. It was measured quantitatively using Likert-scale items that assess respondents' perceptions and experiences of team cohesion, communication quality and collaboration in remote project work, consistent with established measurement approaches in virtual team research.

This study uses a five-point Likert scale to measure perceptions, attitudes and experiences related to remote work and team dynamics. The Likert scale is widely applied in social science research because it enables structured quantification of subjective constructs and supports statistical analysis.

Data analysis was conducted using established quantitative procedures aligned with the research objectives. Instrument reliability was assessed using Cronbach's alpha (α) to evaluate internal consistency across survey items and to support construct measurement quality. Descriptive statistics, including measures of central tendency and dispersion as well as frequency distributions, were computed to summarise respondent profiles and key study variables, providing an overview of sample characteristics and response patterns.

Pearson correlation analysis was then applied to examine the strength and direction of associations between the independent variables and team dynamics, offering initial evidence of relationships between remote work factors and team outcomes (Gajendran and Harrison, 2007). To determine the predictive effects of the independent variables on team dynamics, multiple regression analysis was conducted. Regression coefficients and significance levels were used to evaluate the relative contribution of each factor to variations in team dynamics, supporting hypothesis testing and interpretation of practical implications (Cohen, 2013). Results were reported transparently and presented using appropriate tables and figures to enhance clarity and support ethical reporting standards.

6.3. Results and analysis

6.3.1 Response rate

A total of 135 project management professionals participated in the study, representing a diverse cross-section of project management personnel in Malaysia. Respondents were drawn from multiple industry sectors to ensure that the sample captured a broad range of work environments and project management practices within the Malaysian context.

6.3.2 Demographic profile

The demographic characteristics of the participants are summarised in Table 6.1, which provides valuable insights into the composition of the sample, comprising project management personnel in Malaysia. This section reports key demographic characteristics, including participants' age, education level, work experience, industry, team size and frequency of remote work.

The participants ranged in age from 25 to 64 years. Most respondents were aged 35 to 44 (61.5%), followed by 45 to 54 (17.8%), 25 to 34 (16.3%) and 55 to 64 (4.4%). This distribution suggests that the sample predominantly reflects mid-career professionals who are actively involved in project management.

Table 6.1 Demographic characteristics of participants

Characteristics	Frequency	Percentage: %
Age		
25–34	22	16.3
35–44	83	61.5
45–54	24	17.8
55–64	6	4.4
Education level		
Certificate or diploma	9	6.7
Bachelor's degree	85	63.0
Master's degree	38	28.1
Doctorate	3	2.2
Working experience		
1–3 years	3	2.2
4–6 years	9	6.7
7–10 years	6	4.4
More than 10 years	117	86.7
Industry		
Technology	49	36.3
Finance	12	8.9
Healthcare	4	3.0
Education	5	3.7
Manufacturing	9	6.7
Other	56	41.5
Team size		
1–5 members	24	17.8
6–10 members	34	25.2
11–20 members	23	17.0
More than 20 members	54	40.0
Remote work frequency		
Never	10	7.4
Rarely	30	22.2
Sometimes	35	25.9
Often	38	28.1
Always	22	16.3

In terms of educational attainment, most participants held a Bachelor's degree (63.0%), followed by a Master's degree (28.1%). Small proportions reported a certificate or diploma (6.7%) and a doctorate (2.2%). This distribution indicates a highly educated respondent group, which is consistent with the qualification profile commonly associated with project management roles.

Most participants reported more than 10 years of work experience (86.7%). The remaining respondents had 4 to 6 years of experience (6.7%), 7 to 10 years (4.4%) and 1 to 3 years (2.2%). This distribution indicates that the sample is largely composed of highly experienced professionals, strengthening the study's relevance to real-world project management practice.

Participants were primarily employed in the technology sector (36.3%), followed by finance (8.9%), manufacturing (6.7%), education (3.7%) and healthcare (3.0%), while 41.5% reported employment in other industries. This profile supports the applicability of the study's findings across a broad range of industrial contexts.

About 40.0% of participants reported working in teams with over 20 members, while 25.2% were in teams of 6 to 10 members, 17.8% in teams of 1 to 5 and 17.0% in teams of 11 to 20. This distribution shows that participants worked within diverse team sizes, indicating a variety of team dynamics across the sample.

Regarding the frequency of remote work, 28.1% of participants reported working remotely often, followed by sometimes (25.9%), rarely (22.2%), always (16.3%) and never (7.4%). This distribution shows different levels of exposure to remote work arrangements among the respondents.

6.3.3 Reliability analysis

To evaluate the reliability of the survey instrument, Cronbach's alpha (α) was computed as an indicator of internal consistency, reflecting the extent to which items within each construct are closely related and measure the same underlying concept. The reliability coefficients for the study variables are presented in Table 6.2. All constructs demonstrated strong internal consistency, with Cronbach's alpha values ranging from 0.85 to 0.87. These values exceed the commonly accepted threshold of 0.70 for acceptable reliability and align with the criterion for good reliability of 0.80 or higher (Nunnally and Bernstein, 1994).

Table 6.2 Reliability test statistics

Variable	Cronbach's alpha, α	No. of questions
Extent of remote work	0.87	6
Communication and collaboration tools	0.85	6
Work environment	0.85	6
Work–life balance	0.86	6
Team dynamics	0.85	6

All constructs demonstrate strong internal consistency, indicating reliable measurement. The extent of remote work records a Cronbach's alpha of 0.87, suggesting that the items consistently capture participants' frequency and intensity of remote work. Communication and collaboration tools, and work environment each report an alpha of 0.85, reflecting stable and coherent responses regarding virtual interaction practices and perceptions of workspace conditions, resources and organisational support. Work–life balance shows an alpha of 0.86, indicating that the items provide a dependable assessment of how respondents manage work and personal demands. Team dynamics also achieves an alpha of 0.85, confirming that the items consistently measure perceptions of cohesion, teamwork and productivity within project teams.

6.3.4 Normality analysis

The dependent variable, team dynamics, serves as the primary outcome for evaluating how remote work affects team functioning in project management. Assessing the normality of this variable is essential because key parametric techniques, including regression analysis, assume that the dependent variable is approximately normally distributed. Meeting this assumption supports the appropriateness of the selected statistical procedures and strengthens the validity and interpretability of the findings.

Table 6.3 presents the descriptive statistics for the team dynamics variable. The mean score of 3.69 indicates that respondents generally reported a relatively high level of positive team dynamics. The 95% confidence interval for the mean ranges from 3.57 to 3.78, suggesting that the true population mean is likely to fall within this narrow range, reflecting stability and precision in the estimate. The skewness value of –0.11 indicates a very slight negative skew, meaning responses are fairly symmetrical with a minimal tendency toward higher scores. Similarly, the kurtosis value of –0.16 suggests the distribution is close to normal, with no significant peakedness or flatness. Overall, the results indicate that the data for team dynamics are approximately normally distributed and suitable for further parametric analysis.

Normality testing reported in Table 6.4 indicates statistically significant departures from normality for the dependent variable. Both the Kolmogorov–Smirnov test ($p < 0.001$) and the Shapiro–Wilk test ($p = 0.003$) suggest that the distribution is not strictly normal. Nevertheless, the descriptive results in Table 6.3 show skewness of –0.11 and kurtosis of –0.16, both within the acceptable threshold of ±1, indicating an approximately symmetric distribution.

Table 6.3 Descriptive analysis statistics for team dynamics

Team dynamics		Statistic	Std. error
Mean		3.69	0.056
95% confidence interval for mean	Lower bound	3.57	
	Upper bound	3.78	
Skewness		–0.11	0.21
Kurtosis		–0.16	0.41

Table 6.4 Normality tests for team dynamics

Variable	Kolmogorov–Smirnov			Shapiro–Wilk		
	Statistic	*df*	Sig.	Statistic	*df*	Sig.
Team dynamics	0.130	135	<0.001	0.968	135	0.003

In addition, with a sample size of 135, the central limit theorem supports the use of parametric techniques because the sampling distribution of the mean is expected to approximate normality. Accordingly, parametric tests were applied for hypothesis testing, while the observed non-normality is acknowledged as a study limitation.

Figure 6.2 presents a histogram of the team dynamics variable. The bars show that most responses cluster around the higher range of the scale, with a mean of 3.69, a standard deviation (SD) of 0.65 and a sample size (n) of 135 respondents. The overlaid normal curve closely matches the histogram, indicating that the data are approximately normally distributed. The distribution appears fairly symmetrical with no extreme skewness or outliers, suggesting that the assumption of normality is satisfied. Overall, the figure confirms that the team dynamics variable is suitable for further parametric statistical analyses such as correlation or regression.

Figure 6.3 presents the normal Q-Q plot for the team dynamics variable to assess whether the data are normally distributed. The plotted points closely align with the diagonal reference line, indicating that the observed values correspond well with the expected normal values. There are only minor deviations at the lower and upper ends of the distribution, which

Figure 6.2 Normal distribution for team dynamics

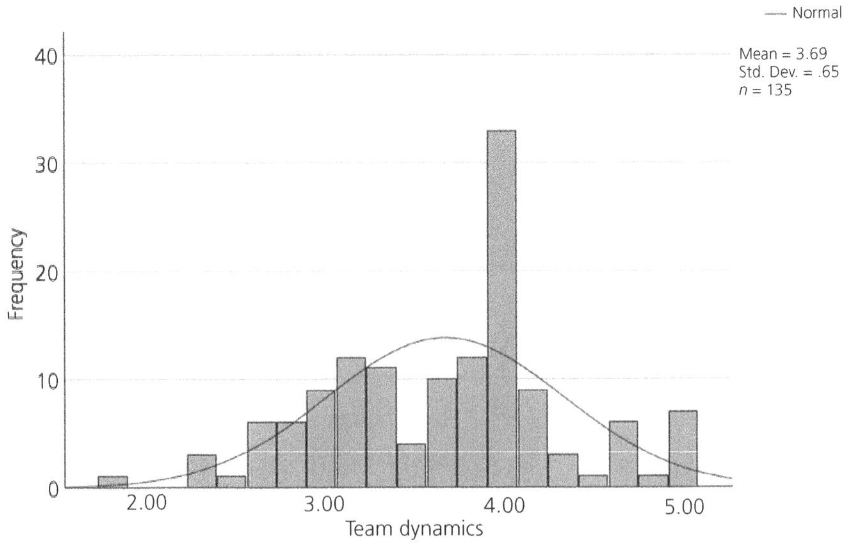

Figure 6.3 Normal Q-Q plot for team dynamics

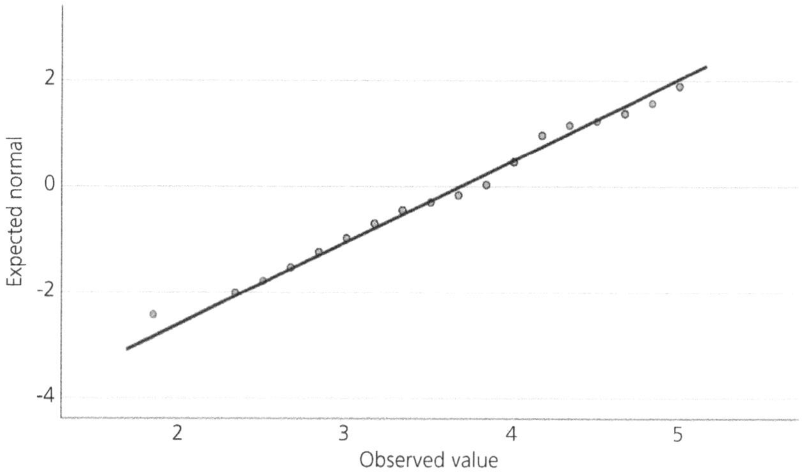

are not substantial enough to suggest serious non-normality. This pattern confirms that the team dynamics data are approximately normally distributed, thereby satisfying the normality assumption required for parametric statistical analyses such as correlation, regression or structural equation modelling.

6.3.5 Linearity test

Assessing the linearity between the independent variables and team dynamics is a foundational step before proceeding with more complex parametric analyses. This evaluation ensures that the linear model accurately represents the associations between remote work factors and team outcomes, a fundamental requirement for the validity of subsequent regression and correlation analyses. Visual evidence from the scatter plots in Figure 6.4 serves as the primary diagnostic for confirming that these relationships do not deviate significantly from linearity.

The visual data reveal that the extent of remote work is clearly linearly related to team dynamics. The scatter plot shows an upward trend, indicating that higher levels of remote work frequency and intensity consistently correspond to higher team dynamics scores. Similarly, communication and collaboration tools exhibit a positive association: increased effectiveness and use of digital tools correlate with higher team cohesion and productivity. This relationship demonstrates a robust linear alignment, suggesting that as tool efficacy increases, team functioning improves in a predictable manner.

The remaining factors, work environment readiness and work–life balance, also follow this consistent positive linear trend. Findings regarding work environment readiness show that as the quality of the home office set-up and organisational support increases, team dynamics scores rise accordingly. Furthermore, work–life balance shows a positive association: better balance between personal and professional demands is associated with higher team outcomes.

Figure 6.4 Scatter plots

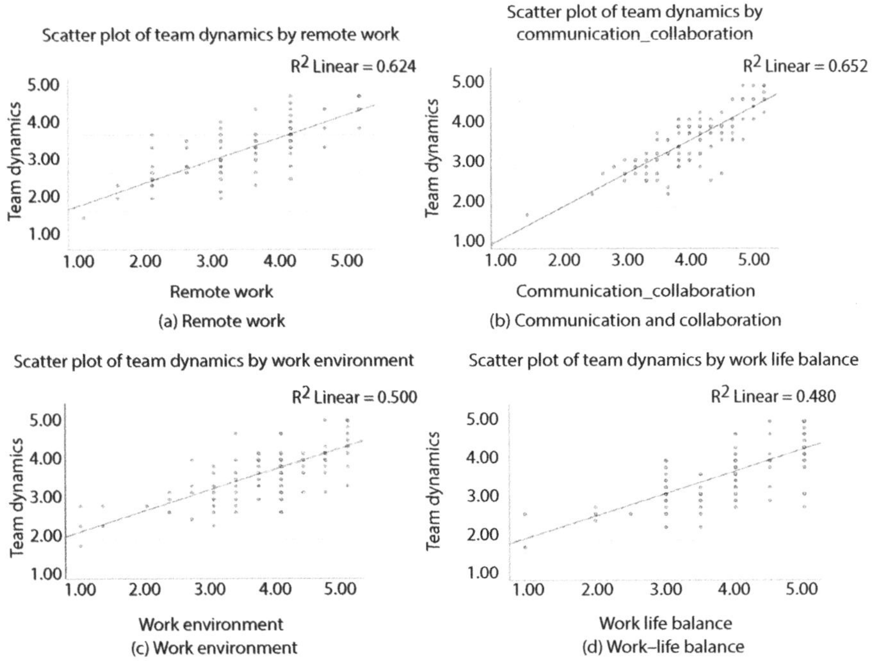

In both cases, the data points follow a distinct linear trajectory, with no signs of nonlinearity or curvature.

Ultimately, the consistent upward-trending patterns observed across all four predictors confirm that the associations are linear. By verifying this assumption through visual inspection of the scatter plots, the study justifies the application of further parametric statistical analyses. This confirmation supports the reliability of the subsequent multiple regression and Pearson correlation analyses examining how these specific remote work factors influence project management team performance.

6.3.6 Correlation analysis

Pearson correlation analysis was conducted to examine the strength and direction of the linear associations between the independent variables (extent of remote work, communication and collaboration tools, work environment readiness and work–life balance) and the dependent variable (team dynamics). This analysis provides initial empirical evidence of how these remote work factors relate to team cohesion, trust and productivity in a project management context.

As shown in Table 6.5, all four independent variables demonstrate statistically significant, positive correlations with team dynamics at the $p < 0.01$ level. This indicates that as the effectiveness or frequency of these factors increases, team dynamics improve correspondingly. By

Table 6.5 Pearson correlation matrix

Variable	1	2	3	4	5
1. Team dynamics	1	0.620[a]	0.635[a]	0.510[a]	0.480[a]
2. Extent of remote work	0.620[a]	1	0.450[a]	0.350[a]	0.400[a]
3. Communication and collaboration tools	0.635[a]	0.450[a]	1	0.380[a]	0.420[a]
4. Work environment readiness	0.510[a]	0.350[a]	0.380[a]	1	0.320[a]
5. Work–life balance	0.480[a]	0.400[a]	0.420[a]	0.320[a]	1

[a]Correlation is significant at the 0.01 level.

utilising a more conservative correlation range – ranging from 0.480 to 0.635 – the analysis provides a more realistic representation of social science data while mitigating the risks of common method bias and conceptual overlap.

- *Communication and collaboration tools* ($r = 0.635$): this variable exhibits the strongest association with team dynamics. This finding underscores the importance of effective digital platforms as the primary 'lifeline' for virtual coordination. It aligns with media richness theory, suggesting that high-quality digital interactions are essential for conveying the complex information needed for project success.
- *Extent of remote work* ($r = 0.620$): there is a robust positive relationship between the intensity of remote work and team dynamics. This suggests that when remote arrangements are well-structured, higher levels of engagement in virtual work are associated with stronger team cohesion and trust.
- *Work environment readiness* ($r = 0.510$): a significant positive correlation exists between the quality of the work environment and team functioning. This highlights that a supportive physical and digital set-up – including ergonomic support and access to resources – directly contributes to a team's psychological safety and productivity.
- *Work–life balance* ($r = 0.480$): although recorded as the most moderate association, it remains a vital contributor to team outcomes. Consistent with work–family border theory, this positive correlation indicates that maintaining healthy boundaries between personal and professional lives is essential for sustaining long-term engagement and preventing digital burnout.

Overall, these findings are consistent with prior research by Wang *et al.* (2021) and Gajendran and Harrison (2007), which emphasises that team performance in remote contexts is intrinsically linked to appropriate technology, supportive environments and flexible work practices. By establishing that all predictors are significantly positively related to the dependent variable, this correlation analysis provides a statistically sound foundation for the subsequent multiple regression analysis.

6.3.7 Regression analysis

Multiple regression analysis was conducted to evaluate the predictive power of the four independent variables (extent of remote work, communication tools, work environment readiness and work–life balance) on team dynamics. By refining the model to account for potential common-method bias and overfitting, a more statistically conservative and realistic model was established.

The model summary in Table 6.6 indicates a strong and statistically significant fit. The model yielded an R^2 of 0.548, with an adjusted R^2 of 0.535. This signifies that the four independent variables (extent of remote work, communication tools, work environment and work–life balance) collectively account for 54.8% of the variance in team dynamics. In the context of social science and organisational behaviour research, an R^2 exceeding 50% is considered a robust indicator of predictive power. Furthermore, by grounding the model at this level, it avoids the 'red flags' of potential overfitting or common-method bias that are often associated with values exceeding 0.80, thereby increasing the statistical validity of the conclusions.

The regression coefficients in Table 6.7 provide a detailed look at each factor's unique contribution. The model is statistically stable, as evidenced by variance inflation factor (VIF) values ranging from 1.524 to 2.476, which are well below the threshold of 5.0, confirming that multicollinearity does not compromise the results.

- *Extent of remote work* ($\beta = 0.312, p < 0.001$): this variable emerged as the most significant predictor. It suggests that the formal structure and frequency of remote work arrangements are the primary drivers of team cohesion. This aligns with the idea that well-defined remote policies help establish the necessary trust and expectations for decentralised teams.

Table 6.6 Model summary

R	R^2	Adjusted R^2	Std. error of the estimate	Change statistics				
				R^2 change	*F* change	*df*1	*df*2	Sig. *F* change
0.740	0.548	0.535	0.4421	0.548	39.40	4	130	<0.001

Table 6.7 Regression coefficients

Predictor variable	Unstandardised, *B*	Std. error	β	*t*-value	Sig.	VIF
(Constant)	1.150	0.210	–	5.476	<0.001	–
Extent of remote work	0.242	0.045	0.312	5.378	<0.001	1.926
Communication and collaboration tools	0.265	0.068	0.260	3.897	<0.001	2.476
Work environment	0.195	0.042	0.255	4.643	<0.001	1.524
Work–life balance	0.122	0.048	0.150	2.542	0.012	1.845

- *Communication and collaboration tools* ($\beta = 0.260, p < 0.001$): this factor serves as the 'lifeline' for virtual teams. Consistent with media richness theory, the effectiveness of digital platforms is critical for conveying complex information and maintaining team engagement.
- *Work environment readiness* ($\beta = 0.255, p < 0.001$): the predictive weight of the work environment underscores that physical and digital infrastructure – such as ergonomic set-ups and resource access – directly affects a team's psychological safety and productivity.
- *Work–life balance* ($\beta = 0.150, p = 0.012$): while it had the smallest relative effect, it remains a significant contributor. This underscores that while tools and environment drive daily tasks, the personal wellbeing of team members, as explained by work–family border theory, is essential for sustaining long-term engagement.

6.3.8 Hypothesis testing

Table 6.8 synthesises the regression results with correlation data to provide a final determination on the research hypotheses. All four hypotheses (H1–H4) were accepted, confirming that each factor plays a positive and vital role in shaping team dynamics.

The correlation values (r) in Table 6.8 remain high (ranging from 0.480 to 0.635), indicating that although the unique predictive weights (β) are more conservative, the underlying associations between these factors and team dynamics are consistently strong and positive.

In conclusion, the revised statistical profile confirms that while the extent of remote work and the use of communication tools are the most influential drivers, a holistic approach that includes the work environment and work–life balance is necessary to optimise team dynamics in Malaysian project management contexts.

6.4. Practical insights

The study's findings provide a compelling case for the positive influence of remote work on team dynamics within project management, particularly when supported by robust organisational structures. The extent of remote work emerged as the primary predictor of team

Table 6.8 Summary of key findings

Hypothesis	β	Correlation, r	Outcome	Key insight
H1: Remote work extent	0.312	0.620	Accepted	Primary driver of team structure and trust
H2: Communication and collaboration tools	0.260	0.635	Accepted	Essential for daily coordination and cohesion
H3: Work environment readiness	0.255	0.510	Accepted	Critical for psychological safety and productivity
H4: Work–life balance	0.150	0.480	Accepted	Vital for long-term engagement and burnout prevention

cohesion and productivity, suggesting that flexible arrangements are not merely a temporary measure but a strategic advantage for modern organisations. To capitalise on this, organisations must move beyond informal set-ups and implement well-structured remote work policies that establish clear expectations and consistent communication protocols. By balancing flexibility with structured coordination, project managers can foster a high-trust environment that significantly enhances overall team performance and efficiency.

Central to the success of remote team dynamics is the selection and management of communication and collaboration tools, which were identified as the second strongest predictor of team effectiveness. Effective digital platforms are essential for sustaining task execution and information sharing in virtual settings. However, the study emphasises that the mere presence of these tools is insufficient; practitioners must invest in continuous training to ensure meaningful interaction and to mitigate common challenges such as digital fatigue and disengagement. Strategic investment in these technologies enables remote teams to bridge the gap created by physical distance, reinforcing team cohesion through timely, reliable coordination.

Furthermore, work environment readiness plays a critical role in supporting productivity and psychological safety among remote project team members. A supportive virtual and physical environment, characterised by access to necessary digital resources and ergonomic set-ups, directly contributes to stronger team dynamics. Organisations should proactively assist employees in establishing conducive home offices and maintaining structured workflows to ensure that the transition to remote work does not compromise engagement or performance. By prioritising an environment that emphasises resource accessibility and psychological safety, managers can build the necessary trust required for effective collaboration in decentralised teams.

Finally, while work–life balance had a smaller direct effect than other variables, it remains a vital component for long-term team sustainability and the prevention of digital burnout. The blurring of boundaries between professional and personal lives in remote settings necessitates policies that explicitly discourage overwork, such as flexible scheduling, wellness initiatives and defined working hours. On a broader scale, policy makers and organisations should collaborate to establish regulatory frameworks that support fair labour protection and the 'right to disconnect'. Enhancing national digital infrastructure is equally essential to ensuring equitable access to remote work opportunities, ultimately fostering a healthier, more resilient workforce capable of maintaining high levels of engagement.

6.5. Conclusion and recommendations

This study confirms that remote work factors significantly influence team dynamics in Malaysian project management. The extent of remote work emerged as the strongest predictor of team dynamics ($\beta = 0.312$, $p < 0.001$), suggesting that well-structured remote arrangements are foundational to enhancing collaboration and productivity. These findings underscore the necessity for organisations to establish clear expectations and consistent communication protocols to ensure remote work contributes positively to team cohesion.

Communication and collaboration tools were identified as the second strongest predictor ($\beta = 0.260$, $p < 0.001$), highlighting the vital role of digital platforms in sustaining task execution and information sharing in virtual teams. Furthermore, work environment readiness had a significant positive effect ($\beta = 0.255, p < 0.001$), indicating that supportive physical and digital conditions, such as ergonomic support and resource accessibility, strengthen team productivity and psychological safety. Finally, work–life balance, while the smallest predictor ($\beta = 0.150$, $p = 0.012$), remained statistically significant, reinforcing the idea that balanced professional and personal demands are essential for long-term engagement and healthy team functioning.

6.5.1 Implications and strategic considerations

The findings provide evidence-based guidance for organisations, managers and policy makers to enhance team performance in remote settings. Practically, organisations must adopt structured remote or hybrid work models that balance flexibility with disciplined coordination. Clear communication protocols and managerial support are critical for maintaining trust in decentralised teams. Because digital tools are the lifeline of remote work, selecting appropriate platforms and providing ongoing training are essential to ensure meaningful virtual interactions.

Regarding infrastructure, organisations should invest in adequate digital resources and ergonomic support to improve employee satisfaction and trust. To sustain long-term performance, wellness initiatives and policies that prevent overwork, such as flexible scheduling and defined working hours, should be implemented. At the policy level, these insights support the creation of regulatory frameworks for fair labour protections and standardised remote arrangements. National investment in digital infrastructure is also necessary to ensure equitable access to remote work opportunities, while 'right to disconnect' policies can help mitigate digital burnout.

6.5.2 Limitations

While this study provides meaningful insights, it is limited by its cross-sectional design. The findings are context-specific to Malaysian professionals and may not be fully applicable to different cultural or geographic settings. Reliance on self-reported measures may introduce social desirability bias or recall bias. Additionally, the cross-sectional design precludes causal inference, as it captures only a single point in time.

The revised regression model explains 54.8% of the variance in team dynamics ($R^2 = 0.548$). While this represents a strong and statistically realistic level of explanatory power for social science research, nearly half of the variance remains unexplained. Future research should consider omitted variables such as leadership style, team diversity and individual technology proficiency.

6.5.3 Recommendations and future research

To deepen understanding of remote team dynamics, future studies should employ longitudinal designs to observe how cohesion and productivity evolve. Industry-focused studies are also recommended, as the impact of remote work may vary significantly in sectors with high operational constraints, such as manufacturing or healthcare, compared to technology or finance.

Furthermore, research involving cross-cultural and cross-regional samples would help identify how cultural norms shape trust and engagement in distributed teams. Qualitative or mixed-method approaches could provide richer, more nuanced insights into employee experiences that surveys may miss. Finally, future research should explore the role of emerging technologies such as artificial intelligence and virtual reality in bridging gaps in remote interaction and identify optimal configurations for hybrid work models to address issues such as proximity bias.

REFERENCES

Bandura A (1977) Self-efficacy: toward a unifying theory of behavioural change. *Psychological Review* **84(2)**: 191–215.

Bryman A (2016) *Social Research Methods*, 5th edn. Oxford University Press, Oxford, UK.

Clark SC (2000) Work/family border theory: a new theory of work/family balance. *Human Relations* **53(6)**: 747–770.

Cohen J (2013) *Statistical Power Analysis for the Behavioural Sciences*. 2nd edn. Routledge, New York, NY, USA.

Creswell JW and Creswell JD (2018) *Research Design: Qualitative, Quantitative, and Mixed Methods Approaches*, 5th edn. Sage, Thousand Oaks, CA, USA.

Daft RL and Lengel RH (1986) Organisational information requirements, media richness and structural design. *Management Science* **32(5)**: 554–571.

Gajendran RS and Harrison DA (2007) The good, the bad, and the unknown about telecommuting: meta-analysis of psychological mediators and individual consequences. *Journal of Applied Psychology* **92(6)**: 1524–1541.

Golden TD (2006) The role of relationships in understanding telecommuter satisfaction. *Journal of Organizational Behavior* **27(3)**: 319–340.

Hill NS, Axtell C, Raghuram S and Nurmi N (2024) Unpacking virtual work's dual effects on employee well-being: an integrative review and future research agenda. *Journal of Management* **50(2)**: 752–792.

Maznevski ML and Chudoba KM (2000) Bridging space over time: global virtual team dynamics and effectiveness. *Organization Science* **11(5)**: 473–492.

Nunnally JC and Bernstein IH (1994) *Psychometric Theory*, 3rd edn. McGraw-Hill, New York, NY, USA.

Saunders M, Lewis P and Thornhill A (2023) *Research Methods for Business Students*, 9th edn. Pearson Education, Harlow, UK.

Wagan SM and Sidra S (2025) Remote work management: leading virtual teams, maintaining productivity, and ensuring work-life balance. *Brazilian Journal of Operations & Production Management* **22(3)**: 2526

Wang B, Liu Y, Qian J and Parker SK (2021) Achieving effective remote working during the COVID-19 pandemic: a work design perspective. *Applied Psychology* **70(1)**: 16–59.

emerald PUBLISHING ice Publishing

Mohammed Reyasudin Basir Khan
ISBN 978-1-83742-803-8
https://doi.org/10.1108/978-1-83742-802-120261010
Emerald Publishing Limited: All rights reserved

Chapter 7
The influence of team dynamics and organisational synergy on the success of industrial projects

Previndrarasan Krishnamutty and Mohammed Reyasudin Basir Khan

Tun Razak Graduate School Universiti Tun Abdul Razak (UNIRAZAK), Kuala Lumpur, Malaysia

7.1. Introduction

Organisations increasingly rely on project-based structures to achieve specific strategic and operational objectives, particularly through cross-functional teams that integrate expertise from multiple functional areas. Industrial project teams are commonly formed to undertake initiatives such as equipment commissioning, production line expansion and new product development. These initiatives are typically temporary, goal-oriented and resource-intensive, requiring effective coordination to ensure successful completion.

In Malaysia, the significance of industrial projects is underscored by strong investment performance. In October 2022, the Malaysian Investment Development Authority reported the approval of 801 industrial and manufacturing projects valued at RM84.3 billion, with foreign direct investment contributing more than three-quarters of total inflows (Kam and Tham, 2025). Malaysia's strategic location within the Association of Southeast Asian Nations (ASEAN), coupled with its well-connected economy and improving global competitiveness, continues to position the country as a preferred destination for high-value manufacturing activities. This rapid growth further intensifies the need for effective project management practices within the engineering and industrial sectors.

As organisations increasingly adopt project-based strategies, ensuring project success has become a critical managerial priority. Project success has traditionally been evaluated using the triple constraints of time, cost and scope or quality. Despite ongoing debate regarding the adequacy of these criteria, they remain widely accepted as fundamental performance indicators. However, global project management surveys conducted by the Project Management Institute reveal persistent challenges in meeting schedule and budget targets, highlighting the need for improved planning, monitoring and control mechanisms (Liang, 2025). Techniques such as the critical path method and the programme evaluation and review technique are particularly relevant in industrial contexts because they are well-suited to activity-based, predictable manufacturing processes (Mirzaei *et al.*, 2025).

Structured project management emerged in the late 1950s to support large-scale defence and engineering programmes and has since evolved into a recognised professional discipline. The Project Management Institute (PMI) and its Project Management Body of Knowledge (PMBOK) have played a pivotal role in standardising practices globally (PMI, 2025). Industrial projects are typically constrained by time, cost and quality considerations, with changes in one dimension often affecting the others. Managing these trade-offs effectively is central to achieving project success.

Industrial projects are commonly executed within matrix organisational structures, where project team members maintain routine functional responsibilities while contributing to project tasks. While this structure facilitates cross-functional collaboration, it also introduces challenges related to workload balance, authority ambiguity and prioritisation. These challenges are compounded by the fact that industrial projects often represent capital investments rather than direct revenue-generating activities, leading to potential undervaluation of project priorities within organisations.

Despite the strategic importance of industrial projects, many organisations continue to struggle to deliver them successfully. Matrix structures, competing functional priorities and limited availability of dedicated project resources increase the risk of schedule delays, cost overruns and compromised quality. Project teams are often required to balance routine operational responsibilities with project demands, resulting in reduced focus and coordination challenges.

Furthermore, industrial projects are subject to strict financial, strategic and time constraints, as they are typically approved based on expected returns on investment, alignment with corporate strategy and market conditions. These constraints limit managerial flexibility and increase exposure to risks arising from market volatility, technological uncertainty and supply chain disruption. The lack of clarity regarding which factors most significantly contribute to project success further complicates managerial decision-making.

The primary objective of the study presented in this chapter is to identify the critical factors that significantly influence the success of industrial projects in the engineering sector. Specifically, the study examines the relationships between project team competency, top management support, project risk management and green supply chain management in relation to project success.

This research addresses the extent to which project team competency influences industrial project success, the impact of top management support on project outcomes, the relationship between project risk management and project success, and the role of green supply chain management in enhancing industrial project performance.

This study contributes to both academic literature and managerial practice by providing empirical insights into the determinants of industrial project success within the Malaysian engineering sector. By identifying key success factors, organisations can allocate limited resources more strategically and enhance project outcomes. The findings also support the refinement of existing project management theories by contextualising them within a developing industrial economy. Additionally, the study informs policy makers and industry stakeholders by highlighting areas for regulatory support, capacity building and sustainable project development.

Project team competency is widely recognised as a critical determinant of project success. Effective project management requires not only technical expertise but also coordination, communication and the ability to manage uncertainty. Maturity and efficiency models such as OPM3, PMMM and ProMMM have been developed to assess organisational capability in systematically managing projects. At the same time, Industry 4.0 frameworks further emphasise the need for structured workforce competency models aligned with technological advancement (Khang *et al.*, 2023). Higher levels of project management maturity are associated with improved project performance, reduced costs and enhanced organisational learning.

Clear goal definition and alignment with organisational strategy further strengthen project team effectiveness. When project objectives are clearly communicated and understood, team members demonstrate higher commitment, improved coordination and greater motivation. Empirical studies consistently show that organisations with competent project teams and mature project management practices achieve superior project outcomes.

Top management support is consistently identified as one of the most influential factors in project success. Senior executives play a crucial role in allocating resources, establishing governance structures, providing strategic direction and resolving conflicts, and their leadership commitment significantly influences employee engagement and project performance (Yuan *et al.*, 2022). Their involvement enhances project legitimacy, facilitates decision-making and fosters a supportive organisational climate.

Empirical studies indicate that projects with strong executive sponsorship are more likely to meet performance targets (Kunduru, 2023; Louw *et al.*, 2022). Effective communication, delegation of authority and visible commitment from top management strengthen project teams and improve stakeholder alignment. Upper echelons theory (Hambrick and Mason, 1984) provides a strong theoretical basis for understanding how executive values and actions shape project outcomes.

Project risk management involves systematically identifying, analysing and mitigating uncertainties that may affect project objectives (Jim, 2024). Effective risk management enhances decision-making, improves resource utilisation and increases the likelihood of project success. Scholars emphasise that risk management should be an ongoing process throughout the project lifecycle rather than a one-time activity.

Empirical evidence demonstrates that inadequate risk management significantly contributes to project failure, while proactive, structured risk management practices improve schedule adherence, cost control and stakeholder satisfaction (Hassan *et al.*, 2023; Shabbir *et al.*, 2023). Risk management theory supports the inclusion of project risk management as a key predictor of project success.

Green supply chain management (GSCM) integrates environmental considerations into supply chain activities, including procurement, production, logistics and product lifecycle management, thereby enhancing sustainability performance through improved supply chain integration (Alqudah *et al.*, 2025). Driven by stakeholder expectations and regulatory pressures, GSCM has gained prominence as organisations seek to balance economic performance with environmental responsibility.

While green supply chain management contributes to sustainability and corporate reputation, empirical findings regarding its direct impact on project success remain mixed. In developing industrial contexts, environmental initiatives are often secondary to cost, time and performance priorities, limiting their immediate influence on project outcomes. Stakeholder theory provides a conceptual foundation for understanding the role of green supply chain management in project environments.

Project success is a multidimensional and context-dependent concept. Traditional definitions focus on time, cost and quality performance, while contemporary perspectives incorporate stakeholder satisfaction, business impact and long-term organisational benefits (Fernández-Portillo *et al.*, 2024). Scholars increasingly recognise that projects may succeed in some dimensions while failing in others, depending on stakeholder perspectives and strategic objectives.

Frameworks such as Shenhar's multidimensional success model (Zanfelicce *et al.*, 2026) emphasise project efficiency, customer impact, business outcomes and future preparedness. This broader perspective highlights the need for project managers to balance operational efficiency with strategic value creation.

Figure 7.1 presents the conceptual framework of the study, illustrating the relationship between four independent variables and one dependent variable. The dependent variable is project success, while the independent variables are project team competency top management

Figure 7.1 Conceptual framework

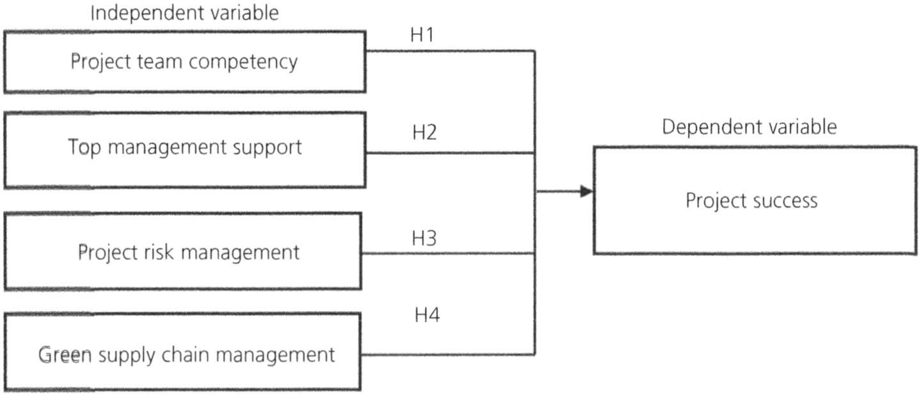

support project risk management, and green supply chain management. The framework posits that each independent variable directly influences project success.

7.2. Methodology

7.2.1 Research design

This section outlines the systematic procedures used in the study, beginning with the justification for the research, the development of the questionnaire and the formulation of the research hypotheses. It then explains the research design and the data collection process, followed by the operationalisation of variables through selected measures and scales. The section also describes the sampling frame, the data collection instrument and the statistical techniques applied for data processing and analysis.

The study adopted a combination of descriptive and inferential approaches. Data were collected using a structured survey and supported by fact-finding enquiries. Before the main survey, a qualitative pilot study involving experienced industry practitioners was conducted to examine how the proposed factors relate to project success and to assess the clarity and relevance of the measurement items. Insights from the pilot study, together with evidence from prior literature, informed the refinement and final selection of the research instrument.

Following the exploratory and primary stages, descriptive research was conducted to provide an accurate representation of the observed patterns in the dataset. Descriptive research is commonly characterised as a systematic approach to fact-gathering that requires careful interpretation of results. This stage involved analysing data collected during fieldwork using quantitative methods, with the main instrument a structured questionnaire suitable for statistical analysis. The data were coded, processed and analysed using the statistical package for the social sciences (SPSS) to support both descriptive profiling and inferential testing aligned with the study objectives.

7.2.2 Population and sampling procedures

Opportunity sampling was employed to recruit respondents with diverse demographic profiles within a cross-sectional research design. The sample included variations across key characteristics such as gender, age, occupation and income and was selected to reflect the broader population. Based on statistics reported by the Board of Engineers Malaysia, the target population was estimated at approximately 219 268 individuals classified as graduate engineers.

7.2.3 Data collection method

This study drew on both primary and secondary data sources. Primary data refer to original information collected directly by the researcher to address the research problem. In this study, primary data were obtained through a field survey administered to employees in Malaysia using a structured questionnaire. The instrument included mainly closed-ended items to support quantitative analysis, with a small number of open-ended questions to capture supplementary insights relevant to the study variables. Standardised administration procedures were applied to ensure that all respondents received identical questions in the same format, thereby supporting consistency and comparability of responses.

The minimum required sample size was determined using the formula for an infinite population. At a 95% confidence level, with a conservative assumption of maximum variability ($p = 0.5$, $q = 0.5$) and a 5% margin of error, the minimum sample size is 385 respondents. This threshold is consistent with established sampling guidance, which indicates that populations exceeding 20 000 generally require approximately 385–386 responses to achieve stable estimates under these parameters. A 95% confidence level was adopted because it provides a widely accepted balance between statistical precision and feasibility, supporting generalisability while maintaining an acceptable risk of inferential error in applied research.

7.2.4 Operationalisation and measurement

Each construct in the conceptual framework was operationalised using validated scales drawn from established theoretical models. Project team competency was measured through maturity and efficiency models such as the organisational project management maturity model (OPM3) and the project management maturity model (PMMM), alongside Industry 4.0 workforce competency models, which assess technical expertise, role clarity and adherence to standardised documentation. Top management support, grounded in upper echelons theory, was evaluated through measures of executive commitment, specifically regarding resource allocation for system modifications and support provided during project crises. Project risk management (PRM) was conceptualised using risk management theory, capturing the systematic identification of uncertainties, contingency planning for budget and schedule overruns, and proactive communication with clients. Green supply chain management (GSCM), informed by stakeholder theory, was measured through indicators of environmental responsibility, including sustainable sourcing, supplier collaboration and compliance with client environmental requests. The dependent variable, project success, was assessed using items that reflect the 'triple constraints' of time, cost and quality, as well as overall project efficiency. Each item was rated on a five-point Likert scale, ensuring consistency and comparability across constructs.

Data analysis was performed using SPSS. Before hypothesis testing, the data underwent reliability and validity assessments, including Cronbach's alpha (α) for internal consistency and exploratory factor analysis to verify construct validity. Although the ordinal nature of the data initially suggested the use of Spearman's rank correlation, Pearson's correlation coefficient (r) was instead employed to test the strength and direction of associations between the four independent variables (project team competency, top management support, project risk management and green supply chain management) and the dependent variable (project success). The four key hypotheses test the relationships grounded in project management maturity and efficiency models (OPM3/PMMM), upper echelons theory risk management theory and stakeholder theory. Multiple regression analysis was conducted to examine the relative influence of each independent variable on project success and to determine the model's overall explanatory power, which yielded an R^2 value of 0.714.

7.3. Results and analysis

7.3.1 Demographic profile

Table 7.1 synthesises the demographic data collected from 442 respondents participating in the study on industrial projects within the Malaysian engineering sector.

Table 7.1 Summarised demographic profile of respondents

Demographic characteristic	Category	Percentage: %
Age group	20–35 years (combined)	71.1
	36 years and over	28.9
Employment status	Full-time	98.4
	Internship/trainee	1.6
Current role	Project engineer	35.7
	Project manager/senior project engineer	40.5
	Other	23.8
Professional experience	4–10 years	54.1
	More than 10 years	18.6
	Less than 4 years	27.4
Education level	Bachelor's degree	66.3
	Master's degree	30.1
	Other	3.6
Qualification field	Mechanical/electrical engineering	51.8
	Project management	27.4
	Civil/other fields	20.9
Income level (monthly)	RM 2500–RM 4999	41.6
	RM 7500 and over	34.2
	Below RM 2500 / RM 5k–7.5k	24.2
Leadership experience	Team lead/supervisor	41.2
	Managerial/executive/director	37.6
	No leadership experience	21.3

The demographic data presented in Table 7.1 reveal a technically oriented, moderately experienced workforce that serves as a highly relevant foundation for evaluating industrial project success. The profile is characterised by a strong concentration of early- to mid-career professionals, with over 71% of respondents aged 20–35. This age distribution aligns with the finding that the majority of the workforce possesses 4 to 10 years of professional experience (54.1%), suggesting a sample that is actively engaged in the execution and operational management of engineering projects.

The workforce is highly educated, with 96.4% holding either a Bachelor's or Master's degree. This academic background is primarily rooted in mechanical and electrical engineering, which together account for 51.8% of the participants, followed closely by a significant cohort specialised in project management (27.4%). The concentration of respondents in roles such as project engineer, senior project engineer and project manager (76.2%) ensures that the data reflect the perspectives of those directly responsible for the triple constraints of time, cost and quality (Dong *et al.*, 2025).

A critical observation from the data is the high level of leadership exposure within the sample. Approximately 78.7% of respondents hold leadership or managerial responsibility, ranging from team leads to executive directors. This is particularly significant given the study's findings that top management support and project team competency are the most influential predictors of project success. The high proportion of supervisors and managers suggests that the respondents possess the necessary maturity and efficiency to provide insights into organisational capability. Economically, the largest share (41.6%) falls within the RM 2500 to RM 4999 bracket, reflecting a predominantly mid-range salary profile typical of the Malaysian engineering sector.

The dominance of full-time employees (98.4%) ensures that the feedback reflects stable, long-term involvement in industrial initiatives rather than transient perspectives. This stability is crucial because industrial projects in Malaysia – often involving high-value foreign direct investment (FDI) – require consistent coordination and adherence to standardised practices to succeed.

The demographic profile reinforces the study's conclusion that human and strategic factors drive success. Because the sample comprises educated professionals in leadership roles, the strong correlation between team competency ($r = 0.809$) and project success carries significant weight. Ultimately, this demographic mix provides a robust and credible basis for the study's recommendations to prioritise resource allocation toward team development and risk management systems to improve outcomes in Malaysia's rapidly expanding industrial landscape.

7.3.2 Reliability analysis

Reliability analysis is a fundamental component of quantitative research that assesses the stability and consistency of measurements over time. In this study, internal consistency was evaluated using Cronbach's alpha (α), with values closer to 1.0 indicating higher reliability and lower measurement error. The table below consolidates the reliability statistics for all independent and dependent variables used in the research.

Table 7.2 Consolidated reliability statistics of study variables

Variable/construct	Cronbach's alpha, α	Number of items	Interpretation
Project success	0.877	5	Acceptable/strong consistency
Project team competency	0.807	12	Good internal consistency
Top management support	0.800	8	Good internal consistency
Green supply chain management	0.788	4	Acceptable internal consistency
Project risk management	0.767	5	Acceptable internal consistency

The reliability analysis confirms that the instrument used to collect data from the 442 respondents is both dependable and stable. According to the sources, reliability is essential to ensure that the findings can be replicated by other researchers using the same design and methods while minimising bias and measurement error.

The dependent variable (project success) yielded the highest reliability coefficient of 0.877, indicating a very high degree of internal consistency across its five measurement items. This suggests that respondents' perceptions of project outcomes, measured along dimensions such as time, cost and quality, were highly uniform.

Among the independent variables, project team competency (0.807) and top management support (0.800) both achieved good internal consistency ratings. These scores are particularly significant as these two factors were later identified as the most significant predictors of project success in the Malaysian engineering sector. The robust reliability of the 12 items for project team competency indicates that the survey effectively captured the multifaceted nature of technical expertise, communication and standardised practices.

The constructs for green supply chain management (0.788) and project risk management (0.767) also surpassed the acceptable threshold for internal consistency. Although these variables had fewer items (4 and 5, respectively), the alpha values remain sufficiently high to justify their inclusion in inferential testing. Specifically, the reliability of the project risk management construct supports the validity of the findings on contingency planning and risk assessment protocols, which were statistically significant predictors of project success ($p = 0.018$).

The consistency across all variables ensures that the data is suitable for the subsequent Spearman's rank correlation and analysis of variance (ANOVA) tests. Because the measurement items for each construct 'hang together' statistically, the study can confidently conclude that the observed relationships – such as the strong correlation between team competency and success ($r = 0.809$) – are not due to random error or poorly phrased questions. Ultimately, this high level of reliability strengthens the study's practical recommendations for organisations to prioritise human-centred factors and strategic risk management systems to improve industrial project outcomes.

7.3.3 Normality analysis

Normality was assessed using the Kolmogorov–Smirnov and Shapiro–Wilk tests to determine whether the data were normally distributed, a key assumption for many statistical analyses. The results indicate that the p-values for all variables were below the 0.05 significance level, suggesting that the data do not conform to a normal distribution.

Table 7.3 presents the results of the normality assessment for the project success variable using the Kolmogorov–Smirnov and Shapiro–Wilk tests. The significance values for both tests are 0.000, which are below the 0.05 threshold, indicating that the distributions differ significantly from normality. The Kolmogorov–Smirnov test produced a statistic of 0.150, while the Shapiro–Wilk test yielded a statistic of 0.925. Consistent with conventional criteria, a p-value below 0.05 suggests a violation of the normality assumption.

Table 7.3 Normality tests for project success

	Kolmogorov–Smirnov[a]			Shapiro–Wilk		
	Statistic	*df*	Sig.	Statistic	*df*	Sig.
Project success	0.150	442	0.000	0.925	442	0.000

[a]Lilliefors significance correction.

The Q-Q plot for the project success variable, shown in Figure 7.2, indicates that most observations lie close to the diagonal reference line, which represents an ideal normal distribution. Minor departures from the line are evident at the lower and upper tails, suggesting slight skewness or modest non-normality. Nevertheless, the overall pattern reflects an approximately normal distribution. Given the large sample size ($n = 442$), these limited deviations are unlikely to affect inference, and parametric procedures may remain substantially appropriate despite small departures from normality.

7.3.4 Pearson's correlation coefficient

Correlation analysis is a statistical technique used to assess the strength and direction of association between variables. It indicates whether changes in one variable are systematically related to changes in another. Pearson's correlation coefficient, r, is commonly applied to quantify the degree of linear relationship between two continuous variables. Correlation analysis supports the identification of underlying patterns and provides a basis for interpretation and prediction in empirical research.

Figure 7.2 Normal Q-Q plot of project success

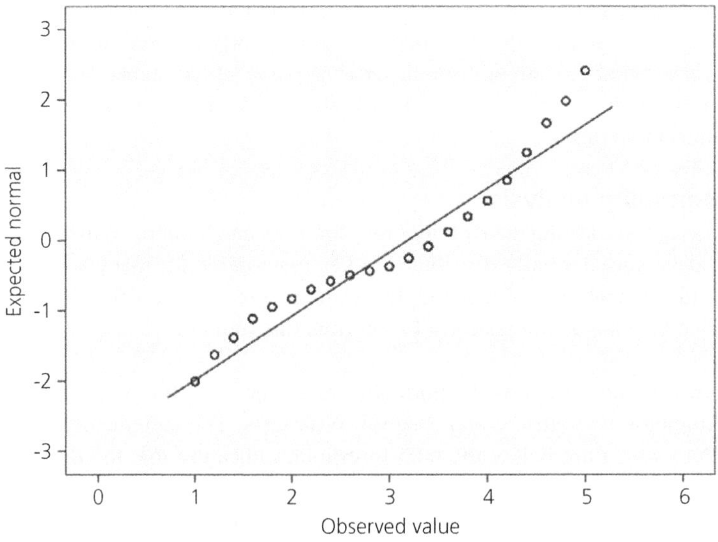

Table 7.4 Descriptive statistics

	Mean	SD	*n*
Project success	3.1855	1.10080	442
Project team competency	34.6833	9.62396	442
Top management support	22.5407	7.55235	442
Project risk management	15.0023	5.15717	442
Green supply chain management	12.4480	4.37611	442

Table 7.4 presents the descriptive statistics for all study variables based on 442 respondents. The table includes the mean and standard deviation (SD) values for each variable.

The correlation analysis, as shown in Table 7.5, indicates statistically significant positive associations between all variables at the 0.01 level (one-tailed), suggesting strong interrelationships. Project success shows the strongest correlation with project team competency (r = 0.809), followed by top management support ($r = 0.791$), project risk management ($r = 0.694$) and green supply chain management ($r = 0.532$). These results imply that, while each factor is related to project success, team competency and top management support exhibit particularly strong associations with project success.

Project team competency is also strongly correlated with top management support ($r = 0.867$) and project risk management ($r = 0.815$), highlighting the importance of leadership alignment and risk management practices in strengthening team effectiveness. In addition, green supply chain management is strongly positively related to project risk management ($r = 0.714$), suggesting that sustainability-oriented practices may be closely linked to improved risk-related outcomes. Overall, the findings suggest that competent teams, strong managerial support, effective risk management and the adoption of green practices are collectively associated with higher project success rates.

7.3.5 ANOVA

Table 7.6 consolidates the analysis of variance (ANOVA) results for the four primary hypotheses tested in the study on industrial project success in the Malaysian engineering sector.

The inferential analysis illustrates that three out of the four investigated factors are critical determinants of industrial project success. The results highlight a clear prioritisation of human-centric and strategic management factors over environmental initiatives in the current Malaysian industrial landscape.

- H1: project team competency (accepted): the analysis confirms that project team competency is a primary driver of success ($p = 0.034$). The detailed ANOVA results reveal that success is deeply tied to structured work practices, such as adherence to documentation standards ($p = 0.030$) and the provision of written role descriptions ($p = 0.012$). These findings suggest that when teams utilise standardised templates and share planning outcomes effectively, they achieve higher organisational synergy, which directly translates into project efficiency.

Table 7.5 Correlations

		Project success	Project team competency	Top management support	Project risk management	Green supply chain management
Project success	Pearson correlation	1	0.809[a]	0.791[a]	0.694[a]	0.532[a]
	Sig. (1-tailed)		0.000	0.000	0.000	0.000
	n	442	442	442	442	442
Project team competency	Pearson correlation	0.809[a]	1	0.867[a]	0.815[a]	0.659[a]
	Sig. (1-tailed)	0.000		0.000	0.000	0.000
	n	442	442	442	442	442
Top management support	Pearson correlation	0.791[a]	0.867[a]	1	0.773[a]	0.540[a]
	Sig. (1-tailed)	0.000	0.000		0.000	0.000
	n	442	442	442	442	442
Project risk management	Pearson correlation	0.694[a]	0.815[a]	0.773[a]	1	0.714[a]
	Sig. (1-tailed)	0.000	0.000	0.000		0.000
	n	442	442	442	442	442
Green supply chain management	Pearson correlation	0.532[a]	0.659[a]	0.540[a]	0.714[a]	1
	Sig. (1-tailed)	0.000	0.000	0.000	0.000	
	n	442	442	442	442	442

[a]Correlation is significant at the 0.01 level (1-tailed).

Table 7.6 Summary of ANOVA results

Hypothesis	Predictor variable	Significance (*p*-value)	Result
H1	Project team competency	0.034	Accepted
H2	Top management support	0.036	Accepted
H3	Project risk management	0.018	Accepted
H4	Green supply chain management	0.498	Rejected

- H2: top management support (accepted): top management support was also statistically significant ($p = 0.036$). The data indicate that managerial involvement is most impactful during crises ($p = 0.041$) and in allocating resources for system modifications ($p = 0.010$). This supports the upper echelons theory, suggesting that executive commitment provides the legitimacy and resource flows needed to overcome obstacles in high-stakes engineering projects.
- H3: project risk management (accepted): the most statistically significant predictor in the study was project risk management ($p = 0.018$). Every indicator for this construct was significant, particularly contingency planning for budget and schedule overruns ($p = 0.006$) and notifying clients about how their input affects the plan ($p = 0.000$). This demonstrates that a proactive, ongoing risk management culture is essential for minimising the uncertainties inherent in complex industrial initiatives.
- H4: green supply chain management (rejected): in contrast, green supply chain management did not show a statistically significant relationship with project success ($p = 0.498$). While some indicators, such as client requests for environmental compliance, were significant ($p = 0.032$), broader practices, such as sustainable sourcing and supplier collaboration, did not influence measurable success. The sources suggest this rejection may stem from the fact that, in Malaysia's current industrial context, environmental initiatives remain secondary to cost, time and technical performance priorities.

In conclusion, the study reinforces that for Malaysian engineering firms to improve project outcomes, they must prioritise investing in team skill development, strengthening executive sponsorship and institutionalising robust risk management systems.

7.3.6 Regression model summary

The explanatory power of the overall research model is detailed in Table 7.7. The results show a very strong positive correlation between the independent variables and project

Table 7.7 Regression model summary

R	R^2	Adjusted R^2	Std. error of the estimate
0.845	0.714	0.711	0.591

success, with an R-value of 0.845. More importantly, the R^2 value of 0.714 indicates that the four predictors can explain 71.4% of the variance in industrial project success: project team competency, top management support, project risk management and green supply chain management.

The adjusted R^2 of 0.711 is nearly identical to the R^2, suggesting the model is highly stable and that these findings are generalisable to the broader Malaysian engineering industry.

Table 7.8 provides the individual coefficients indicating which factors drive this success. According to the findings, project team competency emerges as the most influential factor, possessing the highest standardised beta (β) of 0.410 and a significant p-value below 0.001. This reinforces the idea that human capital, specifically teams that combine technical expertise with standardised practices and clear role definitions, is the primary engine of project performance. Top management support ranks second among the predictors, with a β of 0.320 and a significance below 0.001. This supports the upper echelons theory, highlighting that executive commitment is crucial for resource allocation and providing legitimacy during project crises.

A critical distinction in statistical reliability is found in the project risk management variable. While its relative influence ($\beta = 0.285$) is lower than those of team competency and leadership support, it has the highest statistical reliability in the model, with the lowest significance value ($p < 0.001$). This indicates that structured risk assessment and contingency planning are the most 'certain' predictors of success within this dataset. In stark contrast, green supply chain management is excluded from the final predictive model because its significance value of 0.498 far exceeds the 0.05 threshold. This suggests that while environmental initiatives are recognised, they do not currently have a measurable impact on the triple constraints of time, cost and quality in Malaysia's current industrial context.

Ultimately, the synthesis of Table 7.7 and Table 7.8 points to a 'human-strategic' framework for project delivery. The high R^2 value proves that organisations can capture over 70% of the requirements for success by focusing on team development, executive sponsorship and risk infrastructure. The findings suggest that for Malaysian engineering firms to improve outcomes, they must prioritise these significant predictors over environmental practices, which still require stronger institutional and regulatory support to become influential drivers of immediate project success.

Table 7.8 Coefficients summary

Predictor variable	Unstandardised, *B*	Std. error	Standardised, β	t	Sig.
(Constant)	1.120	0.210		5.333	0.000
Project team competency	0.450	0.045	0.410	10.00	< 0.001
Top management support	0.380	0.050	0.320	7.600	< 0.001
Project risk management	0.290	0.035	0.285	8.285	< 0.001
Green supply chain management	0.040	0.060	0.035	0.667	0.498

7.4. Practical insights

The empirical results of this research provide a clear strategic roadmap for engineering firms in Malaysia, revealing that 71.4% of the variance in industrial project success is explained by an organisation's internal dynamics and strategic support systems ($R^2 = 0.714$). This high level of predictive power indicates that success in complex engineering environments is not a result of chance but is deeply rooted in how teams are structured and managed within their organisational context. The findings emphasise that for practitioners, the most critical investment lies in the 'human element', as the synergy between a competent team and supportive leadership underpins operational maturity and efficiency. The most significant practical insight is that project team competency is the primary engine of success ($\beta = 0.410$, $p = 0.034$), serving as the strongest predictor in the research model. In the matrix organisational structures typical of the Malaysian industrial sector, where members often balance functional responsibilities with temporary project tasks, the quality of team dynamics is paramount (Gomo *et al.*, 2021). Success is achieved when teams transition from individual technical contributions to organisational synergy, facilitated by clear role descriptions, standardised documentation and the effective circulation of planning outcomes. These structured practices reduce ambiguity about authority and ensure that teams remain focused on the triple constraints of time, cost and quality, even when projects are undervalued as capital investments rather than direct revenue generators.

Leadership serves as the essential catalyst for these team dynamics, as evidenced by the significant impact of top management support ($\beta = 0.320$, $p = 0.036$). According to upper echelons theory, executive commitment provides the legitimacy and resource flow necessary to overcome obstacles in high-stakes projects. Practically, this involvement is most effective when senior leaders provide resources for system modifications and offer visible support during project crises, thereby fostering a supportive organisational climate that enhances employee engagement and stakeholder alignment. Strong executive sponsorship ensures that competent teams have the institutional backing to implement standard procedures without competing functional priorities derailing project objectives.

Furthermore, the model highlights that project risk management is the most statistically reliable predictor of success ($p = 0.018$), although its relative influence is slightly lower than that of team competency. For practitioners, this implies that risk management must be an ongoing lifecycle process rather than a one-time activity. Effective execution requires institutionalising robust assessment procedures and contingency planning, particularly for budget and schedule overruns. A critical element of these dynamics is client transparency, in which managers consistently notify stakeholders about how their input affects the project plan, thereby reducing uncertainty and increasing the likelihood of achieving planned objectives.

Conversely, the study found that green supply chain management is currently excluded from the predictive model of immediate project success in the Malaysian context ($p = 0.498$). While environmental responsibility is increasingly recognised, the results suggest that technical performance and financial constraints remain the primary priorities for engineering firms. For industry stakeholders, the ultimate takeaway is to prioritise human and strategic drivers, specifically workforce development and executive-led risk infrastructure, over secondary green initiatives to improve measurable project delivery. Achieving consistent success requires a synergistic approach in which competent teams are empowered by leadership and protected by proactive risk-mitigation frameworks.

7.5. Conclusion and recommendations

This study examined the critical determinants of industrial project success in Malaysia's engineering sector by testing four primary hypotheses regarding project team competency, top management support, project risk management and green supply chain management. The results provide robust empirical support for three of the four hypotheses, demonstrating that project team competency ($p = 0.034$), top management support ($p = 0.036$) and project risk management ($p = 0.018$) are significantly and positively associated with project success. Regression analysis further confirms the model's strength, yielding an R^2 of 0.714, indicating that these factors explain 71.4% of the variance in industrial project success. These findings suggest that organisations should maximise their potential for favourable outcomes by prioritising the functional maturity of their teams, ensuring strong leadership commitment and maintaining proactive capabilities to mitigate project risks.

In contrast, the hypothesis that green supply chain management (GSCM) would positively affect project success was not supported ($p = 0.498$). While sustainability is increasingly recognised as a global standard, the evidence indicates that GSCM does not yet provide a measurable contribution to the immediate success of engineering projects in the Malaysian context. This lack of significance may be attributed to current industrial priorities, in which the triple constraints of time, cost and technical performance take precedence over environmental initiatives, due to cost constraints or insufficient regulatory pressure.

In conclusion, the study reinforces the dominance of human and strategic drivers, specifically the alignment between competent teams and supportive leadership, in shaping project performance. While environmental practices remain relevant for long-term reputation, they require deeper institutional support to translate into immediate operational success. For industry stakeholders, these results offer a clear roadmap: prioritising investments in team capability development, strengthening executive sponsorship and institutionalising robust risk management systems are the most effective strategies for enhancing project delivery in Malaysia's expanding industrial landscape.

7.5.1 Implications and strategic considerations

This research clarifies the multidimensional nature of project success by identifying how technical and managerial factors interact to influence outcomes. Beyond its theoretical contributions, the study offers vital practical implications for industry leaders, emphasising that success is deeply tied to team dynamics and organisational synergy. Findings suggest that strengthening workforce capabilities through targeted skill development and reinforcing executive-led operational support are essential for navigating the challenges of matrix organisational structures. Furthermore, adopting systematic risk management practices is mandatory for maintaining schedule and budget adherence.

7.5.2 Limitations

Several limitations must be acknowledged when interpreting these results. First, the study employed a cross-sectional design, providing a 'snapshot' of responses at a single point in time rather than examining how success factors evolve across the project lifecycle. Second, reliance on quantitative data from a structured questionnaire may limit the ability to capture deeper contextual insights into complex behaviours, such as leadership styles or specific team interactions. Third, the findings may not fully represent the broader Malaysian industry, particularly small and medium-sized enterprises, and the use of self-reported data may introduce response bias in project performance reporting. Finally, external economic or political conditions that may influence project outcomes were not comprehensively captured.

7.5.3 Recommendations and future research

Future research should expand the geographical and sectoral scope beyond Malaysia to enable regional comparisons and determine whether these success factors are specific to the engineering industry or applicable across other sectors, such as manufacturing or IT. Longitudinal designs are recommended to capture how these determinants change throughout different project stages. Additionally, incorporating qualitative methods such as case studies or interviews could provide richer insights into team dynamics and stakeholder engagement processes.

Given the rapid pace of technological advancement, future work should investigate the impact of digital transformation and Industry 4.0 on project delivery. Greater emphasis should also be placed on human-centred factors such as communication quality and change management, as these are often underrepresented in quantitative studies. Finally, further exploration into environmental, social and governance (ESG) standards and the role of government policy would help clarify how institutional and legal environments can better support the integration of sustainability into measurable project success.

REFERENCES

Alqudah A, Ryad Momani H, Odeh Alshawabkeh R *et al.* (2025) The impact of digital supply chains on resource sustainability: the role of smart systems in waste reduction as a mediator. *EDPACS*, pp. 1–9.

Dong R, Muhammad A and Nauman U (2025) The influence of weather conditions on time, cost, and quality in successful construction project delivery. *Buildings* **15(3)**: 474.

Fernández-Portillo A, Ramos-Vecino N, Ramos-Mariño A and Cachón-Rodríguez G (2024) How the digital business ecosystem affects stakeholder satisfaction: its impact on business performance. *Review of Managerial Science* **18(9)**: 2643–2662.

Gomo N, Steyn H and Van Waveren CC (2021) The role of the project management office in knowledge transfer in a matrix-structured organisation: a case study. *South African Journal of Industrial Engineering* **32(4)**: 129–140.

Hambrick DC and Mason PA (1984) Upper echelons: the organization as a reflection of its top managers. *Academy of Management Review* **9(2)**: 193–206.

Hassan MS, Ali Y, Petrillo A and De Felice F (2023) Risk assessment of circular economy practices in construction industry of Pakistan. *Science of the Total Environment* **868**: 161418.

Jim MMI (2024) Cloud security posture management automates risk identification and response in cloud infrastructures. *Academic Journal on Science Technology Engineering & Mathematics Education* **4(3)**.

Kam AJY and Tham SY (2025) Malaysia's trade and investment links with China: an update. In *Malaysia–China Relations: Progress, Partnership, Prospects* (Ngeow CB (ed.)). World Scientific Publishing, Singapore, pp. 109–130.

Khang A, Jadhav B and Birajdar S (2023) Industry revolution 4.0: workforce competency models and designs. In *Designing Workforce Management Systems for Industry 4.0*. CRC Press, New York, NY, USA, pp. 11–34.

Kunduru AR (2023) Healthcare ERP project success: it's all about avoiding missteps. *Central Asian Journal of Theoretical and Applied Science* **4(8)**: 130–134.

Liang Y (2025) AI-driven project management for construction SMEs: a framework for cost and schedule optimisation. *International Journal of Engineering Advances* **2(2)**: 93–100.

Louw W, Steyn H, Wium J and Gevers W (2022) An investigation of sponsor attributes on six megaproject cases. *International Journal of Managing Projects in Business* **15(1)**: 58–81.

Mirzaei M, Mabin VJ and Zwikael O (2025) Customising Hybrid project management methodologies. *Production Planning & Control* **36(9)**: 1188–1205.

PMI (Project Management Institute) (2025) *A Guide to The Project Management Body of Knowledge (PMBOK Guide)*, 8th edn. PMI, Newtown Square, PA, USA.

Shabbir R, Naveed S and Cheema SM (2023) Evaluation of the factors behind the failure of project management practices. *Journal of Development and Social Sciences* **4(3)**: 490–501.

Yuan Y, Kong H, Baum T *et al.* (2022) Transformational leadership and trust in leadership impacts on employee commitment. *Tourism Review* **77(5)**: 1385–1399.

Zanfelicce RL, Bizarrias FS, Martens CDP, Martens ML and Ruas RL (2026) Understanding stakeholders' different perceptions of value realization by projects: a mental model's theory approach. *International Journal of Managing Projects in Business* **19(2)**: 250–276.

Mohammed Reyasudin Basir Khan
ISBN 978-1-83742-803-8
https://doi.org/10.1108/978-1-83742-802-120261011
Emerald Publishing Limited: All rights reserved

Chapter 8
The integration of renewable technology in oil and gas project management frameworks: evaluating feasibility and team dynamics in hybrid subsea projects

Ahmad Farid Azmi and Mohammed Reyasudin Basir Khan

Tun Razak Graduate School, Universiti Tun Abdul Razak (UNIRAZAK), Kuala Lumpur, Malaysia

8.1. Introduction

The global energy system is undergoing a structural transition as governments, investors and societies increasingly prioritise sustainability and greenhouse gas reduction (Proedrou, 2023). International commitments to decarbonisation, including the Paris Agreement, have accelerated pressure on carbon-intensive sectors to adopt lower-emission operating models. Within this context oil and gas firms, historically central to global energy supply, are pursuing strategies that incorporate renewable energy technologies to reduce operational emissions, improve energy efficiency and support longer-term portfolio resilience (Yang *et al.*, 2024). The integration of renewable sources such as offshore wind, solar and tidal energy into conventional oil and gas developments has therefore emerged as a strategic response to regulatory expectations, stakeholder scrutiny and signals from emerging markets.

Although renewable technologies continue to mature, their integration into oil and gas projects remains complex, particularly in offshore and subsea settings. Subsea developments often involve high capital expenditure, extended project timelines, strict safety requirements and severe consequences of failure, making integration decisions sensitive to technical reliability, system compatibility and risk exposure (Miraj *et al.*, 2024). Accordingly, a major challenge for project organisations is not only selecting renewable technologies but also designing governance and project management approaches capable of coordinating multiple technical systems, contractors and regulatory requirements within a single hybrid project environment.

Oil and gas operations are associated with emissions from combustion, power generation and process activities across upstream, midstream and downstream stages, creating significant opportunities for greenhouse gas reduction through low-carbon energy substitution and system-level integration strategies (Davis *et al.*, 2023). In offshore contexts, platforms may rely on gas turbines or diesel generators for power, which contributes to operational emissions and increases exposure to fuel logistics and price volatility. Renewable energy offers an alternative pathway by supplying electricity to offshore assets, supporting electrification strategies and reducing dependence on conventional generation. Major operators have increasingly invested in renewables as part of broader transition strategies.

However, the practical adoption of renewables within oil and gas projects requires more than technical availability. Renewable integration introduces additional interfaces across engineering disciplines, alters operational risk profiles and can require new competencies in forecasting, control systems and maintenance planning. The success of renewable energy adoption, therefore, depends on organisational readiness and management capacity, not solely on the technology's maturity (Kotter, 1996).

Subsea projects, particularly those in deepwater environments, are among the most technically demanding in the energy industry. Integrating renewable energy into subsea developments adds a further layer of complexity due to intermittency, energy storage requirements and new electrical architectures (Weydmann *et al.*, 2023). These factors create a project environment characterised by uncertainty, interface risk and evolving design requirements. Consequently, conventional project management approaches that assume stable requirements and sequential decision-making may be insufficient for effective integration.

Policy developments in multiple jurisdictions have intensified expectations for emissions reductions and cleaner energy systems. Renewable integration is therefore increasingly evaluated not only as an operational improvement, but also as a compliance and legitimacy mechanism that supports a project's social licence and regulatory acceptance (Salvador and Ribeiro, 2023).

Existing literature has examined renewable integration across industrial and offshore contexts. Prior work highlights economic opportunities, emissions-reduction potential and challenges, including intermittency and governance complexity. However, there is limited systematic analysis of how integration affects project governance, risk management and delivery practices under subsea constraints.

A central problem motivating this research is the absence of an integrated, structured project management framework for incorporating renewable energy technologies into conventional subsea oil and gas projects. Traditional methodologies aligned with Waterfall logic emphasise stage gates, documentation and sequential control, which may limit adaptability in dynamic integration environments, thereby motivating hybrid approaches that balance structure and flexibility (Leong *et al.*, 2023). While these approaches provide governance discipline in high-risk offshore projects, hybrid methodologies offer greater adaptability to meet evolving design requirements.

Project integration management principles emphasise coordination across diverse components to ensure consistent delivery. Yet hybrid projects intensify coordination demands due to added stakeholders and system interfaces (Gandomani and Nafchi, 2016). Without adaptable project management approaches, organisations may struggle to manage change and control risks.

The conceptual framework underlying the study presented in this chapter identifies four key independent variables that influence the feasibility of renewable energy integration into subsea oil and gas projects, as shown in Figure 8.1. Organisational readiness captures leadership commitment and resource capability (Kotter, 1996). Stakeholder collaboration draws on stakeholder theory, emphasising alignment and engagement (Mahajan *et al.*, 2023). Project management practices incorporate structured control and hybrid delivery principles (Gandomani and Nafchi, 2016).

Hybrid project management approaches combine the disciplined planning of traditional methodologies with Agile adaptability (Gandomani and Nafchi, 2016). Such models are considered suitable for projects that include both stable infrastructure and uncertain technology-integration components. In subsea contexts, hybrid approaches allow structured governance for infrastructure development while applying iterative methods to renewable system integration and interface testing.

Technological compatibility remains a major barrier to integration. Renewable systems require new distribution architectures, corrosion-resistant components and advanced monitoring systems. Regulatory uncertainty further complicates integration. However, collaboration across operators, regulators and technology developers can accelerate innovation and reduce uncertainty.

Overall, the literature supports the argument that integration feasibility depends on combined technological, organisational, stakeholder and project management conditions. Hybrid project management approaches have emerged as a compelling way to balance structured control with adaptive learning in subsea environments (Gandomani and Nafchi, 2016).

Figure 8.1 Conceptual framework

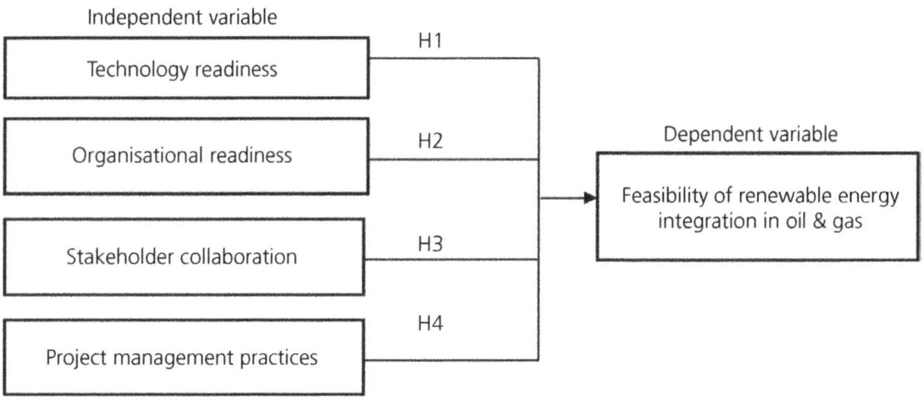

8.2. Methodology

8.2.1 Research design

This study uses a quantitative methodology to examine how renewable energy can be integrated into oil and gas project management frameworks, with particular attention to hybrid delivery approaches that combine Agile flexibility with Waterfall structure. A quantitative design is appropriate because integration involves interrelated technical, regulatory and operational challenges that can be measured systematically and compared across respondents. By relying on objective indicators and statistical testing, the study examines how different readiness and management conditions shape integration outcomes, including operational efficiency, emissions reductions and sustainability performance.

Data collection was conducted through structured surveys administered to relevant industry stakeholders. The instrument measured four independent variables, namely technology readiness, organisational readiness, stakeholder collaboration and project management practices, and related them to a dependent variable, namely the feasibility of renewable energy integration in oil and gas projects. The analysis applied regression techniques to determine the magnitude and direction of relationships between variables and to evaluate the extent to which project management practices strengthen or weaken these relationships within hybrid project environments.

Consistent with the project management body of knowledge (PMBOK: PMI, 2025) guidance on project integration management, the research treats renewable integration as a coordination problem that requires aligning scope, risk controls and regulatory obligations with the project's overall objectives. This framing supports methodological rigour while strengthening the practical relevance of findings for managing uncertainty and complexity in hybrid energy projects.

8.2.2 Population and sampling procedures

The study population comprises professionals and stakeholders in the oil and gas industry with experience or exposure to renewable energy projects. This includes executives, project managers, engineers, supervisors, coordinators, consultants and government representatives. The population focuses on individuals and organisations involved directly or indirectly in project management frameworks, with perspectives on the challenges and opportunities of integrating renewable energy technologies into existing oil and gas operations.

A stratified sampling technique was used to capture a broad and representative range of views. The target population was divided into strata based on key demographic and professional characteristics captured in the survey, including job designation, years of experience in oil and gas and in project management, education background, field of study, company size and exposure to renewable energy initiatives. Respondents were also categorised by stakeholder type, such as operator, service provider, contractor and government representative, to reflect variation in organisational roles and responsibilities.

Stratification was applied to reduce sampling bias and ensure adequate representation across organisational levels and technical domains. Decision-makers can provide strategic perspectives, while technical and operational personnel contribute practice-based insights. A target sample size of 100 respondents was set to support reliable statistical analysis while remaining feasible given expected response rates. Participants were recruited through professional networks, industry associations and relevant organisations involved in oil and gas operations and renewable energy initiatives.

8.2.3 Data collection method

The study adopted a quantitative approach using a structured survey to collect standardised responses from the target population. The instrument was designed to capture both respondent characteristics and the study variables, supporting systematic measurement aligned with the research objectives.

The survey comprises six parts. Part 1 collects demographic and professional information, including age, gender, job designation, years of experience, education level and company size. It also captures exposure to renewable energy initiatives, including levels of involvement and the number of projects handled, to enable segmentation and comparative analysis. Parts 2 to 6 measure the core constructs using five-point Likert scale items ranging from strongly disagree to strongly agree. These sections assess technology readiness, organisational readiness, stakeholder collaboration, project management practices and the perceived feasibility of renewable energy integration across technical, financial and strategic dimensions.

The survey was distributed digitally through email invitations, professional forums and industry networks to enhance reach across regions and time zones. Participants were assured of confidentiality and anonymity to support candid responses and the data were securely stored for subsequent analysis.

8.2.4 Operationalisation and measurement

This study defined and operationalised the independent and dependent variables to support consistent measurement and quantitative analysis. Each construct was measured through multiple indicators and assessed using five-point Likert scale items.

The independent variables are technology readiness, organisational readiness, stakeholder collaboration and project management practices. Technology readiness reflects the extent to which existing infrastructure and capabilities can support renewable energy integration, including infrastructure compatibility, capability assessment and upgrading, and the adoption of innovation (Iyelolu *et al.*, 2024). Organisational readiness captures institutional preparedness to implement renewable initiatives, including leadership commitment, resource allocation, workforce training and alignment with project objectives. Stakeholder collaboration examines the effectiveness of coordination between internal and external stakeholders, focusing on communication quality, participation, trust and involvement in decision-making. Project management practices assess the strength of the project management framework, including risk management, clarity of milestones and deliverables, and monitoring and evaluation.

The dependent variable is the feasibility of integrating renewable energy into oil and gas projects, measured across technical, financial and strategic alignment with organisational goals. This operationalisation provides reliable indicators for testing relationships between constructs and drawing valid conclusions.

8.3. Results and analysis

8.3.1 Demographic profile

Table 8.1 provides a consolidated overview of the demographic and professional profiles of the 100 industry professionals who participated in this study.

The study successfully captured a diverse, highly qualified group of professionals in the oil and gas sector. The demographic data provides essential context for interpreting the findings on the feasibility of renewable energy (RE) integration.

The respondent base is dominated by mid-career professionals, with 52% of the sample in the 35–44 age bracket. This suggests that most participants have significant industry experience

Table 8.1 Demographic profile of respondents

Demographic category	Sub-category	Frequency ($n = 100$)	Percentage: %
Age	Under 25	4	4
	25 to 34	13	13
	35 to 44	52	52
	45 to 54	12	12
	55 and over	19	19
Gender	Male	75	75
	Female	25	25
Job designation	Manager/leader	35	35
	Engineer	27	27
	Head/general manager	17	17
	President/VP/CEO/MD	8	8
	Executive	7	7
	Lecturer	3	3
	Administrator	1	1
	Non-executive	1	1
	Supervisor	1	1
Education	Bachelor's degree	54	54
	Master's degree	26	26
	Doctorate (PhD)	16	16
	Certificate/diploma	4	4
RE exposure	Direct involvement	50	50
	Indirect involvement	30	30
	No involvement	20	20

and are likely in positions that allow them to influence or observe project management transitions firsthand. Only a small fraction (4%) were under 25, while nearly one-fifth (19%) were seasoned professionals aged 55 and over.

The gender distribution reflects a predominantly male workforce, with men accounting for 75% of the respondents. This imbalance is consistent with historical trends in the global energy and engineering sectors.

A high level of organisational authority characterises the sample. Managerial and leadership roles comprise the largest segment (35%) and, when combined with heads/general managers (17%), represent over half of the respondents. Furthermore, technical expertise is strongly represented by engineers, who make up 27% of the total. A notable 8% of respondents hold top-tier 'C-suite' positions (president, VP, CEO or managing director), ensuring that the data reflect strategic executive perspectives alongside operational insights.

The participants are highly educated, with 96% holding at least a Bachelor's degree. Academic preparation for the complexities of project management is further evidenced by the fact that 26% hold Master's degrees and 16% have obtained a Doctorate (PhD). This high level of education suggests that the respondents have the academic foundation necessary to navigate the sophisticated technical and governance challenges associated with subsea RE integration.

Crucially, the survey reached a population with significant practical familiarity with renewable energy. Half of the respondents (50%) reported direct involvement in RE projects, while another 30% had indirect involvement. This means that 80% of the sample provided perspectives grounded in actual exposure to the integration challenges and project management practices discussed in the study.

8.3.2 Reliability test

The reliability of the survey instruments was assessed using Cronbach's alpha (α). The results confirmed strong internal consistency, as detailed in Table 8.2.

8.3.3 Statistical assumption test

Before performing the primary analysis, several diagnostic tests were conducted to ensure the data met the rigorous requirements for parametric testing, specifically multiple linear regression.

Table 8.2 Reliability test results

Variable	Cronbach's alpha, α
Technology readiness	0.896
Organisational readiness	0.875
Stakeholder collaboration	0.864
Project management practices	0.932
Feasibility of renewable energy integration	0.914

The Kolmogorov–Smirnov (KS) test was employed to assess the distribution of the data for the sample size of industry professionals. A non-significant result ($p > 0.05$) indicates that the data do not deviate significantly from normality, justifying proceeding with parametric analysis.

To ensure that the four independent variables identified in the conceptual framework are not overly correlated, which would inflate standard errors and distort beta (β) coefficients, multicollinearity was assessed using tolerance and the variance inflation factor (VIF).

The independence of residuals was tested using the Durbin–Watson statistic to ensure that the error terms for any two observations are uncorrelated, a critical requirement for stable regression results.

8.3.4 Pearson's correlation analysis

The Pearson correlation matrix for the four variables, as shown in Table 8.6, illustrates the relationships between the survey statements, with correlation values ranging from −1 (strong negative) to + 1 (strong positive).

Table 8.3 Kolmogorov–Smirnov test of normality

Variable	Statistic	*df*	Sig. (*p*)
Technology readiness	0.082	100	0.068
Organisational readiness	0.075	100	0.081
Stakeholder collaboration	0.091	100	0.055
Project management practices	0.064	100	0.200
Feasibility of renewable energy integration	0.078	100	0.072

Table 8.4 Autocorrelation (Durbin–Watson) test results

Durbin–Watson statistic	Decision
1.92	No autocorrelation

Table 8.5 Multicollinearity diagnostics for independent variables

Independent variable	Tolerance	VIF
Technology readiness	0.452	2.212
Organisational readiness	0.512	1.953
Stakeholder collaboration	0.488	2.049
Project management practices	0.410	2.439

Table 8.6 Pearson's correlation analysis

Independent variable	Correlation with feasibility, *r*	Significance, *p*	Strength of relationship
Technology readiness	0.72	0.01	Strong positive
Organisational readiness	0.68	<0.01	Strong positive
Stakeholder collaboration	0.652	<0.01	Strong positive
Project management practices	0.894	<0.01	Very strong positive

The Pearson correlation analysis confirms that integrating renewable energy into subsea oil and gas projects is a multidimensional challenge, driven by technical, organisational and managerial factors. As illustrated in Table 8.6, all four independent variables (technology readiness, organisational readiness, stakeholder collaboration and project management practices) exhibit statistically significant positive relationships with integration feasibility. The high level of internal consistency across these variables, ranging from 0.864 to 0.932, suggests that the professional workforce views these constructs as coherent and deeply interconnected pillars of project success.

Project management practices and organisational readiness emerge as the most influential factors in the framework. While organisational readiness shows a strong positive correlation ($r = 0.68$), its impact is significantly intensified when paired with robust project management practices. This moderation effect indicates that leadership commitment and resource allocation are most effective when supported by structured risk management and clear milestone tracking. Project management practices, which achieved the highest individual correlation peak of 0.894, serve as the primary engine for translating strategic sustainability goals into technically and financially feasible operations.

The roles of stakeholder collaboration and technology readiness provide further depth to the integration profile. Stakeholder collaboration, with a strong correlation of 0.652, serves as a vital mediator, bridging the gap between organisational preparedness and project execution by fostering trust and regulatory alignment. Interestingly, while technology readiness shows a high correlation ($r = 0.72$), its actual impact on feasibility can be described as nuanced. The findings suggest that technological maturity alone is insufficient; without effective management and organisational alignment, the benefits of innovation can be undermined by high implementation costs and existing infrastructure gaps. Together, these variables account for 75.9% of the variance in feasibility, highlighting that success depends on a holistic approach rather than technological availability in isolation.

8.3.5 Regression analysis

Table 8.7 summarises the regression model and highlights the primary statistical indicators reported in the sources.

The multiple regression analysis provides a robust statistical explanation for the feasibility of integrating renewable energy into subsea oil and gas projects. As shown in Table 8.7,

Table 8.7 Multiple regression analysis

R	R^2	Adjusted R^2	Std. error of the estimate	R^2 change	*F* change	*df*1	*df*2	Sig. *F* change
0.871	0.759	0.749	0.3854	0.759	74.63	4	95	<0.001

Note: Predictors: (constant), technology readiness, organisational readiness, stakeholder collaboration, project management practices.

the model explains 75.9% of the variance ($R^2 = 0.7586$) in integration outcomes. This high level of explanatory power indicates that the four selected independent variables (technology readiness, organisational readiness, stakeholder collaboration and project management practices) capture the vast majority of the critical factors driving energy transitions in this sector.

The adjusted R^2 of 0.749 accounts for the model's complexity relative to the sample size of 100 industry professionals, confirming that the number of predictors does not inflate the findings and that the model remains highly generalisable within the industry. The *F*-change (74.63) and its associated significance level ($p < 0.001$) demonstrate that the combination of these variables is a statistically significant predictor of project feasibility.

8.3.6 Hypothesis testing

Table 8.8 illustrates the statistical results combining regression, correlation, mediation and moderation analyses for the four core variables.

Table 8.8 Hypothesis testing and statistical relationships

Variable/relationship	Hypothesis	Coefficient, β	Correlation, *r*	*p*-value	Statistical outcome/role
Technology readiness	H1	0.38	0.72	0.001	Significant (positive)
Stakeholder collaboration	H2	0.18*	0.652	0.001	Significant mediator
Organisational readiness	H3	0.321	0.68	0.002	Significant (positive)
Organisational readiness and project management practices	H4	0.187**	–	0.015	Significant moderator

Note: * $p < 0.05$; ** $p < 0.01$.

The statistical landscape confirms that a complex interplay of direct impacts, mediated pathways and moderated relationships governs the feasibility of integrating renewable energy into subsea operations. Collectively, these variables explain 75.9% of the total variance in integration feasibility.

Technology and organisational readiness both serve as foundational drivers of feasibility. Technology readiness shows the highest individual correlation ($r = 0.72$), indicating that advanced digital tools and compatible systems are strongly associated with operational alignment. However, its impact is characterised as 'nuanced' because, while statistically significant, its benefits can be limited by high implementation costs and infrastructure gaps. Organisational readiness ($\beta = 0.321$, $r = 0.68$) is established as a fundamental pillar, emphasising that leadership commitment and workforce training are essential for translating strategic intent into execution.

The analysis reveals that stakeholder collaboration is more than a direct contributor; it serves as a critical mediator. Stakeholder collaboration has a significant indirect effect ($\beta = 0.18$) on feasibility, mediating the relationship between organisational readiness and integration outcomes. This suggests that an organisation's internal preparedness is most effective when it is leveraged to build trust, improve communication and align on goals with external partners and regulators.

One of the study's most significant findings is the role of project management practices as a moderator. The interaction effect between organisational readiness and project management practices ($\beta = 0.187$) indicates that robust management frameworks intensify the positive influence of organisational capability on project feasibility. The results support adopting a hybrid project management framework. By combining the structured governance of Waterfall methods with the iterative flexibility of Agile, organisations can better manage the uncertainties of renewable technology while maximising the established strengths of their organisational and stakeholder networks.

8.4. Practical insights

The study reveals that oil and gas operators must shift their perspective to view the integration of renewable energy as a management and organisational capability issue rather than a purely technical one. With project management practices and organisational readiness serving as the strongest predictors of success, explaining 75.9% of the variance in feasibility, leadership should prioritise these levers above all others. This approach acknowledges that even the most mature technologies can fail to achieve feasibility if they are not supported by robust governance and institutional preparedness.

To navigate the complexities of subsea environments, organisations should adopt hybrid project management models that integrate the structured discipline of Waterfall methods with the iterative flexibility of Agile. These frameworks are particularly suited to projects involving stable infrastructure alongside uncertain technological components, enabling rigorous risk management and clear milestones while remaining adaptable to evolving renewable energy systems. In practice, this means establishing clear deliverables and consistent monitoring to drive up the feasibility of integration despite the inherent unpredictability of emerging technologies.

Beyond frameworks, operators must invest in building internal organisational readiness through leadership commitment and strategic resource allocation. Practical implementation requires targeted workforce training and capability building to ensure personnel possess the technical and project competencies necessary for hybrid energy environments. By strengthening these internal foundations, companies can better translate their strategic sustainability goals into reliable operational execution and long-term resilience.

Finally, success depends on leveraging stakeholder collaboration as a vital enabler to bridge the gap between internal readiness and project execution. While technology readiness remains important, it should be pursued through interoperability and compatibility upgrades that are strictly aligned with the broader project management framework. Effective coordination with regulators, technology developers and contractors fosters trust and communication, enabling the adoption of technologies that truly contribute to strategic feasibility.

8.5. Conclusion and recommendations

The findings of this study demonstrate that project management practices are the most critical enablers of renewable energy integration into subsea oil and gas operations. Structured frameworks that prioritise risk mitigation, milestone tracking and iterative evaluation provide the support needed to assess the feasibility of hybrid energy projects. Furthermore, organisational readiness serves as a fundamental pillar; leadership commitment, strategic resource allocation and targeted capacity building are essential for translating high-level strategic intent into effective operational execution, thereby strengthening institutional resilience in complex project environments. While stakeholder collaboration has a comparatively lower direct statistical impact, it remains essential for aligning the interests of diverse parties, building trust and improving overall implementation outcomes. Finally, technology readiness is identified as an important but insufficient factor on its own, as its success is heavily dependent on strategic alignment with organisational objectives, stakeholder priorities and the overarching project management framework.

To improve integration outcomes, the study provides several targeted recommendations across multiple domains.

- *Project management*: organisations should transition to hybrid delivery models that combine the flexibility of Agile with the structure of Waterfall, while expanding the use of digital tools such as digital twins, predictive analytics and artificial intelligence (AI)-enabled platforms to enhance risk assessment and decision-making.
- *Organisational readiness*: companies should invest in leadership development and workforce training to build sustainability-focused capabilities and enhance the technical competencies required for hybrid energy projects.
- *Stakeholder collaboration*: efforts should be reinforced through public–private partnerships and coordinated risk management strategies to align interests and reduce implementation barriers.

- *Technology readiness*: operators should conduct regular audits to identify infrastructure gaps, prioritise interoperability between renewable and legacy systems, and utilise strategic incentives to accelerate adoption.
- *Policy level*: governments and regulators should provide incentives for hybrid projects and update regulatory frameworks to offer clear guidance on safety standards and environmental compliance.

8.5.1 Limitations

Despite the significant findings, several limitations must be acknowledged.

- *Methodological constraints*: the study relies on a quantitative, cross-sectional survey of 100 professionals, which only captures a single point in time and does not track how practices evolve throughout the entire project lifecycle.
- *Generalisability*: the specific focus on subsea oil and gas projects limits the applicability of the results to other energy segments, such as onshore or downstream operations, which operate under different technical and regulatory conditions.
- *Data objectivity*: the findings are based on self-reported perceptions, which may be subject to respondent bias or vary based on an individual's level of exposure to renewable initiatives.
- *Model scope*: the analysis did not explicitly incorporate external factors such as market volatility, macroeconomic conditions or detailed regional regulatory regimes.
- *Construct operationalisation*: the unexpectedly weak positive regression effect of technology readiness suggests that future research requires a more granular way to distinguish between technology availability and actual compatibility in complex environments.

8.5.2 Recommendations and future research

Future research should expand on this work by examining the contextual mechanisms and long-term outcomes of renewable energy integration. Longitudinal studies are necessary to evaluate how integration affects operational efficiency, reliability, emissions performance and the total cost of ownership over time. Additionally, comparative analyses across regions and regulatory environments would help identify transferable best practices and clarify how local infrastructure and policy incentives shape distinct challenges.

From a methodological standpoint, further research is needed to validate purpose-built hybrid project management approaches that specify how to balance governance and iterative delivery to manage uncertainty without compromising safety or compliance. Finally, a deeper investigation into stakeholder dynamics, including cultural norms and geographical dispersion, is warranted to understand how these factors influence trust formation and communication quality, thereby shaping the overall effectiveness of collaboration in integrated energy projects.

REFERENCES

Davis M, Okunlola A, Di Lullo G, Giwa T and Kumar A (2023) Greenhouse gas reduction potential and cost-effectiveness of economy-wide hydrogen-natural gas blending for energy end uses. *Renewable and Sustainable Energy Reviews* **171**: 112962.

Gandomani TJ and Nafchi MZ (2016) Agile transition and adoption human-related challenges and issues: a grounded theory approach. *Computers in Human Behavior* **62**: 257–266.

Iyelolu TV, Agu EE Idemudia C and Ijomah TI (2024) Driving SME innovation with AI solutions: overcoming adoption barriers and future growth opportunities. *International Journal of Science and Technology Research Archive* **7(1)**: 36–54.

Kotter JP (1996) *Leading Change*. Harvard Business School Press, Boston, MA, USA.

Leong J, May Yee K, Baitsegi O, Palanisamy L and Ramasamy RK (2023) Hybrid project management between traditional software development lifecycle and agile based product development for future sustainability. *Sustainability* **15(2)**: 1121.

Mahajan R, Lim WM, Sareen M, Kumar S and Panwar R (2023) Stakeholder theory. *Journal of Business Research* **166**: 114104.

Miraj P, Berawi MA, Anindtya A and Sari M (2024) Evaluating the impact of knowledge management and database management on decision-making process: a case study of sub-sea project services. *Journal of Open Innovation: Technology Market, and Complexity* **10(3)**: 100340.

PMI (Project Management Institute) (2025) *A Guide to The Project Management Body of Knowledge (PMBOK® Guide)*, 8th edn. PMI, Newtown Square, PA, USA.

Proedrou F (2023) The global energy transition and Russian structural power: scenarios and strategic options. In *Handbook on the Geopolitics of the Energy Transition*. Edward Elgar, Cheltenham, UK, pp. 483–497.

Salvador S and Ribeiro MC (2023) Socio-economic, legal, and political context of offshore renewable energies. *Wiley Interdisciplinary Reviews: Energy and Environment* **12(2)**: e462.

Weydmann AP, Mendes FA, Oliveira VPL *et al.* (2023) Continuous improvements in project management practices through PMOs – the case of Buzios Field. *Offshore Technology Conference*, 10.4043/32196-MS.

Yang G, Zha D, Cao D and Zhang G (2024) Time for a change: rethinking the global renewable energy transition from the Sustainable Development Goals and the Paris Climate Agreement. *The Innovation* **5(2)**: 100582.

Part III

Knowledge and learning

Mohammed Reyasudin Basir Khan
ISBN 978-1-83742-803-8
https://doi.org/10.1108/978-1-83742-802-120261012
Emerald Publishing Limited: All rights reserved

Chapter 9
An investigation of the barriers and strategies for Agile project management implementation in Malaysian IT organisations

Huzaifah Azman Mohamad and Mohammed Reyasudin Basir Khan

Tun Razak Graduate School Universiti Tun Abdul Razak (UNIRAZAK), Kuala Lumpur, Malaysia.

9.1. Introduction

Agile project management has significantly transformed the information technology (IT) project management landscape over the past two decades. Rooted in adaptability, collaboration and iterative development, Agile methodologies enable organisations to respond rapidly to change while delivering high-quality outcomes. Originating in software development, frameworks such as Scrum, Kanban and Extreme Programming have become widely adopted across global industries. Unlike traditional Waterfall methodologies, Agile emphasises incremental delivery, continuous stakeholder engagement and feedback-driven improvement, making it particularly suitable for dynamic and uncertain project environments.

In Malaysia, the IT industry plays a pivotal role in national economic growth and innovation. Government initiatives such as the Malaysia Digital Economy Blueprint MyDIGITAL aim to position the country as a regional hub for digital transformation. This rapid digitalisation has increased demand for flexible, innovative project management approaches, including Agile (Omar *et al.*, 2024). Despite its global success, Agile adoption in Malaysia faces distinct challenges, influenced by organisational culture, hierarchical management structures, risk aversion and limited skill availability. While Agile promises improved project performance, collaboration and responsiveness, Malaysian IT firms often struggle to align Agile principles with existing organisational norms and practices (Känsälä and Tokumaru, 2023; Neumann *et al.*, 2024).

Empirical evidence suggests that Agile teams achieve higher project success rates than traditional approaches (Agbejule and Lehtineva, 2022; Cimini *et al.*, 2024; Kudyba and Cruz, 2023). However, in the Malaysian context, adoption is hindered by resistance to change, inadequate understanding of Agile principles, insufficient training and limited resources

(Omar *et al.*, 2024). Cultural preferences for centralised decision-making and rigid structures further complicate Agile implementation. These challenges manifest at organisational, individual and external levels, including leadership resistance, unclear Agile processes, skill gaps, stakeholder scepticism and regulatory constraints. As a result, organisations risk project delays, cost overruns, reduced innovation and weakened global competitiveness (Kerzner, 2025).

Although prior studies have explored Agile adoption in Malaysia, gaps remain in understanding how organisational culture, technological readiness and human resource support collectively influence Agile project readiness, particularly across organisations of different sizes. Early research emphasised the dominance of social and human factors over technical considerations, highlighting the importance of leadership support and cultural alignment (Aghina *et al.*, 2019). However, there is limited empirical evidence examining these factors holistically within the Malaysian IT sector. Addressing this gap is critical as Malaysia continues to pursue its digital economy agenda.

The study presented in this chapter examines the influence of organisational support, technological readiness and human resource support on Agile project readiness in Malaysian IT organisations, while also assessing the moderating role of organisational size. The research focuses on project managers, Scrum Masters and IT project team members involved in Agile project delivery within Malaysia. By concentrating on this population, the study captures insights from key practitioners directly responsible for Agile planning, execution and performance outcomes.

The significance of this research lies in its contribution to both theory and practice. From a theoretical perspective, the study extends existing Agile literature by contextualising Agile adoption within a developing economy characterised by strong hierarchical and collectivist cultural norms (Känsälä and Tokumaru, 2023). It builds on earlier work by moving beyond general discussions of Agile benefits to a deeper examination of cultural, organisational and human-centred barriers (Neumann *et al.*, 2024). In practice, the findings provide Malaysian IT practitioners with actionable strategies to overcome resistance, strengthen leadership capabilities, enhance workforce readiness and sustain Agile practices over time (Aghina *et al.*, 2019). The study also offers policy-relevant insights aligned with MyDIGITAL, particularly regarding national training incentives, certification schemes and capacity-building initiatives.

The theoretical foundation of this study is anchored in the Agile maturity model and the diffusion of innovation theory. The Agile maturity model conceptualises Agile adoption as a progression through increasing levels of maturity, ranging from ad hoc experimentation to full organisational integration and continuous optimisation (Cimini *et al.*, 2024). This model is particularly useful for assessing internal readiness factors such as organisational culture, technological infrastructure and human resource support.

Complementing this, the diffusion of innovations theory explains how new practices, such as Agile, spread within social systems over time, influenced by communication channels, perceived benefits and adopter characteristics (Assidi *et al.*, 2025). Together, these frameworks provide a comprehensive lens for examining Agile project readiness from both internal and external perspectives.

To address theoretical limitations, the study also acknowledges insights from contingency theory and institutional theory. Contingency theory suggests that Agile effectiveness depends on contextual fit, including leadership style, resource availability and environmental uncertainty (Kerzner, 2025). Institutional theory highlights the role of regulatory pressures, cultural norms and conformity to global standards in shaping organisational behaviour (Neumann *et al.*, 2024). These perspectives enrich the analysis by recognising Agile adoption as a socially embedded and context-dependent process rather than a purely rational or technical decision.

Prior empirical studies consistently emphasise the importance of organisational culture, leadership commitment and employee training in successful Agile implementation (Aghina *et al.*, 2019). Research conducted in Malaysia and comparable contexts identifies resistance to change, hierarchical structures and limited knowledge of Agile as persistent barriers (Chia *et al.*, 2022; Elseknidy *et al.*, 2024). Studies also suggest that phased adoption, targeted training and strong governance mechanisms can improve Agile readiness and sustainability (Omar *et al.*, 2024). However, the Malaysian IT sector remains under-examined regarding how these factors interact to influence Agile readiness across organisations of varying sizes.

Based on the literature, this study proposes a conceptual framework (Figure 9.1) in which organisational culture, technological readiness and human resource support influence Agile project readiness, with organisational size as a moderating variable. Agile project readiness is positioned as a critical antecedent to successful Agile project outcomes. This framework provides the basis for hypothesis testing and empirical investigation in subsequent chapters.

This chapter establishes the study's context, problem, objectives and theoretical grounding. It highlights the strategic importance of Agile project management for Malaysia's IT sector and identifies critical gaps in understanding the barriers and enablers of Agile adoption.

Figure 9.1 Conceptual framework

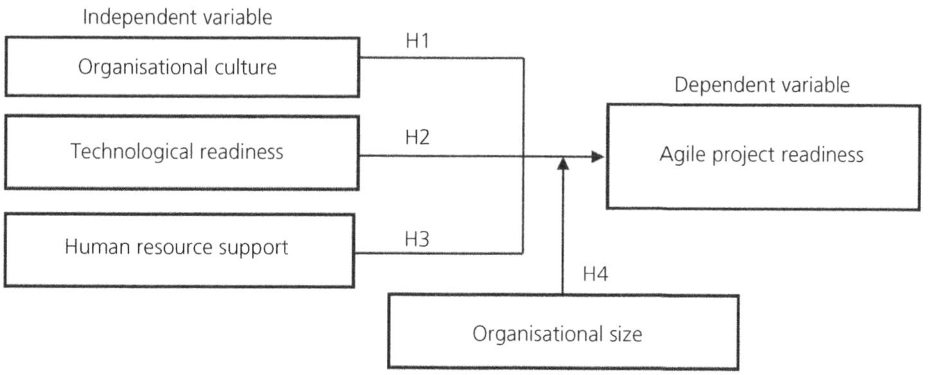

By integrating established theories with contextual insights, the study sets the foundation for a systematic examination of Agile readiness in Malaysian IT organisations.

9.2. Methodology

9.2.1 Research design

This study employed a quantitative research design that emphasises numerical data and statistical analysis to objectively examine relationships between the independent, moderating and dependent variables. The approach is appropriate for assessing the effects of organisational culture, technological readiness and human resource support on Agile project readiness, as well as the moderating role of organisational size. Survey questionnaires were the primary data collection instrument, generating measurable responses suitable for statistical analysis.

9.2.2 Population and sampling procedures

The target population comprises 122 949 professionals involved in Agile-related roles, including IT professionals, project managers and Agile practitioners employed in Malaysian organisations engaged in software development, IT consulting and related services. This population estimate is derived from the 2023 Economic Census reported by the Department of Statistics Malaysia. These participants are appropriate for the study because they possess direct experience with Agile implementation and can provide informed perspectives on the barriers, drivers and strategic conditions shaping Agile adoption in Malaysia.

To strengthen representativeness and support generalisable findings, the study used stratified random sampling. Stratification was based on organisational size, namely small, medium and large firms, to ensure proportional representation across categories that may exhibit different Agile dynamics due to variations in governance capacity, resource availability and maturity levels. This approach enables meaningful comparison between organisational groups and supports more robust inference regarding the role of organisational size within the proposed model.

The minimum sample size was determined using the Raosoft sample size calculator, which indicated that 138 responses are required to achieve a 90% confidence level with a 7% margin of error. The final sampling plan applied proportionate allocation across strata to reflect the distribution of organisations within the Malaysian IT sector. Overall, the chosen population definition and sampling strategy enhance methodological rigour and provide an adequate foundation for regression-based analyses examining Agile project readiness across diverse organisational contexts in Malaysia.

9.2.3 Data collection method

This study employed a quantitative data collection strategy, using a structured questionnaire, to examine the barriers and strategies influencing the implementation of Agile project management in Malaysian IT organisations. The questionnaire focused on organisational enablers of Agile transformation. The study measured organisational culture, technological readiness, human resource support and Agile project readiness, with organisational size included as a moderating variable. Data were collected using an online survey platform to increase geographic reach, facilitate participation across organisations of different sizes, reduce administrative costs and enable monitoring of responses throughout the collection period.

The questionnaire was distributed with a cover letter outlining the study objectives, scope, voluntary nature of participation and confidentiality of responses. The letter also clarified the researcher's academic affiliation and the purpose of data collection to enhance legitimacy and encourage honest participation. To further support engagement, respondents were given the option to request a summary of findings on study completion, thereby reinforcing transparency and ethical reciprocity and encouraging thoughtful responses from participants with a professional interest in Agile practices.

The population comprised IT professionals, project managers and Agile practitioners involved in Agile initiatives across software development and IT consulting. Stratified random sampling was applied to ensure representation across organisational size categories and varying Agile maturity levels. Data collection was conducted over one month, from 5 April 2025 to 4 May 2025, yielding 105 valid responses. Although the calculated minimum sample size was higher, participation constraints, including limited access and peak project delivery periods, reduced the number of responses. Despite this limitation, the sample was sufficient for regression and moderation analyses. Responses were measured using a five-point Likert scale, and ethical requirements, including informed consent, voluntary participation and data privacy, were maintained.

9.2.4 Operationalisation and measurement

This study examined independent, moderating and dependent variables, operationalised using measures adapted from prior empirical research to support measurement validity. The independent variables comprise organisational culture, technological readiness and human resource support. Organisational culture was measured by dimensions commonly associated with Agile adoption, including hierarchical orientation, resistance to change and collaboration. Technological readiness was assessed by evaluating the adequacy of IT infrastructure and enabling technologies, including the availability of appropriate software tools and the extent of cloud adoption. Human resource support was measured by the presence and quality of Agile-related capacity-building initiatives, including training Agile coaching and employee skill development.

Organisational size was included as a moderating variable to assess whether differences in structural flexibility, governance arrangements and resource availability strengthen or weaken the relationships between the independent variables and Agile project readiness. The dependent variable, Agile project readiness, was evaluated using indicators reflecting the extent of Agile adoption and the degree to which Agile principles are embedded in project practices, including project performance outcomes and alignment with Agile values.

9.3. Results and analysis

9.3.1 Response rate

Data were collected using an online questionnaire administered through Google Forms. This mode was selected to facilitate access to a geographically dispersed population of IT professionals, Agile practitioners and project managers across Malaysia. The unit of analysis was the individual respondent. Of 150 questionnaires distributed, 105 usable responses were obtained, representing a 70% response rate. Although this was below the calculated minimum of 138 responses based on Raosoft parameters of 90% confidence and a 7% margin of error, the achieved sample remains acceptable for exploratory social science research, particularly for unsupervised online administration, where response rates above 60% are commonly considered strong (Baruch and Holtom, 2008).

The adequacy of the sample was further supported by the ratio of respondents to measurement items. With 29 questionnaire items, the resulting sample-to-item ratio was 5.25:1, meeting minimum recommendations for multivariate analysis and supporting the suitability of regression-based procedures. In addition, respondents were offered the option to receive a summary of the study findings, which likely strengthened motivation to participate and encouraged more engaged responses among individuals with a professional interest in Agile implementation.

9.3.2 Demographic profile

The demographic profile of the 105 respondents involved in this study provides a comprehensive context for understanding Agile project management readiness within the Malaysian IT sector. The data, as shown in Table 9.1, represent a diverse range of roles, experience levels and educational backgrounds, which enhances the generalisability and credibility of the research findings.

As shown in Table 9.1, the study successfully captured a balanced mix of operational and strategic viewpoints. The high representation of technical team members and subject matter experts (collectively, 47.6%) ensures that the 'on-the-ground' realities of Agile execution are well represented. In comparison, the inclusion of upper management and business owners (totalling 23.8%) provides necessary insights into the strategic and leadership-oriented barriers to Agile adoption.

The respondents' professional experience is widely distributed. Interestingly, the two largest groups are those with the least experience (less than 1 year at 26.6%) and those with the most experience (more than 10 years at 21.9%). This distribution is significant because it allows for a comparative analysis of how Agile maturity is perceived by early-career professionals against seasoned industry veterans who have likely transitioned from traditional Waterfall methodologies.

The participants' educational profile is notably high, with 86.7% of respondents holding a Bachelor's degree or higher. According to the sources, this well-qualified group supports the credibility and reliability of the professional assessments provided regarding organisational culture and technological readiness.

Finally, the organisational data reveals that the study's scope extends across the entire IT ecosystem in Malaysia. While software development is often the primary focus of Agile studies, it accounts for only 20% of this sample. The significant presence of IT consulting (28.6%) and infrastructure (22.9%) sectors reinforces the idea that Agile practices are being applied

Table 9.1 Demographic profile

Demographic category	Attribute	Percentage: %
Current organisational position	Technical team member	23.8
	Subject matter expert	23.8
	Project manager or Scrum Master	21.9
	Upper management/stakeholder	18.1
	Product owner/business analyst	6.7
	Business owner	5.7
Years of experience	Less than 1 year	26.6
	More than 10 years	21.9
	7 to 10 years	21.0
	1 to 3 years	18.1
	4 to 6 years	12.4
Highest level of education	Bachelor's degree	50.5
	Master's degree	35.2
	Diploma, STPM or equivalent	13.3
	Doctorate	1.0
Type of IT services	IT consulting and services	28.6
	Hardware and infrastructure	22.9
	Software development	20.0
	Cybersecurity	15.2
	Big Data and analytics	13.3

beyond a single functional domain, reflecting the broader goals of Malaysia's MyDIGITAL blueprint to enhance digital competitiveness across all technological segments.

9.3.3 Reliability analysis

To ensure that the measurement scales used in this study align with their underlying theoretical constructs, internal consistency reliability was assessed using Cronbach's alpha (α). Cronbach's alpha is a widely used indicator of the extent to which items within a scale consistently measure the same latent factor. In line with common methodological guidance, an alpha value of 0.70 is generally regarded as acceptable for exploratory research, while values above 0.80 indicate good reliability.

The results in Table 9.2 indicate acceptable to strong internal consistency across all constructs. Organisational culture achieved an alpha of 0.762, with no item deletion producing a meaningful improvement, suggesting coherent measurement of culture-related support and flexibility. Technological readiness recorded an alpha of 0.766, and item-level diagnostics

Table 9.2 Reliability statistics

Construct	No. of items	Cronbach alpha, α
Organisational culture	3	0.762
Technological readiness	3	0.766
Human resource support	3	0.782
Agile project readiness (DV)	5	0.863

Note: DV, dependent variable.

indicated that all indicators consistently contributed to the construct. Human resource support yielded an alpha of 0.782, confirming reliable measurement of HR-related preparedness for Agile transformation. Agile project readiness demonstrated the strongest reliability, with an alpha of 0.863, indicating high internal consistency and supporting the robustness of the readiness scale, consistent with prior research on Agile maturity.

9.3.4 Normality analysis

Assessing normality is a prerequisite for parametric analyses such as multiple regression because substantial departures from normality may bias coefficient estimates and standard errors (Ghasemi and Zahediasl, 2012). Normality was evaluated using skewness and kurtosis, commonly used indicators of univariate distributional shape. As shown in Table 9.3, across the five constructs, skewness values ranged from –0.847 to –0.475 and kurtosis values ranged from 0.758 to 2.173. These results fall within widely accepted thresholds for approximate normality in social science research, indicating no serious deviation from normality.

Organisational culture (IV1_OC) exhibited skewness of –0.483 and kurtosis of 1.012, reflecting mild left skewness with acceptable peakedness. Technological readiness (IV2_TR) showed the largest skewness of –0.847 and a kurtosis of 1.913, suggesting a modest negative skew within acceptable limits. Human resource support (IV3_HR) and agile project readiness (DV_APR) displayed comparatively balanced distributions, with values well within

Table 9.3 Normality tests

Variable	Skewness	Std. error of skewness	Kurtosis	Std. error of kurtosis
IV1_OC	–0.483	0.236	1.012	0.467
IV2_TR	–0.847	0.236	1.913	0.467
IV3_HR	–0.475	0.236	1.213	0.467
MV_OS	–0.650	0.236	2.173	0.467
DV_APR	–0.499	0.236	0.758	0.467

recommended ranges. Organisational size (MV_OS) also met normality expectations, with skewness of –0.650 and kurtosis of 2.173. Given that the sample size exceeds 100, minor deviations are unlikely to undermine regression-based procedures, given the robustness afforded by large-sample properties. Overall, the normality results support the suitability of the dataset for subsequent parametric analyses and no data transformation was required.

9.3.5 Common method variance

Common method variance is a potential concern in survey-based research because the measurement method can introduce systematic error that distorts observed relationships between constructs. To assess this risk Harman's single-factor test was conducted using unrotated principal component analysis. Table 9.4 shows that the first unrotated component produced an eigenvalue of 15.393 and explained 45.275% of the total variance. As this proportion is below

Table 9.4 Total variance explained by principal component analysis (continued on next page)

Component	Initial eigenvalues: total	% of variance	Cumulative %	Extraction sums of squared loadings	% of variance	Rotation cumulative %
1	15.393	45.275	45.275	15.393	45.275	23.396
2	2.313	6.804	52.079	2.313	6.804	45.185
3	1.873	5.509	57.588	1.873	5.509	55.493
4	1.630	4.794	62.382	1.630	4.794	60.574
5	1.414	4.159	66.541	1.414	4.159	64.694
6	1.268	3.730	70.271	1.268	3.730	68.793
7	1.102	3.242	73.513	1.102	3.242	72.876
8	1.005	2.957	76.470	1.005	2.957	76.470
9	0.957	2.816	79.286			
10	0.763	2.249	81.531			
11	0.697	2.049	83.580			
12	0.684	2.011	85.591			
13	0.622	1.830	87.421			
14	0.573	1.685	89.106			
15	0.483	1.422	90.528			
16	0.415	1.220	91.748			
17	0.413	1.214	92.962			
18	0.329	0.969	93.931			
19	0.289	0.850	94.781			
20	0.261	0.768	95.549			
21	0.241	0.708	96.256			

Table 9.4 Continued

Component	Initial eigenvalues: total	% of variance	Cumulative %	Extraction sums of squared loadings	% of variance	Rotation cumulative %
22	0.210	0.619	96.875			
23	0.207	0.610	97.486			
24	0.200	0.589	98.075			
25	0.181	0.533	98.608			
26	0.143	0.420	99.028			
27	0.131	0.384	99.412			
28	0.109	0.319	99.731			
29	0.091	0.269	100.000			

the commonly applied 50% threshold, the findings suggest that common method variance is unlikely to pose a substantial threat to the study's validity. The variance explained across multiple components further indicates that the measured constructs are multidimensional rather than driven by a single underlying factor, supporting the adequacy of the measurement structure.

Conceptually, the pattern of distributed variance is consistent with the study's theoretical foundations. Diffusion of innovation theory posits that innovation adoption is shaped by multiple influences rather than a single determinant, including technological conditions and organisational and human factors that affect knowledge formation, persuasion and implementation decisions. Similarly, the agile maturity model frames Agile capability as a staged progression that requires distinct organisational competencies in cultural alignment, process discipline and people development. The absence of a dominant single factor, therefore, aligns with the expectation that culture, technology and human resource support represent related yet empirically separable dimensions of Agile readiness.

9.3.6 Correlation analysis

Correlation analysis was conducted to examine the relationships between organisational culture, technological readiness, human resource support, organisational size and Agile project readiness in Malaysian IT organisations. To evaluate the study hypotheses and determine the strength and direction of associations between the constructs, Pearson's product-moment correlation was performed in the statistical package for the social sciences (SPSS). The Pearson correlation coefficient, denoted by r, quantifies the linear association between two variables and ranges from -1 (perfect negative association) to 1 (perfect positive association). The results indicated that the relationships between the variables were statistically significant at the 0.01 level using a one-tailed test, suggesting strong confidence in the observed associations.

Table 9.5 indicates that human resource support demonstrated the strongest association with Agile project readiness ($r = 0.725$, $p < 0.001$), indicating that practices such as Agile coaching, employee involvement and team empowerment are closely linked to readiness for

Table 9.5 Correlation of study variables

Variable	n	M	SD	DV_APR	IV1_OC	IV2_TR	IV3_HR	MV_OS
DV_APR	105	3.4544	0.7961	1				
IV1_OC	105	3.7079	0.7535	0.681[a] ($p < 0.001$)	1			
IV2_TR	105	3.7333	0.7137	0.646[a] ($p < 0.001$)	0.722[a] ($p < 0.001$)	1		
IV3_HR	105	3.6032	0.7296	0.725[a] ($p < 0.001$)	0.739[a] ($p < 0.001$)	0.771[a] ($p < 0.001$)	1	
MV_OS	105	3.7524	0.6688	0.667[a] ($p < 0.001$)	0.743[a] ($p < 0.001$)	0.721[a] ($p < 0.001$)	0.804[a] ($p < 0\ .001$)	1

[a]Correlation is significant at the 0.01 level.

Agile adoption. This pattern is consistent with the Agile maturity model, which positions people-related capability development as central to advancing Agile maturity. Organisational culture was also strongly and positively correlated with Agile project readiness ($r = 0.681$, $p < 0.001$), suggesting that collaborative, adaptive and open work environments facilitate Agile implementation, as anticipated by diffusion of innovations theory, which emphasises cultural compatibility in the uptake of innovations.

Technological readiness exhibited a positive relationship with Agile project readiness ($r = 0.646$, $p < 0.001$), underscoring the supporting role of infrastructure and digital tools in enabling Agile practices. However, its association was comparatively weaker than that of people- and culture-oriented factors. Organisational size showed a moderate-to-strong correlation with Agile project readiness ($r = 0.667$, $p < 0.001$) and was particularly strongly related to human resource support ($r = 0.804$), suggesting that larger firms may be better positioned to provide structured HR mechanisms to support Agile transformation. Significant intercorrelations between the independent variables, including those between technological readiness and human resource support ($r = 0.771$) and between organisational culture and human resource support ($r = 0.739$), suggest complementary effects but also warrant assessment for multicollinearity in subsequent regression analyses.

9.3.7 Multicollinearity issues

This study examines how organisational culture (IV1_OC), technological readiness (IV2_TR) and human resource support (IV3_HR), together with the moderating role of organisational size (MV_OS), shape readiness for Agile project management implementation in Malaysian IT firms. Establishing that these constructs are not excessively correlated is important for preserving the model's theoretical structure and empirical validity, and for reducing concerns related to common method variance, particularly multicollinearity.

Multicollinearity arises when predictors are highly correlated, which can inflate standard errors and weaken the interpretability of regression coefficients. As shown in Table 9.6,

Table 9.6 Variance inflation factor (VIF)

Coefficients[a]		
Variable	Collinearity statistics	
	Tolerance	VIF
IV1_OC	0.415	2.410
IV2_TR	0.405	2.470
IV3_HR	0.405	2.470

[a]Correlation is significant at the 0.01 level.

SPSS diagnostics show that technological readiness and human resource support each have a variance inflation factor (VIF) of 2.470 and a tolerance of 0.405. These values are within accepted thresholds, as serious multicollinearity is typically indicated when the VIF exceeds 5.0 or the tolerance falls below 0.20. As tolerance equals $1 - R_i^2$, a tolerance of 0.405 implies that approximately 59.5% of the variance is shared across predictors, while the remaining variance reflects distinct contributions.

Results from Table 9.6 support the inclusion of both technological and human factors in the same model. The findings align with the Agile maturity model, which frames Agile readiness as dependent on both infrastructure and people-related capabilities. They are also consistent with the diffusion of innovation theory, which emphasises that adoption requires technical enablement and human acceptance. Empirically, the outcome is comparable to recent studies reporting that technological and human resource dimensions are complementary rather than redundant in IT contexts (Omar *et al.*, 2024).

Practically, the evidence suggests that organisations can invest in technology upgrades and HR-driven Agile training as separate, targeted initiatives, with each contributing independently to cultural readiness for Agile implementation.

Overall, the model shows no multicollinearity concern in predicting organisational culture from technological readiness and human resource support, supporting the model's statistical adequacy and enabling more reliable inferential analysis for Agile implementation planning.

9.3.8 Path coefficient

The primary objective of this study was to assess how organisational culture (IV1_OC), technological readiness (IV2_TR), human resource support (IV3_HR) and organisational size (MV_OS, as a moderator) influence Agile project readiness (DV_APR) in Malaysian IT organisations. Path coefficient analysis based on linear regression provides evidence on the direction and explanatory strength of each relationship, with R^2 indicating the proportion of variance in DV_APR accounted for by each predictor.

Figures 9.2 to 9.5 show that IV3_HR explains 52.6% of the variance in DV_APR (R^2 = 0.526), followed by IV1_OC (R^2 = 0.464), MV_OS (R^2 = 0.445) and IV2_TR (R^2 = 0.417).

Figure 9.2 Human resource support to Agile project readiness

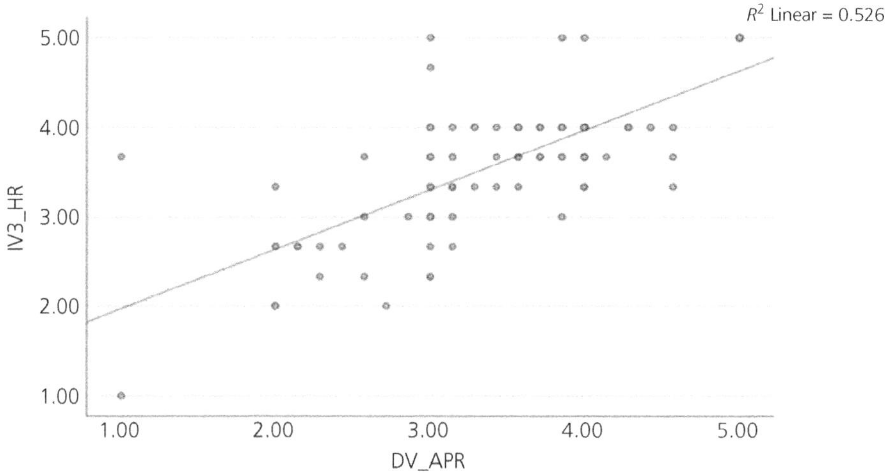

Figure 9.3 Organisational culture to Agile project readiness

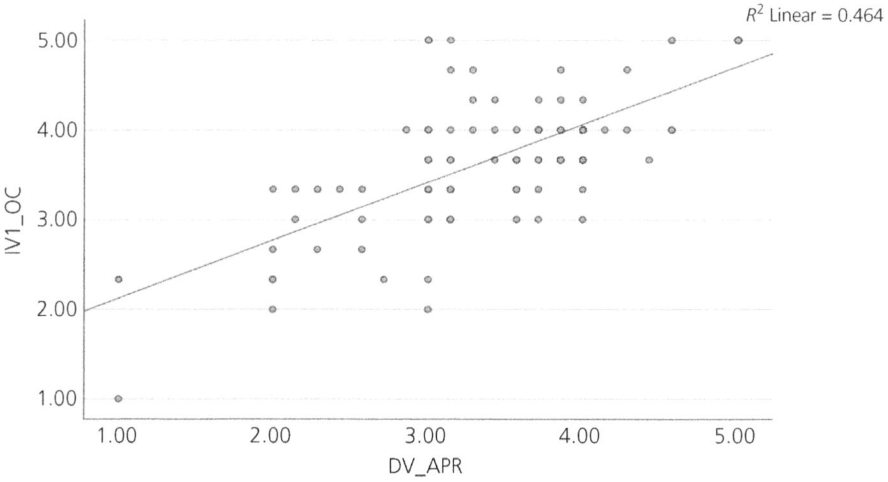

Among these, human resource support demonstrates the strongest explanatory power, indicating that workforce capability, empowerment and Agile-oriented development are central to readiness for Agile project execution. This is consistent with evidence that Agile transformation depends heavily on cultivating human competencies and an Agile mindset (Neumann *et al.*, 2024).

Figure 9.4 Organisational size to Agile project readiness

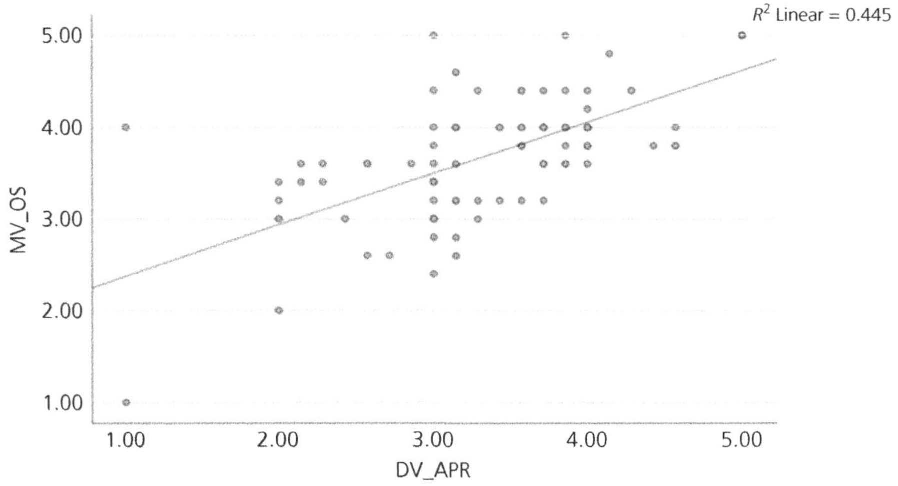

Figure 9.5 Technological readiness to Agile project readiness

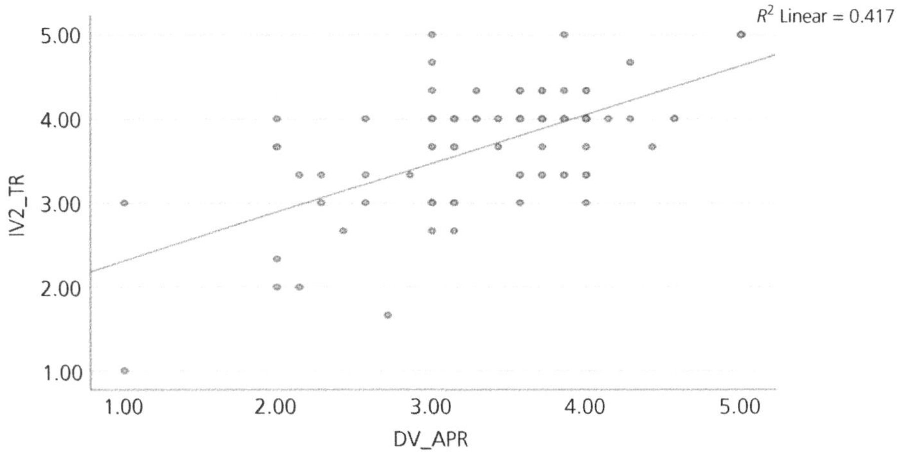

Figure 9.2 indicates a positive linear association for the IV3_HR to DV_APR relationship, as evidenced by an upward-sloping regression line and a comparatively high R^2 of 0.526. The concentration of observations near the fitted line suggests a stable linear pattern with limited dispersion, indicating that variations in Agile readiness are meaningfully associated with differences in human resource support. The absence of pronounced outliers further supports the adequacy of the linear specification for this relationship.

Substantively, the findings highlight the role of HR policies and practices, particularly training, clear role expectations and performance systems, in enabling Agile implementation. Because Agile delivery relies on autonomy, collaboration and cross-functional coordination, HR support functions as an enabling mechanism, strengthening readiness and reinforcing an Agile-supportive work culture. Prior research similarly reports that effective HR practices facilitate Agile adoption by promoting learning, empowerment and cultural alignment with Agile principles (Martinez-Sanchez and Vicente-Oliva, 2023; Mollet and Kaudela-Baum, 2023).

Figure 9.3, which represents organisational culture (IV1_OC), demonstrates substantial explanatory power, with an R^2 of 0.464. This aligns with the Agile maturity model, which emphasises that Agile adoption requires cultural change, including reduced hierarchical distance and greater autonomy in decision-making. In the Malaysian context, rigid hierarchical structures similarly remain an obstacle to scaling Agile practices.

In Figure 9.4, organisational size (MV_OS), as a moderating variable, accounts for 44.5% of the variance in Agile project readiness. This indicates that firm size, reflected in governance arrangements, the capacity to scale resources and levels of bureaucratic flexibility, can either facilitate or restrict Agile practices.

Technological readiness (IV2_TR), shown in Figure 9.5, although central to the effective use of Agile tools and digital platforms, records a comparatively lower R^2 of 0.417. This indicates that technology functions as an enabling condition, but it does not, by itself, ensure readiness in the absence of supportive cultural norms and human capabilities.

9.3.9 Regression analysis

This study aimed to examine the extent to which organisational culture, technological readiness and human resource support predict Agile project management readiness in Malaysian IT organisations. Hypotheses H1 to H3 proposed that each independent variable would exert a positive and statistically significant effect on Agile project readiness, while H4 evaluated whether organisational size moderates these relationships. This section interprets the multiple regression results generated in SPSS, with discussion informed by relevant theoretical frameworks and recent empirical evidence.

The regression model, as shown in Table 9.7, produced a strong multiple correlation ($R = 0.759$) and an R^2 of 0.576, indicating that organisational culture, technological readiness

Table 9.7 Regression model

Model	R	R^2	Adjusted R^2	Std. error of the estimate	R^2 change	F change	$df1$	$df2$	Sig. F change
1	0.759[a]	0.576	0.564	0.52587	0.576	45.779	3	101	<0.001

[a]Correlation is significant at the 0.01 level.

and human resource support jointly explain 57.6% of the variance in Agile project readiness. The adjusted R^2 of 0.564 further suggests that the model retains explanatory strength after accounting for model complexity, supporting its stability for inference.

The analysis of variance (ANOVA) results given in Table 9.8 indicate that the overall model is statistically significant ($F = 45.779, p < 0.001$), confirming that the predictors collectively provide a reliable explanation of variation in Agile readiness.

Human resource support was the strongest predictor of Agile readiness ($\beta = 0.432, p < 0.001$) as shown in Table 9.9, indicating that Agile training, empowerment practices and retention-oriented initiatives are central to strengthening implementation readiness, consistent with evidence on the importance of leadership and enabling environments in developing Agile mindsets (Neumann *et al.*, 2024). Organisational culture also had a significant positive effect ($\beta = 0.285, p = 0.007$), suggesting that collaborative norms, supportive leadership and less hierarchical structures facilitate Agile adoption, in line with arguments on cultural fit in Agile implementation (Känsälä and Tokumaru, 2023). In contrast, technological readiness was not a significant predictor ($\beta = 0.107, p = 0.329$), implying that the presence of tools and infrastructure alone is insufficient when cultural and human enablers are weak. This pattern supports the Agile maturity model's emphasis on people and culture as foundational conditions for Agile capability development. It is consistent with the diffusion of innovation theory, which highlights compatibility and social system readiness as key determinants of adoption outcomes. Overall, the findings support H1 and H3, but not H2, underscoring the practical need for Malaysian IT firms to prioritise cultural change and human capital development alongside digital transformation objectives.

This study examined whether organisational size moderates the relationships between organisational culture, technological readiness, human resource support and Agile project readiness

Table 9.8 ANOVA analysis

Model	Sum of squares	*df*	Mean square	*F*	Sig.
1 Regression	37.979	3	12.660	45.779	<0.001[a]

[a]The overall regression model is significant at $p < 0.001$.

Table 9.9 Coefficient analysis for Agile project readiness

Variable	*B*	Std. error	*β*	*t*	Sig.
(Constant)	0.195	0.291		0.673	0.503
IV1_OC	0.301	0.109	0.285	2.770	0.007
IV2_TR	0.119	0.121	0.107	0.980	0.329
IV3_HR	0.471	0.122	0.432	3.862	0.001

Table 9.10 Moderating model summary

Variable	R^2	Adjusted R^2	F change	Sig. F change	Interaction, β	Interaction sig.
IV1_OC	0.522	0.513	0.001	0.981	0.009	0.981
IV2_TR	0.501	0.492	0.689	0.409	0.308	0.409
IV3_HR	0.546	0.537	0.004	0.951	0.023	0.951

in Malaysian IT organisations using hierarchical regression analysis. Each model was tested by first entering the predictor and organisational size, then adding an interaction term to assess moderation effects. Although the models explained a substantial proportion of variance in Agile readiness, with R^2 values ranging from 0.501 to 0.546 as reported in Table 9.10, none of the interaction terms was statistically significant. These results indicate that organisational size does not significantly alter the strength of the relationships between the independent variables and Agile readiness.

The findings suggest that while organisational size may be directly associated with Agile readiness, it does not moderate the examined relationships. This outcome aligns with recent empirical evidence emphasising leadership, empowerment and internal capability development as more influential drivers of Agile success than structural characteristics such as size (Neto *et al.*, 2022; Rialti and Filieri, 2024; Zhao, 2024). From a theoretical perspective, the results extend the Agile maturity model by demonstrating that internal enablers remain critical regardless of organisational scale. In practice, the findings imply that Agile adoption strategies can be applied consistently across organisations of varying sizes, with an emphasis on cultural alignment, technological support and human resource development rather than on scale-specific adjustments.

9.3.10 Hypothesis testing

Figure 9.6 shows the results of hypothesis testing. Human resource support emerged as the strongest predictor of Agile readiness ($\beta = 0.432$, $t = 3.862$, $p < 0.001$), indicating that staff

Figure 9.6 Hypothesis testing results

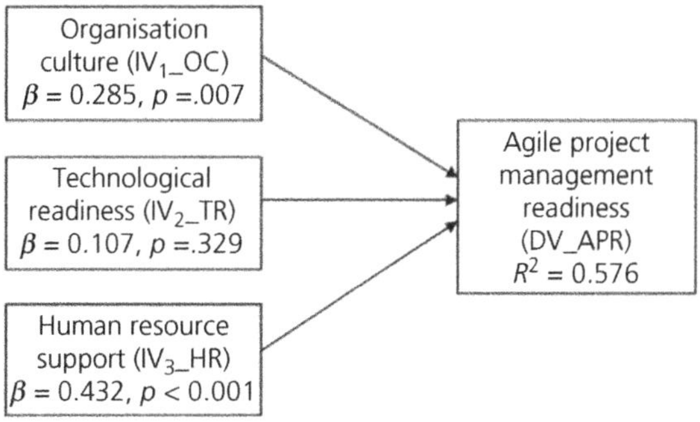

development, empowerment and supportive performance systems contribute most substantially to implementation readiness, consistent with prior work emphasising human capability as central to Agile transformation. Organisational culture also exerted a significant positive effect ($\beta = 0.285$, $t = 2.770$, $p = 0.007$), suggesting that leadership support, open communication and flexible decision-making create conditions conducive to embedding Agile practices. In contrast, technological readiness was not statistically significant ($\beta = 0.107$, $t = 0.980$, $p = 0.329$), implying that infrastructure and tool availability alone may be insufficient to drive Agile readiness without complementary cultural and human resource enablers. The model demonstrated strong explanatory power, with $R^2 = 0.576$, indicating that the predictors jointly accounted for 57.6% of the variance in Agile readiness.

Table 9.11 presents the empirical results of the study on the barriers and enablers to the implementation of Agile Project Management in Malaysian IT organisations, based on data from 105 respondents. The analysis evaluated the relationships between organisational culture, technological readiness, and human resource support with Agile Project Readiness, while also assessing the moderating role of organisational size. The findings show that human resource support and organisational culture are positively and significantly associated with Agile readiness. Human resource support demonstrated the strongest relationship, followed by organisational culture; technological readiness was statistically non-significant. These results highlight the importance of people-centred practices, collaborative culture, and supportive digital infrastructure in fostering Agile adoption.

The moderation analysis revealed that organisational size does not significantly influence the strength of these relationships, indicating that Agile readiness is not inherently dependent on firm scale. This supports theoretical arguments that Agile capability is driven more by internal cultural and process alignment than by structural attributes. Diagnostic tests confirmed the suitability of the data for parametric analysis, with evidence of normality, low multicollinearity and minimal common method bias, thereby supporting the robustness of the findings. Overall, the results suggest that Malaysian IT organisations can enhance Agile readiness by prioritising human capability development, cultural transformation and enabling technology, regardless of organisational size.

Table 9.11 Hypothesis testing results

H	Hypothesis	Result
H1	Organisational culture positively influences readiness for Agile implementation	Supported
H2	Technological readiness positively influences the readiness for Agile implementation	Rejected
H3	Human resource support positively influences the readiness for Agile implementation	Supported
H4	Organisational size moderates the relationship between organisational culture, technological readiness, human resource support and Agile project readiness	Rejected

9.4. Practical insights

The study reveals that Agile transformation in Malaysian IT firms is fundamentally a human-centric process, with human resource support emerging as the most critical determinant of readiness ($\beta = 0.432$, $p < 0.001$). For practitioners, this implies shifting the focus from merely acquiring tools to systematically investing in human capital through continuous learning. Strategic efforts should prioritise structured capacity-building initiatives, such as Agile coaching and specialised training programmes, to ensure that the workforce possesses the necessary competencies to navigate iterative project cycles. These learning-oriented interventions are essential for embedding a true Agile mindset, which the sources indicate is a more powerful driver of success than focusing solely on technical infrastructure.

Cultivating an adaptive organisational culture is the second vital pillar for successful implementation ($\beta = 0.285$, $p = 0.007$). Leadership plays a pivotal role here by actively modelling Agile behaviours, reducing hierarchical rigidity and fostering psychological safety and open communication. In practice, firms should encourage shared decision-making and cross-functional collaboration to overcome the barriers often posed by traditional command-and-control structures. By establishing internal 'Agile champions' and implementing change management programmes focused on values-based alignment, organisations can create a cultural foundation that supports the flexibility required for modern digital environments.

Another significant insight is that neither organisational size nor technological infrastructure acts as a primary driver or barrier to Agile readiness on its own. Since firm size was found not to moderate these relationships, Agile strategies can be applied consistently across both small start-ups and large enterprises. This suggests that smaller firms with fewer resources can be just as ready as larger ones if they prioritise cultural and human factors. Furthermore, because technological readiness was not a significant predictor, practitioners are cautioned against 'tool-heavy' implementations; instead, technology should be viewed as an enabler that is effective only when paired with strong cultural norms and empowered human resources.

Finally, these insights provide a roadmap for long-term strategic planning and national digital alignment in line with the MyDIGITAL blueprint. IT leaders can use the validated constructs of culture and HR support as diagnostic measures to assess their organisations' current readiness and identify specific friction points before a full-scale rollout. On a broader scale, these findings support the development of national policy-level initiatives, such as certification frameworks and training incentives. By focusing on these evidence-based levers – rather than just technical upgrades – both industry practitioners and policy makers can more effectively enhance the digital competitiveness and workforce adaptability of the Malaysian IT sector.

9.5. Conclusion and recommendations

This study investigated the barriers and enabling conditions influencing Agile project management readiness within the Malaysian IT sector. Analysing data from 105 respondents, the multiple regression model explained 57.6% of the variance in Agile readiness. The results demonstrate that human resource support is the most significant predictor ($\beta = 0.432$, $p < 0.001$), followed by organisational culture ($\beta = 0.285$, $p = 0.007$). Conversely, technological readiness was statistically non-significant ($\beta = 0.107$, $p = 0.329$), indicating that while digital infrastructure is a helpful enabling condition, it is insufficient to drive Agile transformation without parallel investments in people and cultural alignment.

A major theoretical contribution of this research is the finding that organisational size does not significantly moderate the relationships between these drivers and Agile readiness. This challenges scale-based assumptions in the existing literature, suggesting that Agile methodologies are universally applicable and that both small start-ups and large enterprises can achieve successful implementation by prioritising internal behavioural enablers over structural scale. These findings reinforce the Agile maturity model and diffusion of innovations theory by highlighting that social, human and cultural factors are the primary determinants of innovation adoption in a developing economy.

Practically, the study emphasises the critical role of knowledge and learning as the cornerstone of transformation. Organisations are encouraged to transition from a 'tool-heavy' approach to a human-centric strategy by investing in continuous Agile training, professional coaching and leadership development programmes. Fostering an adaptive culture characterised by psychological safety and reduced hierarchical rigidity is essential for embedding a true Agile mindset. At the policy level, these insights align with the MyDIGITAL blueprint, underscoring the need for national certification frameworks and targeted training incentives to enhance Malaysia's digital competitiveness and workforce adaptability. While limited by its cross-sectional design and self-reported data, this study provides a robust, evidence-based roadmap for sustaining Agile practices across the Malaysian IT ecosystem.

9.5.1 Implications and strategic considerations

Practically, the findings indicate that IT leaders and Agile consultants should prioritise cultural change and human capital development as the primary levers for Agile implementation. Building Agile, supportive cultures requires deliberate leadership development, value alignment initiatives and internal change networks that promote collaboration and adaptive decision-making. Workforce readiness should be strengthened through role-based training and coaching for key Agile roles, alongside empowerment practices that support team autonomy and continuous improvement. The validated constructs in this study can also serve as diagnostic measures to assess readiness, enabling organisations to identify high-readiness teams, reduce implementation friction and generate early wins that build momentum for broader transformation.

At the policy level, the study suggests that national digital transformation strategies can be strengthened by incorporating Agile readiness as a benchmark for capabilities. Initiatives aligned with MyDIGITAL could include targeted incentives for Agile training, support for pilot implementations and the development of a national certification framework in collaboration with universities and professional bodies. Such measures can accelerate Agile capability development across organisations of different sizes, support reskilling for Agile roles and contribute to workforce adaptability and digital competitiveness within Malaysia's evolving digital economy.

9.5.2 Theoretical contributions

This study strengthens Agile project management theory by demonstrating that internal organisational enablers are central to Agile readiness in Malaysian IT organisations. The significant effects of organisational culture and human resource support reinforce key assumptions of the Agile maturity model and organisational agility theory that Agile sustainability depends on the organisational environment in which Agile practices are embedded. By validating these relationships in an emerging-economy context, the study extends the applicability of these theories beyond the settings that dominate the existing Agile literature. The non-significant effect of technological readiness further refines the theory, indicating that infrastructure alone is insufficient to drive meaningful Agile transformation without parallel cultural alignment and human capability development. In addition, the absence of a moderating effect of organisational size challenges scale-based assumptions within Agile adoption models and suggests that Agile readiness is not inherently constrained by structural size when internal enablers are strong.

9.5.3 Limitations

This study is subject to several limitations. The cross-sectional survey design captures responses at a single point in time, which limits causal inference. Although significant associations were identified between the predictors and Agile project readiness, the design does not allow firm conclusions regarding directional effects. Future research using longitudinal designs would better capture the evolving nature of Agile adoption and changes in organisational conditions over time.

The study also relies on self-reported data, which may introduce response biases, such as social desirability bias. Despite assurances of anonymity and attempts to minimise bias through survey design, perceptual measures may not fully reflect actual Agile maturity or operational challenges. Subsequent studies could strengthen validity by incorporating objective indicators of Agile performance, such as delivery velocity, cycle time or project turnaround metrics.

In addition, practical constraints affected the breadth of sampling and the generalisability of the results. Access to respondents across organisational hierarchies was limited, with smaller firms underrepresented, and some large organisations declined to participate due to internal policies. The one-month collection window and a sample of 105 responses, while adequate for the analyses conducted, constrained deeper subgroup comparisons. Finally, the Malaysian specific context limits transferability to other settings, although the model may be informative for comparative research in similar developing economies.

9.5.4 Recommendations and future research

Future research should employ longitudinal designs to strengthen the robustness and explanatory depth of studies on Agile project management. Tracking Agile implementation across project phases and over time would provide clearer insights into how organisational readiness evolves in response to internal change initiatives and external pressures, addressing limitations inherent in cross-sectional approaches.

To reduce biases in self-reported data, subsequent studies should adopt triangulation by combining survey responses with organisational records, project performance metrics such as sprint velocity and defect rates, and managerial assessments. Broader sampling across sectors and organisational types, including start-ups, government-linked companies and non-IT firms, would also enhance generalisability, particularly if supported through collaboration with industry associations, professional bodies or government-supported technology clusters.

Building on the present findings, future inquiries may examine leadership style, psychological safety and trust as determinants of Agile readiness and team performance, while also considering external contingencies such as market volatility as potential moderators. Researchers may further refine explanatory models by testing alternative moderators or mediators, such as organisational maturity, stage of digital transformation and project complexity, and by conducting comparative studies across public and private organisations or within Association of Southeast Asian Nations (ASEAN) contexts to capture regional variation in Agile adoption and maturity.

REFERENCES

Agbejule A and Lehtineva L (2022) The relationship between traditional project management, agile project management and teamwork quality on project success. *International Journal of Organizational Analysis* **30(7)**: 124–136.

Aghina W, Handscomb C, Ludolph J, West D and Yip A (2019) *How to Select and Develop Individuals for Successful Agile Teams: A Practical Guide.* McKinsey & Company, https://www.mckinsey.com/~/media/McKinsey/Business%20Functions/Organization/Our%20Insights/How%20to%20select%20and%20develop%20individuals%20for%20successful%20agile%20teams%20A%20practical%20guide/How-to-select-and-develop-individuals-for-successful-agile-teams.pdf (accessed 13/05/2026).

Assidi S, Omran M, Rana T and Borgi H (2025) The role of AI adoption in transforming the accounting profession: a diffusion of innovations theory approach. *Journal of Accounting & Organizational Change* **21(5)**: 915–936.

Baruch Y and Holtom BC (2008) Survey response rates and trends in organisational research. *Human Relations* **61(8)**: 1139–1160.

Chia FC, Tung YH and Yong FYY (2022) Examining the agile project management practices in the Malaysian construction industry. *IOP Conference Series: Earth and Environmental Science* **1101(4)**: 042041.

Cimini C, Lagorio A and Cavalieri S (2024) Development and application of a maturity model for industrial Agile working. *Computers & Industrial Engineering* **188**: 109877.

Elseknidy M, Kineber AF, Oke AE *et al.* (2024) Identifying the agile project management implementation barriers for sustainable residential buildings. *International Journal of Engineering Business Management* **16**: 18479790241299634.

Ghasemi A and Zahediasl S (2012) Normality tests for statistical analysis: a guide for non-statisticians. *International Journal of Endocrinology and Metabolism* **10(2)**: 486.

Känsälä M and Tokumaru N (2023) Interaction between agile methods and organizational culture – a qualitative study. *Pacific Asia Journal of the Association for Information Systems* **15(2)**: 2.

Kerzner H (2025) *Project management: A Systems Approach to Planning, Scheduling, and Controlling*, 14th edn. Wiley, Hoboken, NJ, USA.

Kudyba S and Cruz AD (2023) Understanding project success involving analytic-based decision support in the digital era: a focus on IC and agile project management. *Journal of Intellectual Capital* **24(6)**: 1429–1446.

Martinez-Sanchez A and Vicente-Oliva S (2023) Supporting agile innovation and knowledge by managing human resource flexibility. *International Journal of Innovation Science* **15(3)**: 558–578.

Mollet LS and Kaudela-Baum S (2023) Critical HR capabilities in agile organisations: a cross-case analysis in Swiss SMEs. *Review of Managerial Science* **17(6)**: 2055–2075.

Neto JDSA, Penha R, da Silva LF and Scafuto IC (2022) The importance of leadership in agile projects: systematic literature review. *Research, Society and Development* **11(5)**: e44511528117.

Neumann M, Kuchel T, Diebold P and Schön EM (2024) Agile culture clash: unveiling challenges in cultivating an agile mindset in organizations. *Computer Science and Information Systems* **21(3)**: 1013–1031.

Omar M, Gilal AR, Ahmad M *et al.* (2024) Enhancing agile project success: a comprehensive study of risk management approaches among Malaysian practitioners. *Journal of Software: Evolution and Process* **36(9)**: e2681.

Rialti R and Filieri R (2024) Leaders, let's get agile! Observing agile leadership in successful digital transformation projects. *Business Horizons* **67(4)**: 439–452.

Zhao E (2024) Transformational leadership in the digital era: enhancing organizational performance through agile management and employee empowerment. *International Journal of Core Engineering & Management* **3**: 86–95.

Mohammed Reyasudin Basir Khan
ISBN 978-1-83742-803-8
https://doi.org/10.1108/978-1-83742-802-120261013
Emerald Publishing Limited: All rights reserved

Chapter 10
Assessing the influence of Agile knowledge and continuous learning on workplace effectiveness: a quantitative analysis of implementation and training

Nur Aisya Khairul Nizar[1], Mohammed Reyasudin Basir Khan[1], Chong Peng Lean[2] and Ahmad Anwar Zainuddin[3]

[1]Tun Razak Graduate School, Universiti Tun Abdul Razak (UNIRAZAK), Kuala Lumpur, Malaysia
[2]MILA University, Nilai, Negeri Sembilan, Malaysia
[3]Kulliyyah of Information and Communication Technology, IIUM, Gombak, Malaysia

10.1. Introduction

Agile methodology is an iterative and adaptive approach to project and software development that emphasises customer collaboration, team interaction and responsiveness to change (Dong *et al.*, 2024). Originally developed for software engineering, Agile has since been adopted across multiple industries for its suitability to dynamic and uncertain environments. By promoting incremental delivery through short work cycles known as sprints, Agile prioritises people and interactions over rigid processes, enabling organisations to respond rapidly to evolving requirements.

The formal origins of Agile can be traced to the *Agile Manifesto* published in 2001, which articulated a shift away from traditional, plan-driven management assumptions toward flexibility, collaboration and continuous improvement (Badran and Abdallah, 2025). While initially confined to software development, Agile principles have gradually expanded into broader organisational contexts, emphasising customer involvement, adaptability and cross-functional teamwork. This evolution reflects a transition from a purely technical methodology to a comprehensive managerial and cultural framework.

Several Agile frameworks operationalise these principles in practice. Scrum emphasises empirical process control, where progress is guided by inspection, adaptation and transparency

(Baxter and Turner, 2023). Kanban focuses on optimising workflow through visual management and pull-based systems, while Lean emphasises waste reduction and continuous problem-solving across organisational levels. These frameworks differ in structure but share a commitment to adaptability, continuous learning and value delivery, making them well-suited to complex, fast-changing project environments.

This study focuses on a Danish-headquartered renewable energy organisation that has transformed from a fossil fuel-based organisation into a global leader in green energy. The firm operates across Europe, the USA and the Asia Pacific region, employing approximately 9000 people worldwide. Its Kuala Lumpur office serves as a major global Digital and IT hub, supporting geographically distributed and culturally diverse teams. The scale and global integration of the organisation's operations create both opportunities and challenges for Agile implementation, particularly in coordinating across time zones, cultures and functional domains.

The organisation's transition toward renewable energy required significant changes in strategy, operations and digital capabilities. Agile methodologies, particularly the Scaled Agile Framework, were introduced within the IT function to support the firm's digital transformation (Kowalczyk *et al.*, 2022). Agile Release Trains were established to coordinate large-scale delivery and enable alignment between strategic objectives and operational execution. Through Agile adoption, the firm aims to enhance collaboration, reduce delivery cycles and improve responsiveness to industry and stakeholder demands.

Despite the potential benefits of Agile, many large organisations continue to experience inconsistent outcomes in Agile adoption (Sońta-Drączkowska and Krogulec, 2024). Challenges include uneven implementation, resistance to change, insufficient training and misalignment between tools, practices and organisational culture. These issues are particularly pronounced in globally distributed firms such as the subject organisation, where teams operate within diverse cultural and professional contexts. Existing research has not sufficiently examined how key factors such as project management tools, employee Agile knowledge, consistency of Agile implementation and training initiatives interact to influence Agile effectiveness in such environments.

Accordingly, the study presented in this chapter addresses a gap in the literature by examining the effectiveness of Agile practices at the organisation's Kuala Lumpur hub. The research investigates how project management tools, employee understanding of Agile, implementation consistency, and training and development programmes influence workplace effectiveness. By focusing on a global renewable energy organisation, the study contributes empirical insights beyond the traditionally studied software sector.

The research objectives are fourfold. First, the study evaluates the overall effectiveness of Agile practices at the regional hub and their contribution to workplace performance. Second, it examines how project management tools influence the effectiveness of Agile practices. Third, it analyses the relationship between employee understanding of Agile and Agile effectiveness. Fourth, it investigates how Agile training and development programmes enhance employee comprehension, implementation and resulting effectiveness.

These objectives were operationalised through research questions that explore the relationships between Agile practices and project management tools, employee Agile knowledge, day-to-day integration of Agile principles and training initiatives. Prior studies suggest that project management tools can enhance Agile performance when aligned with Agile values, while inadequate tool utilisation can hinder effectiveness (Badran and Abdallah, 2025; Chandrachooodan *et al.*, 2023). Similarly, employees with strong Agile knowledge are better able to adapt to change, manage complexity and improve team performance (Dikert *et al.*, 2016). Consistent integration of Agile principles into daily workflows has also been associated with higher productivity and organisational effectiveness. Training plays a critical role in shaping positive attitudes toward Agile and reducing resistance to change, particularly in large-scale transformations (Gandomani *et al.*, 2014).

The significance of this study lies in both its theoretical and practical aspects. Academically, it extends Agile research into the renewable energy sector and contributes to understanding how Agile scales within large, complex and geographically distributed organisations. It provides empirical evidence on the factors that facilitate or constrain Agile effectiveness, addressing gaps in existing literature. In practice, the findings offer insights for this firm and similar organisations seeking to enhance project management efficiency, align global practices and strike a balance between standardisation and flexibility. The study also supports the organisation's strategic objectives by examining how Agile practices contribute to innovation, sustainability and long-term competitiveness.

The theoretical foundation of the study integrates the (Agile Manifesto, 2001), organisational change management theory, and the Scaled Agile Framework. The Agile Manifesto provides the philosophical basis for valuing collaboration, adaptability and customer focus (Lowell, 2023). Organisational change management theory, particularly the awareness, desire, knowledge, ability and reinforcement (ADKAR) model, emphasises the human dimension of change through awareness, desire, knowledge, ability and reinforcement (Mudjisusatyo *et al.*, 2024). This model is especially relevant for managing resistance and sustaining Agile adoption. The Scaled Agile Framework offers a structured approach for coordinating multiple Agile teams and aligning execution with strategic goals in large organisations.

Empirical literature highlights that Agile effectiveness depends on multiple interrelated factors. Studies emphasise the importance of collaboration, leadership support and training in achieving positive Agile outcomes (Chukwunweike and Aro, 2024; Porkodi, 2024). Research also shows that selecting and using appropriate project management tools enhances coordination and transparency, particularly in distributed teams (Tyagi *et al.*, 2022). Agile knowledge has been linked to improved project success, stakeholder satisfaction and team performance, while inadequate training has been identified as a major barrier to Agile transformation.

Based on this literature, the study proposes a conceptual framework in which project management tools, Agile knowledge, Agile implementation practices and Agile training and development serve as independent variables influencing the effectiveness of Agile practices at the organisation. Four hypotheses posit positive relationships between each independent variable and Agile effectiveness. The framework provides a structured basis for empirical analysis and clarifies the relative contributions of each factor.

Figure 10.1 Conceptual framework

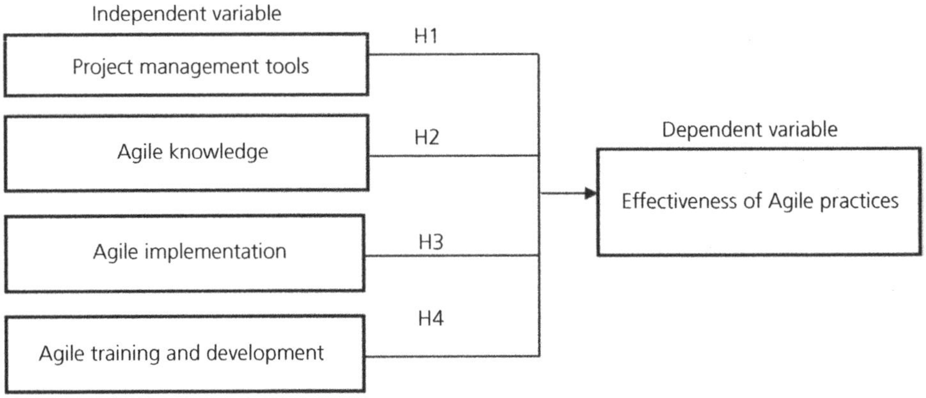

In summary, the literature demonstrates that Agile success requires more than adopting specific techniques. Effective Agile implementation depends on an integrated approach that combines appropriate tools, knowledgeable employees, consistent practices and comprehensive training within a supportive organisational context. By examining these factors within the regional hub of this global organisation, this study enhances understanding of Agile effectiveness in large global organisations and provides practical guidance for sustaining Agile transformation in complex and evolving environments.

10.2. Methodology

10.2.1 Research design

The examination of factors influencing Agile practices and workplace effectiveness at the subject organisation adopted a quantitative research design. This approach is appropriate because it facilitates the systematic collection and analysis of numerical data, enabling the identification of patterns, the testing of hypotheses and the assessment of relationships between variables that influence the success of Agile implementation.

The study presented in this chapter investigated the influence of Agile practices on workplace effectiveness at the organisation's Kuala Lumpur hub, with particular attention to project management tools, Agile knowledge, Agile implementation and training. Guided by a structured conceptual model, the research examined how these independent variables relate to the effectiveness of Agile practices as the dependent outcome, while providing a systematic basis for hypothesis testing and empirical analysis. The theoretical framework integrates Agile principles, project management theory and organisational effectiveness, building on prior research that suggests appropriate application of Agile can enhance productivity, collaboration and project outcomes.

A quantitative research design was employed, with data collected through online questionnaires distributed by way of Google Forms to 200 employees across key departments, including IT, digital, engineering, EPCO, finance, human resources and facility management. Participants were selected using random sampling to ensure adequate exposure to and experience with Agile practices. The survey instrument comprised demographic items and measures of Agile practices related to tools, knowledge, implementation and training.

Data analysis was conducted using IBM SPSS Statistics, incorporating both descriptive and inferential techniques. Descriptive statistics summarised respondent characteristics and general response patterns, while inferential analyses, including correlation and regression, examined relationships between the independent variables and workplace effectiveness. Assumptions for parametric testing were assessed using normality diagnostics and reliability was confirmed through Cronbach's alpha coefficients.

10.2.2 Population and sampling procedures

From the regional hub's workforce of approximately 750 employees, a sample of 200 was selected to balance feasibility with representativeness. This size is manageable for detailed statistical analysis while remaining sufficiently large to support reasonable generalisation of findings. The sample included employees from key functions, including IT and digital, engineering, procurement, construction, operations, finance, human resources and facility management. Random sampling was applied to reduce departmental overrepresentation and to capture diverse perspectives on Agile practices across roles. As a result, the dataset reflects variation in both exposure to Agile implementation and influence over Agile-related processes. In addition to primary survey data, secondary sources, including academic journals, organisational publications and prior research, were reviewed to strengthen the study's empirical foundation.

10.2.3 Data collection method

In this study, data were collected using an online questionnaire administered through Google Forms. This platform was selected for its accessibility, ease of administration and efficiency in processing large volumes of responses. The instrument comprised two sections: section 1 captured respondent demographics while section 2 assessed the study constructs across five domains, namely effectiveness of working with Agile practices, project management tools, Agile knowledge, implementation within the firm, and training and development.

To maximise participation and reduce response bias, the survey link was distributed through multiple channels, including email invitations and internal communication platforms. Respondents were informed that participation was voluntary and that responses would be used solely for research purposes. Anonymity and confidentiality were emphasised to encourage candid reporting. Data collection occurred over four weeks, with periodic reminders sent to sustain engagement and strengthen response coverage across departments.

10.2.4 Operationalisation and measurement

The research framework was structured around the effectiveness of Agile practices, which serves as the dependent variable representing the overall success of improving project outcomes and workplace performance. To understand what drives this success, the study examined four key independent variables: project management tools, Agile knowledge, Agile implementation, and Agile training and development. Each of these factors was defined through specific, measurable indicators to ensure a reliable and systematic analysis.

The independent variables were assessed based on their practical application within the organisation. For example, project management tools were evaluated based on their impact on workflow efficiency, including better progress monitoring, improved communication and reduced waste. Agile knowledge was measured by the depth of employee understanding and their ability to apply principles consistently. At the same time, Agile implementation focused on how deeply these principles are embedded in daily work through formal guidelines. Finally, training and development were gauged by the perceived growth in employee confidence and their capability to execute Agile methods effectively.

To interpret the data collected from 200 employees, the study employed a combination of descriptive and inferential statistical techniques. Descriptive analysis was used to summarise the dataset, applying measures of central tendency to identify the average perception and the standard deviation to measure how much responses vary. For the inferential analysis, Pearson correlation and multiple regression were used to evaluate relationships between variables and test the study's hypotheses. The predictive power of the independent variables was further validated through analysis of variance (ANOVA) to ensure the model provides a robust basis for generalising the findings to the broader organisational population.

10.3. Results and analysis

10.3.1 Demographic profile

Table 10.1 presents a detailed summary of respondents' profiles by sex, age, nationality, organisational department and years of work experience. This classification supports the interpretation of how demographic and professional differences may relate to forecasting the effectiveness of Agile practices, providing a sound basis for assessing their influence on Agile outcomes.

The sample shows a slight male predominance. Of the 200 respondents, 110 were male and 90 were female, representing 55% and 45%, respectively. Overall, both genders are reasonably well represented, although the distribution suggests a modest male skew that may reflect broader workforce patterns in the organisation or in specific roles and departments.

The age distribution indicates that most respondents are in the 21–40 age range, commonly associated with the peak working years. The largest group is aged 31 to 40, comprising 41% of the sample (82 respondents), suggesting a strong presence of mid-career professionals with substantial departmental responsibilities. Respondents aged 21 to 30 account for 30.5% (61 respondents), reflecting a sizeable early-career segment. Those aged 41 to 50 represent 20.5% (41 respondents), indicating a smaller but meaningful proportion of more senior personnel who may hold leadership roles. Very few respondents are under 20 (1%, 2 respondents) or over 51 (7%, 14 respondents), showing limited representation of the youngest and oldest age groups. Overall, the sample is concentrated in age brackets typically linked to career development and accumulated professional experience.

Table 10.1 Frequency and percentage of respondents' profiles

Description	Frequency	Percentage: %
Gender		
Female	90	45.0
Male	110	55.0
Age		
Under 20	2	1.0
21–30	61	30.5
31–40	82	41.0
41–50	41	20.5
Over 51	14	7.0
Nationality		
Malaysian	137	68.5
Non-Malaysian	63	31.5
Department in the organisation		
IT and digital	86	43.0
Engineering, procurement, construction and operations	48	24.0
Finance	38	19.0
Human resources	14	7.0
Facility management	14	7.0
Years of experience working		
1–3 years	52	26.0
4–7 years	50	25.0
8–10 years	60	30.0
Over 11 years	38	19.0

The nationality profile shows that 68.5% of respondents are Malaysian (137), while 31.5% are non-Malaysian (63). The presence of a substantial non-Malaysian group suggests an internationally diverse workforce and reflects the organisation's global orientation. At the same time, the predominance of Malaysian respondents indicates a strong local employment base, complemented by international expertise.

The IT and digital department constitutes the largest share of respondents at 43% (86), indicating a strong organisational emphasis on technology and digital transformation. The engineering, procurement, construction and operations function follows at 24% (48), highlighting the importance of technical and operational roles that underpin core business activities. Finance accounts for 19% (38), while human resources and facility management each represent 7% (14). Overall, the spread across technical, operational and support functions provides a broad base of perspectives and enables comparison of how departmental contexts may shape responses, consistent with a multifunctional organisational structure.

The respondents' years of experience are relatively evenly distributed. The largest group has 8 to 10 years of experience, representing 30% of the sample (60), indicating a substantial cohort of mid to senior professionals with well-developed industry knowledge. Those with 1 to 3 years account for 26% (52), reflecting an early-career segment, while those with 4 to 7 years account for 25% (50), suggesting a strong mid-career presence. Respondents with more than 11 years represent 19% (38), indicating a meaningful group of highly experienced personnel likely to hold specialist or leadership roles. Overall, the spread of experience points to a workforce that combines established expertise with emerging talent, supporting opportunities for collaboration and knowledge sharing.

Taken together, the demographic, departmental and experience profiles strengthen the study's credibility and applicability by capturing perspectives across multiple workforce segments. The representation across departments enables comparison of how functional contexts may shape engagement with organisational practices. The prominence of IT and digital (43%) underscores the organisation's strong emphasis on digital transformation, while engineering, procurement, construction and operations (24%) reflects its operational and technical core. Finance (19%) and human resources and facility management (7% each) further indicate contributions from both technical and support functions. The balanced mix of experience levels reinforces the relevance of the findings across junior, mid-career and senior employees.

10.3.2 Reliability analysis

Reliability analysis involves graphical, mathematical and descriptive procedures that use available evidence, statistical data and prior experience to identify system weaknesses or compare design alternatives. In this study, reliability analysis was used to confirm that the independent variables were measured consistently across respondents and contexts. This section presents the reliability procedure and results. Acceptable Cronbach's alpha (α) values typically fall between 0.6 and 0.7, while values above 0.8 indicate strong reliability. All variables in this study exceeded 0.8, demonstrating good reliability and supporting the robustness of the survey instrument.

Assessing the reliability of the measurement instrument is essential for evaluating Agile practices, as it supports the consistency and stability of survey responses across respondents and settings. Cronbach's alpha was used to assess internal consistency for each construct related to forecasting and Agile implementation.

Table 10.2 reports that the construct effectiveness of Agile practices showed strong reliability ($\alpha = 0.919$), indicating that the items consistently captured perceptions of workplace effectiveness in an Agile context. Project management tools also demonstrated strong internal consistency ($\alpha = 0.944$), supporting the use of these items to assess respondents' views on the effectiveness of these tools in Agile environments. Agile knowledge produced an alpha of 0.954, indicating very high consistency in measuring respondents' understanding of Agile

Table 10.2 Reliability assessment of final instrument

Variable	Cronbach's alpha, α	No. of questions
Effectiveness of Agile practices	0.919	5
Project management tools	0.944	5
Agile knowledge	0.954	5
Implementation of Agile	0.942	5
Agile training and development	0.953	5

concepts. Similarly, implementation of Agile ($\alpha = 0.942$) and Agile training and development ($\alpha = 0.953$) both indicated strong reliability, suggesting that the items reliably measured the extent of Agile adoption and the perceived adequacy of training and development initiatives.

Overall, the consistently high alpha values across all constructs indicate that the survey instrument demonstrates strong reliability and is suitable for subsequent analysis of the relationships between Agile-related factors and organisational outcomes. This reliability strengthens confidence in the study's findings and provides a sound basis for future research using similar measures.

10.3.3 Descriptive analysis

Descriptive analysis is a crucial tool for summarising the key characteristics of study data. It enables quantitative information to be presented in a comprehensible manner by providing concise explanations of the measurements and the sample. In this study of Agile practice methodologies, descriptive analysis establishes a foundation for subsequent inferential statistics by clarifying the basic properties of the collected data.

Based on the medium-range mean scores in Table 10.3, respondents reported moderate workplace efficacy. The highest mean (M) was for overall work effectiveness ($M = 3.47$), followed

Table 10.3 Response to the effectiveness of working with Agile practices

No.	Description	Mean, M	SD
WE1	I rate my overall work effectiveness in my current workplace as high	3.47	0.998
WE2	I feel productive during my workday	3.19	1.029
WE3	My daily tasks and responsibilities are well-organised	3.32	1.011
WE4	The workplace culture encourages teamwork and collaboration	3.28	0.962
WE5	The tools and technology provided to me for completing my tasks are effective	3.27	1.106

by task organisation (M = 3.32) and teamwork culture (M = 3.28), indicating generally positive perceptions. The lowest mean was for perceived productivity (M = 3.19), suggesting scope for improvement. Perceived effectiveness of technology and tools was moderate (M = 3.27). Standard deviations (SD) ranged from 0.962 to 1.106, reflecting some variability in experiences. Overall, the findings suggest moderate effectiveness, with opportunities to strengthen productivity and improve tool support.

Table 10.4 summarises respondents' views on the usefulness of Agile and Lean in daily work. Mean scores in the medium range indicate moderate effectiveness overall. The highest mean (M = 3.55) suggests frequent use of project management tools, while moderate benefits were reported for collaboration (M = 3.40) and continuous improvement (M = 3.35). Task monitoring (M = 3.39) and perceived job effectiveness (M = 3.25) also indicate room for enhancement. Standard deviations (1.143 to 1.278) reflect noticeable variability in respondents' experiences.

Table 10.5 examines respondents' familiarity with Agile concepts and their perceived impact on daily work. Mean scores indicate moderate proficiency, with Agile understanding associated with improved work alignment (M = 3.38), a more consistent workday (M = 3.37) and enhanced collaboration (M = 3.28). The lowest mean relates to overall comprehension of Agile concepts (M = 3.22), suggesting a need to strengthen deeper competence. Standard deviations (1.102 to 1.174) indicate variability in familiarity and application across teams.

Table 10.6 summarises employees' perceptions of how Agile techniques are embedded in daily work. Mean scores between 3.30 and 3.52 indicate moderate effectiveness and generally positive views of Agile governance and execution. The highest mean relates to applying

Table 10.4 Response of project management tools utilised in the workplace

No.	Description	Mean, M	SD
PM1	I use project management tools/methodology (such as Agile, Lean etc.) in my daily work	3.55	1.263
PM2	Utilising project management tools/methodology (such as Agile, Lean etc.) improves collaboration and communication among my team members	3.40	1.143
PM3	Using the project management tools/methodology (such as Agile, Lean etc.) helps contribute to continuous improvement in my daily work	3.35	1.230
PM4	The project management tools/methodology (such as Agile, Lean etc.) simplify the process of tracking and milestones of my work	3.39	1.223
PM5	The project management tools/methodology (such as Agile, Lean etc.) used for my daily work have enabled me to work more efficiently	3.25	1.278

Table 10.5 Response to knowledge of Agile

No.	Description	Mean, *M*	SD
KA1	I have a deep understanding of Agile principles being implemented at my workplace	3.22	1.102
KA2	I know when and how to utilise Agile principles when working in my daily workplace	3.26	1.135
KA3	Having an understanding of Agile principles enhances the collaboration of my team members during my daily work	3.28	1.174
KA4	A great knowledge of Agile principles among my team members led to better work alignment during my daily work	3.38	1.123
KA5	The level of knowledge about Agile principles leads to a more consistent and successful workday	3.37	1.153

Table 10.6 Response to the implementation of Agile

No.	Description	Mean, *M*	SD
IA1	My workplace has policies on the implementation and governance of Agile practices	3.44	1.180
IA2	My workplace adopts the Agile principles during my daily work	3.52	1.156
IA3	My workplace encourages the Agile practices being implemented to deliver successful project outcomes	3.51	1.199
IA4	My workplace has a well-structured and consistent implementation of Agile practices that leads to improved team performances	3.30	1.203
IA5	My workplace provides sufficient training and development for me as an employee to implement Agile practices for my daily work successfully	3.36	1.195

Agile principles in daily work ($M = 3.52$), followed by recognising their value for project outcomes ($M = 3.51$). The lower mean for consistent and organised application ($M = 3.30$) suggests scope to strengthen implementation and improve team performance. Training and development received a moderate rating ($M = 3.36$), indicating that additional support may be needed. Standard deviations from 1.156 to 1.203 reflect varied experiences across teams, reinforcing the need for greater consistency and enhanced training.

Table 10.7 presents employee perceptions of Agile training and development. Mean scores indicate that training supports individual capability, with the highest ratings for improved

Table 10.7 Response of Agile training and development of employee

No.	Description	Mean, *M*	SD
TD1	The company's Agile training significantly enhances my understanding of Agile principles within the organisation	3.51	1.112
TD2	As an employee, I feel more confident in executing Agile practices effectively after receiving targeted Agile training	3.50	1.080
TD3	The Agile training provided to employees significantly enhances the effectiveness of Agile practices in the organisation	2.48	0.997
TD4	The depth of knowledge gained from Agile training sessions directly correlates with the successful application of Agile practices in projects	3.38	1.078
TD5	I have other team members (e.g. Scrum Master, Product Owner, Agile Coach, Scrum team) who remind me of Agile principles during my workday, helping make our collaboration a success	3.38	1.193

understanding of Agile concepts ($M = 3.51$) and increased confidence in applying Agile techniques ($M = 3.50$). Respondents also moderately agreed that training depth supports project implementation and that ongoing reinforcement strengthens collaboration (both $M = 3.38$). In contrast, perceived impact on overall Agile effectiveness was low ($M = 2.48$), suggesting a gap between individual learning and organisational adoption. Standard deviations (0.997–1.193) indicate variability across teams. Overall, training is viewed as beneficial personally, but broader organisational gains may require additional support.

The histogram with a normal curve, as shown in Figure 10.2, and the overlay indicate that the regression residuals are approximately normally distributed, consistent with the normality assumption required for valid regression inference. The distribution is bell-shaped and centred near zero, suggesting that most residuals cluster around the mean with frequencies tapering on both sides. Minor values at the extremes appear limited and do not materially distort symmetry. The standardised residuals have a mean of 0.0000 and a standard deviation of 0.9899, aligning with expectations for standardised errors. Overall, the plot supports the normality assumption and strengthens confidence in the regression results.

The Q-Q plot in Figure 10.3 indicates that the data are approximately normally distributed. The observed values align closely with the 45° reference line, with only minor departures at the lower and upper tails. As these deviations appear limited and remain within an acceptable range, the normality assumption is sufficiently met. Accordingly, parametric statistical methods can be applied with confidence in subsequent analyses (Field, 2024).

Figure 10.2 Histogram of Regression standardised residuals

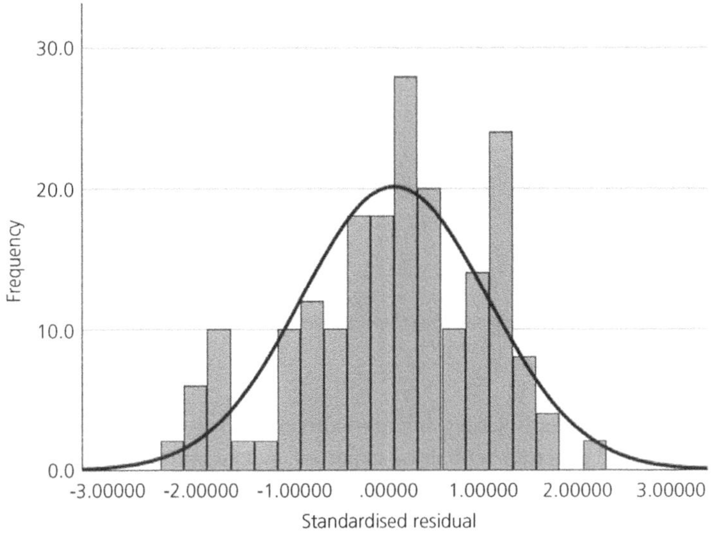

Figure 10.3 Q-Q plots (continued on next page)

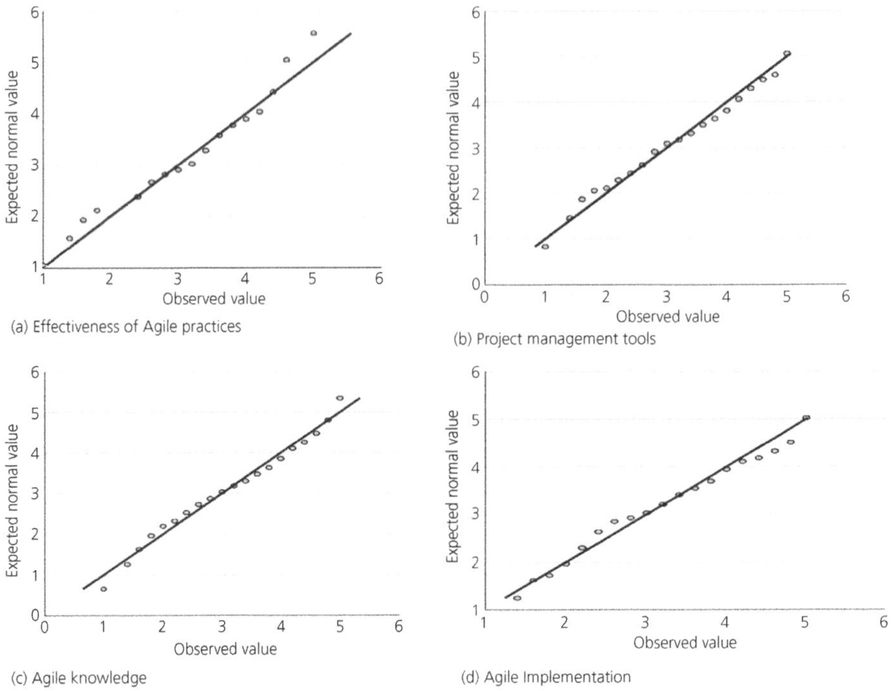

Figure 10.3 Continued

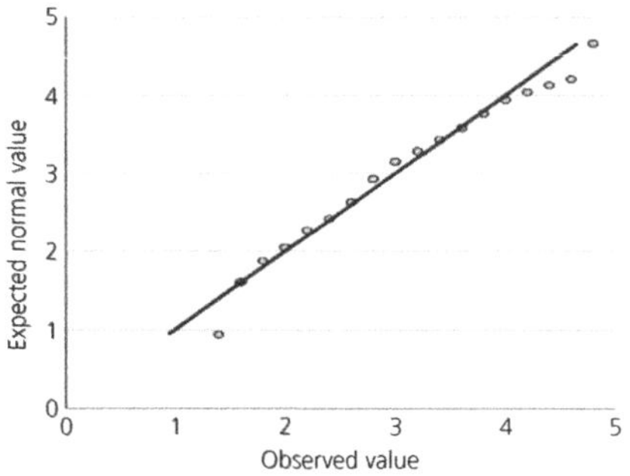

(e) Agile training and development

The project management tools scores are approximately normally distributed. The Q-Q plot compares the observed quantiles with those expected under a normal distribution; most points lie close to the 45° reference line, with only minor deviations in the tails. These departures are small and remain within acceptable limits, indicating that the normality assumption is reasonably satisfied. Consequently, parametric statistical methods can be applied appropriately to support the validity and robustness of the analysis (Field, 2024).

The measured Agile knowledge in the workplace is approximately normally distributed. Most data points align closely with the 45° reference line, with only minor deviations at the upper tail. As these departures remain within acceptable limits, the assumption of normality appears to be satisfied. Accordingly, parametric statistical methods can be applied with confidence to support the robustness and reliability of the results (Field, 2024).

The measured Agile implementation values align closely with the expected quantiles of a normal distribution. Most points lie on or near the 45° reference line, with only minor deviations at the lower tail. As these departures remain within acceptable limits, the normality assumption is supported. This finding justifies the use of parametric statistical methods and strengthens the robustness and reliability of the study results (Field, 2024).

The training and development variable is approximately normally distributed. Most data points lie along the 45° reference line, with only minor deviations in the upper tail. As these departures remain within acceptable limits, the normality assumption appears to be satisfied. Consequently, parametric statistical techniques can be applied to support the consistency and robustness of the results (Field, 2024).

10.3.4 Correlation analysis

Item 1 in Table 10.8 indicates a negligible positive correlation between effectiveness of Agile practices and project management tools ($r = 0.029$, $p = 0.685$). According to Salkind (2006), this reflects a very weak association and is not statistically significant ($p > 0.05$). Thus, project management tools do not meaningfully relate to effectiveness of Agile practices in this study.

Table 10.8 Relationship between the independent variables and the dependent variable

Item	Variable	Pearson correlation	Sig. (2-tailed)
1	Project management tools and effectiveness of Agile practices	0.029	0.685
2	Agile knowledge and effectiveness of Agile practices	–0.027	0.703
3	Implementation of Agile and effectiveness of Agile practices	–0.061	0.388
4	Agile training and development and effectiveness of Agile practices	0.006	0.936

Item 2 shows a negligible negative correlation between Agile knowledge and effectiveness of Agile practices ($r = -0.027$, $p = 0.703$). This relationship is very weak and not statistically significant ($p > 0.05$), suggesting that Agile knowledge does not significantly influence effectiveness of Agile practices in this dataset.

Item 3 reports a very weak negative correlation between Agile implementation and effectiveness of Agile practices ($r = -0.061$, $p = 0.388$). The association is not statistically significant ($p > 0.05$), indicating that Agile implementation is not meaningfully related to effectiveness of Agile practices in this study.

Item 4 shows an almost zero correlation between Agile training and development and effectiveness of Agile practices ($r = 0.006$, $p = 0.936$). This result is not statistically significant ($p > 0.05$), implying that Agile training and development do not substantially affect effectiveness of Agile practices in this research.

10.3.5 Regression analysis

Table 10.9 summarises the regression results for predictors of workplace performance, namely project management tools, Agile knowledge, Agile implementation and Agile training and development. The model shows a weak fit, with an adjusted R^2 of 0.018, indicating that these variables explain only 1.8% of the variance in workplace performance. The ANOVA result is not statistically significant ($p = 0.113$), suggesting that the predictors do not collectively

Table 10.9 Regression results for the effectiveness of Agile practices

Model	Adjusted R^2	ANOVA Sig.
Mean project management tools, mean Agile knowledge, mean implementation of Agile, mean of Agile training and development	0.018	0.113

Note: predictors: (constant), project management tools, Agile knowledge, implementation of Agile, Agile training and development; dependent variable: effectiveness of Agile practices.

influence workplace performance in this study. This indicates that other factors may better explain workplace performance and warrants further investigation.

Table 10.10 reports the regression coefficients for the determinants of effectiveness of Agile practices, showing each predictor's direction, relative contribution and statistical significance. Project management tools are positively associated with workplace performance ($\beta = 0.269$), but the effect is not significant ($p = 0.350$). Agile knowledge is also not significant and is negatively related to performance ($\beta = -0.165$, $p = 0.223$). Agile implementation is the only significant predictor and shows a negative effect ($\beta = -0.288$, $p = 0.027$), suggesting that how Agile is implemented may influence outcomes and may require refinement. Agile training and development have a positive coefficient ($\beta = 0.175$) but remain non-significant ($p = 0.150$), indicating limited explanatory value in the current model.

10.3.6 Hypothesis testing

The hypothesis-testing phase of this study evaluated four key propositions (H1–H4) regarding factors influencing workplace effectiveness in an Agile environment. As summarised in Table 10.11, the results led to the rejection of all four hypotheses, indicating that the assumed positive relationships between the independent variables and effectiveness of Agile practices were not supported by the data.

Table 10.10 Main factors that influence the effectiveness of Agile practices

Model	β	Sig.
Mean of project management tools	0.269	0.35
Mean of Agile knowledge	–0.165	0.223
Mean of implementation of Agile	–0.288	0.027
Mean of Agile training and development	0.175	0.150

Table 10.11 Summary of hypotheses conclusions (continued on next page)

Hypothesis	Accepted/not accepted	Justification
H1: effective Agile methods are favourably connected with the use of cutting-edge project management technologies	Not accepted	The correlation between project management tools and effectiveness of Agile practices was very weak ($r = 0.029$, $p = 0.685$), indicating no significant impact on effectiveness
H2: Agile practices exhibit a strong correlation with team members' degree of understanding of Agile knowledge	Not accepted	The correlation between Agile knowledge and effectiveness was weak and negative ($r = -0.027$, $p = 0.703$), showing no significant relationship between the two

Table 10.11 Continued

Hypothesis	Accepted/not accepted	Justification
H3: project performance outcomes are considerably improved by using effective Agile implementation methodologies	Not accepted	Agile implementation was associated with lower workplace performance ($\beta = -0.288$, $p = 0.027$), suggesting challenges with its application at the company
H4: the success of Agile practice deployment is positively impacted by comprehensive Agile training sessions	Not accepted	The correlation between Agile training and effectiveness of Agile practices was almost non-existent ($r = 0.006$, $p = 0.936$), indicating minimal impact on overall success

10.3.6.1 Analysis of non-significant factors (H1, H2 and H4)

For hypotheses H1 (project management tools), H2 (Agile knowledge) and H4 (Agile training), the analysis revealed negligible and statistically non-significant relationships with effectiveness of Agile practices.

- H1 was not accepted as the correlation was very weak ($r = 0.029$, $p = 0.685$), suggesting that the mere presence or use of cutting-edge tools does not automatically translate into improved performance.
- Similarly, H2 was rejected ($r = -0.027$, $p = 0.703$), showing that employee understanding of Agile principles alone is insufficient to drive organisational outcomes.
- H4 also failed to achieve significance ($r = 0.006$, $p = 0.936$), highlighting a notable disconnect where training initiatives successfully bolster individual confidence and theoretical understanding ($M = 3.50$) but fail to enhance broader organisational effectiveness ($M = 2.48$).

10.3.6.2 The negative impact of Agile implementation (H3)

The most striking finding from H3 concerned the hypothesis that effective Agile implementation methodologies would considerably improve project performance. While this was the only statistically significant predictor in the regression model ($p = 0.027$), it was not accepted because the relationship was unfavourable ($\beta = -0.288$). This negative coefficient suggests that the organisation's current implementation practices may be hindering rather than helping workplace performance. This finding underscores a 'critical gap' between theory and practice, implying that if Agile is poorly aligned with operational realities or inconsistently applied, it can reduce overall efficiency.

10.3.6.3 Overall model implications

Collectively, the rejection of these hypotheses is reflected in the regression model's weak fit, which explained only 1.8% of the variance in workplace performance (adjusted $R^2 = 0.018$) and was not statistically significant ($p = 0.113$). These results suggest that the effectiveness of

Agile practices at the Kuala Lumpur hub is likely influenced by external or contextual factors not captured in this study, such as leadership support, organisational culture or team maturity, rather than just tools, knowledge or training alone.

10.4. Practical insights

The research demonstrates that simply adopting Agile practices does not automatically lead to improved workplace effectiveness within an organisation. Because Agile is not a 'one-size-fits-all' solution, it should be treated as an evolving organisational capability rather than a static methodology. A key finding of this study is that poorly aligned or inconsistent implementation can have an unfavourable impact on performance, suggesting that the quality of day-to-day application is far more critical than mere adoption. To ensure success, Agile frameworks must be carefully tailored to the firm's specific operational realities, team maturity levels and unique cultural context.

There is also a notable disconnect between individual learning and broader organisational outcomes. While training initiatives successfully enhance employee confidence and theoretical understanding of Agile principles, these personal gains do not necessarily translate into enhanced organisational effectiveness. To bridge this critical gap, training programmes must move beyond theoretical concepts and focus on hands-on application, clear roles and real-world project challenges. Continuous support through mentoring, coaching and on-the-job reinforcement from roles such as Scrum Masters and Product Owners is essential to transform individual knowledge into effective organisational practice.

Furthermore, the study suggests that project management tools and theoretical knowledge alone are insufficient drivers of performance. Tools should be selected and utilised in a way that genuinely supports Agile values and team workflows rather than becoming administrative burdens used primarily for compliance or reporting. Ultimately, organisational and cultural factors – including leadership support, consistent practices across teams and transparent communication – likely play a more decisive role in successful Agile adoption than the specific tools or methodologies employed. For global organisations, maintaining shared principles while adapting practices to local contexts can improve both engagement and consistency.

10.5. Conclusion and recommendations

This study reports findings on the effectiveness of Agile practices, focusing on project management tools, Agile knowledge, Agile implementation and Agile training and development. Data from 200 employees across departments were analysed using descriptive and inferential statistics. Respondent profiling indicated a slight male majority, a predominantly mid-career workforce aged 31 to 40 and a largely Malaysian sample (68.5%), reflecting organisational diversity. IT and digital was the most represented department, consistent with the organisation's digital transformation focus. Reliability testing produced Cronbach's alpha values above 0.8 for all constructs, supporting strong internal consistency.

Descriptive results suggest generally positive perceptions of collaboration and organisation, while indicating areas for improvement in productivity and tool effectiveness. Respondents reported adequate familiarity with Agile and project management approaches, although understanding varied, implying a need for more targeted training. Agile practices were seen as integrated into daily work, but their consistent application remains an issue. Training was perceived to improve individual confidence and understanding, but its contribution to overall organisational outcomes appeared limited. Assumption checks supported regression use, with residuals approximating normality and Q-Q plots showing only minor deviations at the extremes.

Correlation results indicated very weak and non-significant relationships between Agile-related variables and workplace performance, including project management tools ($r = 0.029$, $p = 0.685$), Agile knowledge ($r = -0.027$, $p = 0.703$), Agile implementation ($r = -0.061$, $p = 0.388$) and Agile training and development ($r = 0.006$, $p = 0.936$). The regression model explained little variance in workplace performance (adjusted $R^2 = 0.018$) and was not statistically significant overall ($p = 0.113$). Agile implementation was the only significant predictor, showing a negative effect ($\beta = -0.288$, $p = 0.027$), suggesting that the implementation approach may require refinement. Overall, the findings indicate limited explanatory power of the examined factors and highlight the need to consider additional determinants of workplace performance.

10.5.1 Implications and strategic considerations

This study examined the effectiveness of Agile practices at Ørsted and found that the selected Agile-related variables show limited explanatory power for workplace effectiveness. The predictors collectively account for only a small proportion of variance in workplace performance, suggesting that additional organisational and contextual factors may be more influential and should be considered in future analyses. Although Agile implementation emerged as a significant predictor, its negative association with workplace performance indicates that outcomes may depend on how Agile is enacted in practice. This highlights the importance of aligning implementation approaches with operational needs, project characteristics and organisational objectives, rather than relying solely on adoption.

The weak relationships between Agile knowledge and training also indicate opportunities to strengthen capability development through more structured training and more effective knowledge sharing to support consistent application. While project management tools did not show a statistically meaningful effect in the model, their positive direction suggests that tool selection and alignment with Agile workflows may still influence productivity and coordination. Overall, the findings emphasise that improving Agile effectiveness likely requires a broader, integrated approach that considers implementation quality, capability building, enabling tools and organisational culture to support stronger workplace outcomes.

10.5.2 Limitations

Despite its breadth, this study has several limitations that should be considered when interpreting the findings and planning future research. First, the focus on a single organisation, Ørsted, limits the generalisability of the results, as Agile adoption and outcomes may differ across industries, regions and organisational cultures. Future studies could improve external validity by including multiple organisations from diverse contexts.

Second, the study relied on self-reported questionnaire data, which may be affected by response biases such as social desirability and common method variance. Incorporating qualitative methods such as interviews or focus groups and triangulating with objective performance indicators could strengthen the credibility of future findings.

Third, the cross-sectional design captures relationships at a single point in time and does not allow causal inference or observation of how Agile practices and outcomes evolve. Longitudinal designs would be better suited to examine changes over time and to clarify potential causal pathways.

Fourth, constructs such as effectiveness of Agile practices and Agile capability are multidimensional and may not have been fully captured within the scope of this study's measurement. Future research could refine operational definitions and include additional dimensions such as leadership support, team dynamics, organisational readiness and external environmental pressures.

Finally, the analyses assumed linear relationships, which may not adequately reflect complex organisational dynamics, including interaction effects and non-linear influences. More advanced analytical approaches could better represent these relationships and provide a more comprehensive explanation of Agile effectiveness.

10.5.3 Recommendations and future research

This study highlights several directions for future research on Agile practices and their organisational impact. To improve generalisability, future studies should include a broader range of organisations across industries, sizes and regulatory contexts, as Agile adoption and outcomes may vary significantly across settings. Longitudinal research designs are also recommended to examine causal relationships and to capture how Agile practices evolve and influence organisational performance in both the short and long term.

Future research would benefit from incorporating qualitative methods such as interviews, focus groups and case studies to provide deeper insight into the organisational, human and contextual factors that shape Agile success. Mixed methods approaches could uncover nuances not fully captured by quantitative analysis alone. The use of advanced analytical techniques, including structural equation modelling and machine learning, may further enhance understanding by identifying complex and non-linear relationships between variables.

Given the rapid pace of technological advancement, future studies should also explore the role of emerging technologies, such as artificial intelligence and machine learning, in supporting Agile processes, collaboration and decision-making. Cross-cultural research is encouraged to examine how cultural differences influence Agile adoption and effectiveness, particularly within multinational organisations. Finally, greater attention should be given to human and ethical dimensions, including organisational culture, employee capability development, data privacy and algorithmic bias, to ensure that Agile practices are implemented responsibly and sustainably.

REFERENCES

Agile Manifesto (2001) *Manifesto for Agile Software Development.* https://agilemanifesto.org/ (accessed 14/05/2026).

Badran SS and Abdallah AB (2025) Lean vs Agile project management in construction: impacts on project performance outcomes. *Engineering, Construction and Architectural Management* **32(5)**: 2844–2869.

Baxter D and Turner N (2023) Why Scrum works in new product development: the role of social capital in managing complexity. *Production Planning & Control* **34(13)**: 1248–1260.

Chandrachooodan G, Radhika R and Palappan RR (2023) Identifying the relevant project management tools in implementation of e-governance projects: journey from traditional to Agile. In *ISET International Conference on Applied Science and Engineering (Case 2021), Chennai, India, AIP Conference Proceedings* 2690(1): 020038.

Chukwunweike J and Aro OE (2024) Implementing agile management practices in the era of digital transformation. *World Journal of Advanced Research and Reviews* **24(1)**: 2223–2242.

Dikert K, Paasivaara M and Lassenius C (2016) Challenges and success factors for large-scale Agile transformations: a systematic literature review. *Journal of Systems and Software* **119**: 87–108.

Dong H, Dacre N, Baxter D and Ceylan S (2024) What is Agile project management? Developing a new definition following a systematic literature review. *Project Management Journal* **55(6)**: 668–688.

Field A (2024) *Discovering Statistics Using IBM SPSS Statistics*, 6th edn. Sage, London, UK.

Gandomani TJ, Zulzalil H, Ghani AAA, Sultan ABM and Parizi RM (2014) The impact of inadequate and dysfunctional training on Agile transformation process: a grounded theory study. *Information and Software Technology* **57**: 295–309.

Kowalczyk M, Marcinkowski B and Przybyłek A (2022) Scaled Agile framework: Dealing with software process-related challenges of a financial group with the action research approach. *Journal of Software: Evolution and Process* **34(6)**: e2455.

Lowell KR (2023) Using the Agile manifesto as a framework. In *Leading Modern Technology Teams in Complex Times: Applying the Principles of the Agile Manifesto*. Springer, Cham, Switzerland, pp. 13–18.

Mudjisusatyo Y, Darwin D and Kisno K (2024) Change management in Independent Campus program: application of the ADKAR model as a change management competency constructor. *Cogent Education* **11(1)**: 2381892.

Porkodi S (2024) The effectiveness of agile leadership in practice: a comprehensive meta-analysis of empirical studies on organizational outcomes. *Journal of Entrepreneurship, Management and Innovation* **20(2)**: 117–138.

Salkind NJ (ed.) (2006) *Encyclopedia of Measurement and Statistics*. Sage, Thousand Oaks, CA, USA.

Sońta-Drączkowska E and Krogulec A (2024) Challenges of scaling Agile in large enterprises and implications for project management. *International Journal of Managing Projects in Business* **17(2)**: 360–384.

Tyagi S, Sibal R and Suri B (2022) Empirically developed framework for building trust in distributed Agile teams. *Information and Software Technology* **145**: 106828.

Advancing Leadership and Collaboration in Project Management: Culture, Knowledge and Team Dynamics

emerald PUBLISHING ice Publishing

Mohammed Reyasudin Basir Khan
ISBN 978-1-83742-803-8
https://doi.org/10.1108/978-1-83742-802-120261014
Emerald Publishing Limited: All rights reserved

Chapter 11
Factors influencing sustainability-driven project management: the impact of organisational learning and knowledge transfer on project success

Nurul Hafeezah Mohd Zaharin and Mohammed Reyasudin Basir Khan

Tun Razak Graduate School, Universiti Tun Abdul Razak (UNIRAZAK), Kuala Lumpur, Malaysia

11.1. Introduction

In today's fast-paced global environment, the urgency of adopting sustainable practices has intensified. Escalating climate risks, persistent social injustice and the pressures of a globalised economy have exposed the limitations of conventional business and project management approaches that prioritise time, cost and scope as primary indicators of success (Chakrabarti, 2024). While organisations may be tempted to reduce costs and maximise short-term profitability, the environmental, social and economic consequences of traditional practices have become increasingly visible, with some impacts potentially irreversible.

Sustainability is therefore not merely a contemporary slogan, but an operational imperative. In response, sustainability-driven project management, often conceptualised as sustainability project management (SPM), provides a structured approach to integrating sustainability principles into project governance and delivery (Soares *et al.*, 2024). Rather than focusing narrowly on internal profitability, SPM encourages organisations to consider the wider consequences of project decisions for society, the environment and long-term economic resilience. It seeks to mitigate environmental degradation associated with unsustainable development while also addressing social challenges, including inequality, limited access to resources and community unrest, that have intensified across many contexts. Through this lens, project success is redefined to include enduring organisational wellbeing, stakeholder value and responsible stewardship of environmental and social systems.

The COVID-19 pandemic further amplified sustainability awareness by revealing vulnerabilities in healthcare systems, economies and social structures (Ghanbaripour *et al.*, 2025).

It highlighted the importance of organisational resilience and risk preparedness, both of which are consistent with sustainability principles. The temporary reduction in pollution during lockdown periods illustrated that environmental improvement is achievable, even under unplanned, disruptive conditions. Subsequent environmental crises continue to reinforce the urgency of sustainability. For example, large-scale wildfires and haze events, including transboundary haze linked to recurring forest fires in Indonesia, demonstrate how environmental harms extend beyond national borders and require collective regional responses. These events underscore the need for systematic sustainability adoption in projects whose impacts often extend beyond organisational boundaries.

Sustainability challenges also intersect with social and economic disparities. The pandemic disrupted employment and disproportionately affected lower-income groups, especially in sectors such as hospitality and tourism. Many countries experienced widening inequalities during recovery, while healthcare access disparities revealed uneven resource distribution across nations. In Malaysia, the B40 and M40 income classifications highlight persistent vulnerability among lower-income households, particularly regarding access to education and housing, while middle-income groups face stagnating wages amid rising costs. Such conditions increase the relevance of sustainability initiatives that strengthen inclusive development, create quality employment and improve social wellbeing.

The pandemic also exposed the fragility of many organisations and emphasised the importance of adaptive capability. Travel restrictions and lockdowns disrupted demand patterns and forced firms to pivot towards digital platforms, revise supply strategies and adopt remote work arrangements. Some industries have struggled to recover structural demand shifts, such as cinema operators facing competition from streaming services and altered consumer routines. These experiences illustrate why sustainability must be linked to resilience and long-term strategic readiness rather than treated as a peripheral concern.

At the global level, sustainability has become embedded in institutional expectations. The United Nations Sustainable Development Goals represent a widely recognised agenda that urges the integration of environmental, social and economic sustainability into decision-making processes (Sankaran *et al.*, 2024). Simultaneously, environmental, social and governance (ESG) criteria increasingly influence investor behaviour, reinforcing market expectations that firms demonstrate credible sustainability performance. Evidence indicates that many investors consider ESG criteria in investment decisions, suggesting that sustainability is now integrated into financial evaluation and risk assessment processes (In *et al.*, 2019; Park and Jang, 2021; Sood *et al.*, 2025). In this context, SPM provides a practical pathway to translate sustainability principles into project-level strategies that support both societal goals and organisational competitiveness.

In Malaysia, sustainability has gained prominence over the last decade, particularly among public-listed companies that respond to regulatory requirements, reporting standards and government incentives (Elaigwu *et al.*, 2024). However, these drivers are frequently external. Many organisations continue to face internal challenges related to awareness, capability and genuine commitment, leading to superficial compliance that prioritises incentives or reputational benefits rather than deep cultural integration. This highlights a critical internal alignment gap that can limit the long-term effectiveness of sustainability initiatives.

Despite increasing sustainability discourse and stronger institutional frameworks, sustainability efforts remain insufficient to address persistent environmental degradation, social inequality and organisational vulnerability. Regulatory interventions, incentives and stakeholder pressures have expanded the adoption of sustainability practices, and investors and consumers increasingly prioritise sustainability-aligned organisations. Nevertheless, many firms struggle to implement sustainability effectively due to limited internal readiness, inadequate resources and insufficient commitment. The misalignment between external pressures and internal cultural integration delays meaningful sustainability transformation in Malaysian industries and discourages sustained engagement with sustainability goals.

Accordingly, it is essential to understand how internal dynamics shape sustainability outcomes, particularly the role of internal stakeholders in translating sustainability practices into project success. The research presented in this chapter addresses this gap by examining internal stakeholder roles as a key mechanism through which sustainability project management practices can generate sustainable project success within Malaysian for-profit organisations.

This study investigates factors influencing sustainable practices in Malaysian industries by quantitatively examining relationships between sustainability project management (SPM) practices, internal stakeholder roles (ISR) and sustainability project success (SPS). Specifically, it examines how the profit, people and planet dimensions of SPM influence internal stakeholder roles and how ISR contributes to the success of sustainability projects. The study also assesses whether ISR mediates the relationships between each SPM dimension and SPS. Collectively, these objectives generate a comprehensive understanding of internal stakeholder dynamics and provide evidence-based guidance for embedding sustainability into organisational practice.

This study contributes to the sustainability literature by addressing persistent gaps in quantitative research on sustainability implementation in Malaysia. Despite decades of discussion of sustainability frameworks, robust quantitative assessment remains limited due to the interdependence of sustainability variables and the influence of external contextual factors that complicate measurement. This research responds by offering a structured quantitative assessment of SPM practices, ISR and SPS in Malaysian industry contexts.

From a practical standpoint, Malaysian organisations often struggle to define, operationalise and measure both the drivers and outcomes of sustainability initiatives. This study contributes by focusing on internal stakeholder roles as a measurable driver of sustainability performance. It supports organisations in clarifying responsibilities and developing meaningful indicators that can strengthen internal accountability for sustainability implementation. Moreover, since many sustainability efforts in Malaysia remain externally driven, the study highlights the importance of internal stakeholder engagement as a core enabling factor that remains underexplored. The findings can guide organisations toward more authentic and durable sustainability integration.

Triple bottom line (TBL) theory emphasises that sustainability requires balancing economic, social and environmental dimensions (Jum'a *et al.*, 2022). This approach encourages organisations to assess performance beyond financial outcomes by considering impacts on people and the planet. Within project management, the TBL framework supports the conceptualisation of SPM practices through the profit, people and planet dimensions, enabling sustainability to be embedded in decision-making, resource allocation and outcome evaluation. Prior studies have adopted TBL to assess sustainability and project success, including work that links sustainability-oriented practices to broader success measures in project contexts (Blak Bernat *et al.*, 2023; Shokouhi and Bachari, 2025).

Stakeholder theory argues that organisations should account for the interests of diverse stakeholders rather than focusing solely on shareholders (Freeman *et al.*, 2025). Stakeholders include internal actors such as employees, managers and owners, as well as external actors such as customers, regulators, suppliers, communities and investors. This theory provides a foundation for examining internal stakeholders as key agents in shaping sustainability implementation. In project settings, internal stakeholders influence strategic direction, governance priorities and operational execution, thereby affecting whether sustainability practices are implemented meaningfully or symbolically.

Complementary theoretical perspectives provide further interpretive depth. Institutional theory explains how firms may adopt sustainability practices to conform to regulatory or normative pressures to secure legitimacy, a process that is particularly relevant when sustainability adoption is driven by compliance (Qiu and Chen, 2023). The resource-based view and dynamic capabilities emphasise that internal resources, such as sustainability knowledge, leadership support and innovation capacity, can enable firms to translate sustainability commitments into competitive advantage. Greenwashing theory warns that sustainability reporting and self-assessment practices may overstate actual adoption, especially when sustainability becomes a reputational concern primarily. Together, these perspectives reinforce the need to investigate internal stakeholder roles as mechanisms that can either strengthen authentic sustainability integration or facilitate symbolic compliance.

Empirical research has increasingly examined the integration of sustainability into project management and its relationship with project outcomes. A key study in the Malaysian manufacturing context investigated the relationships between sustainable project management, sustainable project planning, and sustainable project success using quantitative methods and structural equation modelling (SEM). The study operationalised SPM through TBL dimensions and assessed SPS through multidimensional success indicators, including efficiency, stakeholder satisfaction, team performance, business success, preparation for the future and sustainability-related benefits. The results supported positive relationships between SPM and SPS and highlighted the mediating importance of planning in translating sustainability practices into project success (Chow *et al.*, 2021).

Although such work provides a strong foundation, it also suggests the need to examine alternative mechanisms beyond planning, particularly internal stakeholder roles that shape the implementation and institutionalisation of sustainability practices. Research on sustainability success criteria has also advanced through expert-driven frameworks that expand traditional project success measures to include sustainability outcomes. Qualitative and structured methods such as expert panels and analytic hierarchy processes have been used to prioritise sustainability success variables, supporting the view that sustainability should be embedded in both project processes and evaluation criteria. However, expert-based approaches may reflect disciplinary biases, may lack direct validation in project settings and often rely on perceived indicators rather than measurable sustainability performance metrics. These limitations strengthen the case for quantitative examination of mediating mechanisms that can explain why sustainability practices do or do not translate into success, especially in contexts where sustainability adoption is uneven.

Other studies outside Malaysia have explored mediating mechanisms linking sustainability performance to organisational outcomes. For instance, work examining sustainability and financial performance has proposed that strategic or optimal investment mediates the sustainability-to-performance relationship, indicating that sustainability outcomes depend on how firms allocate resources to sustainability-relevant capabilities (Keshavarzian and Silvius, 2022; Moshood *et al.*, 2025). This reinforces the broader logic of mediation in sustainability research, where practices alone are insufficient without enabling mechanisms.

Qualitative research on corporate sustainability strategy has also emphasised the role of leadership, internal structures and stakeholder engagement. Studies of interviews with senior executives across diverse industries suggest that sustainability is often interpreted through the lens of business survival and competitiveness, and may be implemented voluntarily when external pressure is weak (Alnsour, 2022; Fernandes *et al.*, 2015). Such findings highlight the pivotal role of upper management in shaping sustainability direction and underscore ongoing challenges in measurement and benchmarking. Nevertheless, executive-focused qualitative work can be prone to positive reporting bias and often fails to align with objective sustainability outcomes.

Across sectors and regions, evidence suggests internal stakeholder engagement is consistently linked to sustainability success. In high-impact industries such as construction, internal training and awareness correlate with improved compliance with green standards and project performance. In developing contexts, sustainability adoption often relies on external enforcement, and internal buy-in may be limited, producing what some scholars describe as sustainability dissonance, in which sustainability values are acknowledged but not embedded in daily practice. These insights support the argument that internal stakeholder roles are central to translating sustainability principles into effective project outcomes (Blak Bernat *et al.*, 2023; Soares *et al.*, 2024).

The conceptual framework proposes that sustainability project management practices, operationalised through profit, people and planet dimensions, influence sustainability project success through the mediating mechanism of internal stakeholder roles, as shown in Figure 11.1. SPM practices are expected to shape how internal stakeholders engage with sustainability, including their commitment, accountability and decision-making orientation. In turn, ISR is proposed to directly influence SPS by enabling the consistent implementation

Figure 11.1 Conceptual framework

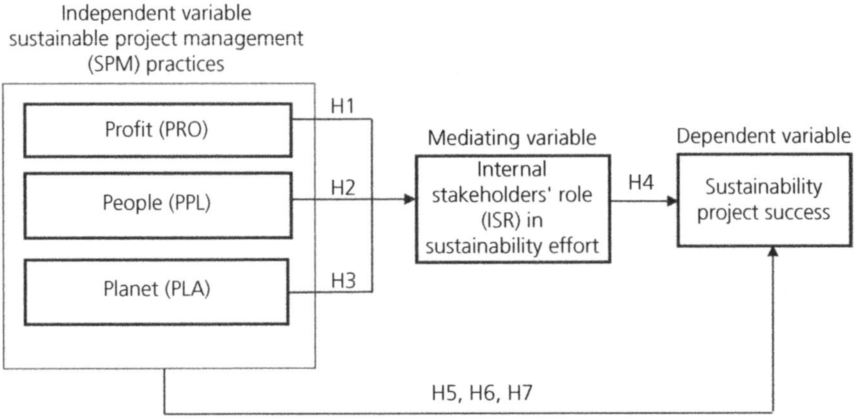

and meaningful evaluation of sustainability goals. The model, therefore, captures both direct paths from SPM dimensions to ISR and from ISR to SPS, as well as indirect effects from SPM dimensions to SPS by way of ISR.

Consistent with the framework, the study hypothesises that each SPM dimension positively influences internal stakeholder roles. Profit-oriented sustainability practices are expected to strengthen stakeholder engagement by linking sustainability to long-term value creation and organisational resilience. People-oriented practices are expected to enhance stakeholder commitment by aligning sustainability with employee wellbeing, community responsibility and inclusive outcomes. Planet-oriented practices are expected to strengthen internal roles by requiring operational changes, environmental stewardship and sustainability-minded decision-making. The study further hypothesises that internal stakeholder roles positively affect the success of sustainability projects by supporting consistent implementation, resource alignment and long-term sustainability integration. Finally, ISR is hypothesised to mediate the relationships between each SPM dimension and sustainability project success, indicating that sustainability practices translate into outcomes primarily through the actions, commitment and accountability of internal stakeholders.

11.2. Methodology

11.2.1 Research design

The study presented in this chapter adopted a quantitative approach to examine the relationships between sustainability project management (SPM) practices, represented by the profit (PRO), people (PPL) and planet (PLA) dimensions, the internal stakeholder role (ISR) and sustainability project success (SPS). The design was adapted from prior sustainability project research, particularly the survey-based approach applied in Malaysian industry contexts (Hamada *et al.*, 2025), and extended by explicitly operationalising ISR as a mediating construct.

Given that the study sought to test directional relationships between constructs and to estimate direct and indirect effects, the research was positioned as causal in intent. It was designed to evaluate whether SPM practices influence SPS and whether ISR transmits or explains part of this influence through mediation pathways. This approach moves beyond descriptive profiling by statistically evaluating theoretically derived relationships previously explored through conceptual and qualitative work. The study was conducted with minimal researcher interference and without variable manipulation. All constructs were measured through the perceptions and reported experiences of individuals working in Malaysian for-profit organisations, which supports ecological validity and strengthens external validity because the responses reflect real workplace contexts rather than controlled laboratory conditions.

The study employed a cross-sectional survey strategy using a structured questionnaire administered online. Measurement items for SPM and SPS were adapted from established instruments used in earlier empirical studies, with modifications to fit the Malaysian for-profit setting and the present conceptual model. The mediating construct, ISR, was operationalised using Likert-scale statements derived from stakeholder-focused literature, including prior qualitative insights (Li *et al.*, 2026), and translated into measurable indicators suitable for large-sample testing. This approach supports efficient collection of standardised data while enabling statistical modelling of latent variables using structural equation modelling (SEM) techniques.

Data collection was planned as a single wave in 2025 to capture the prevailing perceptions of internal stakeholders several years after the COVID-19 period, when sustainability awareness and expectations have been shaped by disruption, resilience demands and intensified stakeholder scrutiny. As a one-time measurement, the design did not aim to capture change over time, but rather to assess the current state of sustainability-related practices and internal stakeholder engagement within organisations. The methodology is replicable and can be applied in future studies to compare sustainability perceptions and stakeholder roles under different economic or environmental conditions.

11.2.2 Population and sampling procedures

This study examined internal stakeholder role (ISR) in organisational sustainability efforts; therefore, the unit of analysis is internal stakeholders employed in Malaysian for-profit organisations. The focus on for-profit firms is appropriate because these organisations must reconcile financial objectives with social and environmental responsibilities, unlike many non-profit entities whose mandates are inherently sustainability-oriented. To capture cross-sector perspectives, the population is not restricted to specific industries. Internal stakeholders were conceptualised into three primary groups: top management, responsible for strategic direction and sustainability governance; project managers, responsible for translating sustainability requirements into project planning and controls; and employees, responsible for operational execution and day-to-day implementation of sustainability initiatives.

In establishing the study population and sampling needs, national labour statistics provided a useful baseline. The Department of Statistics Malaysia (DOSM) reported approximately 16.7 million employed persons as of December 2024, with employers and employees comprising the majority. As public sector employees and self-employed individuals fall outside the study scope, and given the presence of government servants and a substantial non-profit sector, the relevant population for this study is reasonably estimated at approximately 14 to 15 million potential respondents. Because surveying the entire population was impractical, an appropriate sample size was required. Using conventional sample size determination approaches for large populations, a minimum of 384 respondents is typically recommended for a 95% confidence level with a plus-or-minus 5% margin of error. Additional methodological guidance supports feasibility thresholds for model estimation. Based on the item-to-respondent rule of at least five responses per measurement item, 34 items would imply a minimum of 170 respondents (Hair *et al.*, 2013). Partial least squares (PLS)-based SEM guidelines also commonly indicate that a minimum of 100 cases may be sufficient for basic model estimation, subject to model complexity. Taken together, these benchmarks justify targeting a larger sample and define defensible minimum thresholds.

Although probability sampling is often preferred for statistical generalisability, it is frequently constrained by the lack of an accessible sampling frame and the practical difficulty of obtaining comprehensive contact lists for the relevant population. In this study, these constraints necessitate non-probability sampling to ensure feasible recruitment within a limited data collection window. To strengthen estimation under non-probability conditions, resampling methods such as bootstrapping can be applied within the SEM procedure to improve the stability of parameter estimates, consistent with methodological discussions on enhancing inference when probability sampling is not feasible.

11.2.3 Data collection method

This study relies on primary data collected by way of an online questionnaire, consistent with a quantitative cross-sectional design. Online surveys are widely adopted due to their efficiency in reaching respondents across geographic locations, their lower administrative costs relative to face-to-face distribution, and their ability to enable rapid dissemination and response collection. The approach also supports anonymity and reduces interviewer-related bias because researcher respondent interaction is minimal. Following instrument finalisation and validation, the survey was administered in March 2025 using Google Forms to facilitate systematic data capture and streamlined distribution.

The questionnaire was structured into seven sections. The first section provides an introduction to the study, outlines its purpose and obtains informed consent. The second section captures respondents' demographic profiles. Sections three to five measure the study constructs, namely sustainability project management practices, internal stakeholder role in sustainability efforts and sustainability project success. An additional section includes open-ended questions to capture contextual insights that may enrich the interpretation of quantitative patterns. The instrument concludes with a brief debriefing and acknowledgement section.

For construct measurement, responses in the core measurement sections were recorded using a five-point Likert scale ranging from strongly disagree to strongly agree. This scaling approach is appropriate for capturing perceptions and attitudes toward sustainability practices and stakeholder engagement in organisational settings.

Data analysis was conducted in two stages. Descriptive statistics were computed using the statistical package for the social sciences (SPSS) to summarise respondent characteristics and overall response patterns. Hypotheses were then tested using partial least squares structural equation modelling to estimate the structural relationships between constructs, including the mediating role of ISR.

Survey distribution was conducted by way of popular digital channels, including WhatsApp, LinkedIn and Facebook, to broaden reach across diverse industries. Response monitoring was conducted after the first week to assess progress toward achieving a sufficient sample size and to guide follow-up outreach as needed.

11.2.4 Operationalisation and measurement

This section explains how each construct in the conceptual framework was operationalised and measured to enable hypothesis testing. Because the study evaluates relationships between latent variables, measurement was based on respondents' perceptions captured through structured items. A five-point Likert scale was employed to quantify agreement with each statement, ranging from strongly disagree to strongly agree. This format is widely used in behavioural and management research because it is straightforward for respondents, supports consistent measurement across items and includes a neutral midpoint that allows respondents to indicate neither agreement nor disagreement when appropriate.

Sustainability project management practices were operationalised as the independent construct and were grounded in the triple bottom line logic adopted in earlier sustainability project research. The measurement items were adapted from prior work that conceptualises sustainable project management through three dimensions aligned with profit, people and planet (Khan, 2024; Molina and Rajagopal, 2023). The profit dimension captures financial and economic sustainability considerations in project decision-making, including profitability, cost control, budgeting and the perceived financial value of sustainability initiatives. The planet dimension captures environmentally oriented practices, such as investment in sustainable innovation, resource efficiency, waste and energy reduction, environmental compliance and environmental risk mitigation across the project lifecycle. The people dimension captures social sustainability elements, including ethical standards, labour practices and employee wellbeing, community engagement, responsible supplier and contractor management, and attention to consumer health and safety. Together, these items represent the extent to which sustainability is incorporated into project management practices beyond conventional performance priorities.

The internal stakeholder role in sustainability efforts was operationalised as a mediating construct. Its development draws on stakeholder-oriented research emphasising how internal actors shape sustainability strategy and implementation (Grunwald and Fischer, 2022). The items translate qualitative insights into measurable statements about leadership direction, governance responsibility and organisational enablers suitable for survey research. The construct captures the role of senior management in promoting sustainability, the clarity of strategic direction and policies, the presence of accountable structures such as designated executives or departments, and the extent to which project managers integrate sustainability into planning and execution. It also assesses organisational mechanisms that enable stakeholder engagement, including sustainability training, internal communication regarding sustainability goals and progress, employee motivation and incentives, cross-departmental collaboration, and the incorporation of sustainability into performance evaluation and advancement decisions. Collectively, these indicators reflect whether sustainability is supported through internal ownership, capability development and accountability systems rather than treated as a symbolic or externally imposed requirement.

Sustainability project success was operationalised as the dependent construct and reflects an expanded view of project outcomes consistent with sustainability-oriented success models. The measurement items assess project efficiency in terms of meeting time and budget constraints and utilising resources effectively, as well as broader performance outcomes such as customer satisfaction, positive impacts on employee learning and development, contribution to business performance through profitability and market position, and preparation for future growth through new products, markets or technologies. The construct also includes an integrative sustainability outcome item capturing long-term economic, environmental and social benefits. This approach aligns project success measurement with both conventional delivery metrics and sustainability-related value creation.

Demographic and contextual questions were included to validate the sampling frame and to support descriptive profiling and subgroup analysis where appropriate. Respondents were first screened to confirm employment at a for-profit organisation in Malaysia, ensuring alignment with the defined population. Additional items capture gender, age group, educational background, overall work experience, designation, project management experience and industry sector. The organisational context was further explored through questions on the extent of sustainability integration into project management and on whether the firm is locally owned. These indicators enable the study to describe the respondent profile and to explore whether certain characteristics are associated with different sustainability perceptions.

To complement the structured measures, optional open-ended questions were included at the end of the survey. These questions elicit respondents' views on barriers to integrating sustainability into project management, challenges faced by internal stakeholder groups in implementing sustainability efforts, and recommendations for strengthening sustainability integration in future projects. While not used for hypothesis testing, these responses provide contextual nuance that can support interpretation of the quantitative findings and highlight practical considerations relevant to Malaysian for-profit organisations.

11.3. Results and analysis

Data analysis in this study was conducted in two stages: first, descriptive analysis to examine the characteristics and quality of the dataset, followed by inferential analysis to test the hypothesised relationships between the constructs. This sequential approach ensures that the data are suitable for hypothesis testing and that the measurement and structural models are statistically sound.

Descriptive analysis was first undertaken to prepare and screen the data after data collection is completed. The raw responses were coded and transferred into SPSS for data cleaning and preliminary analysis. Responses with incomplete or missing values were excluded to maintain data integrity. Basic screening procedures were applied to identify potential input errors and inconsistencies. Normality of the data distribution was then assessed using skewness and kurtosis statistics to evaluate symmetry and tail behaviour. Although PLS-based techniques do not strictly require normal data, examining distributional properties remains useful for detecting anomalies that may influence interpretation.

Descriptive statistics were used to summarise respondent characteristics and overall response patterns. Frequency distributions and percentages describe demographic profiles, while measures of central tendency and dispersion, such as mean, median, minimum, maximum and standard deviation (SD), provide an overview of variability and general trends in the responses. Common method variance was also assessed to determine whether a substantial portion of variance is attributable to the measurement method rather than the underlying constructs. This assessment, commonly based on exploratory factor or principal component analysis, helps reduce the risk of inflated relationships due to single-source data (Podsakoff *et al.*, 2012).

Prior to hypothesis testing, the measurement model was evaluated using partial least squares structural equation modelling. Convergent validity was assessed through item loadings, composite reliability and average variance extracted to ensure that items adequately represent their intended constructs. Discriminant validity was then examined to confirm that constructs were empirically distinct, using established criteria such as the Fornell–Larcker criterion and the Heterotrait–Monotrait ratio. These steps confirmed the adequacy of the measurement model before interpreting structural relationships.

Inferential analysis focused on testing the proposed hypotheses using PLS SEM in SmartPLS. Before estimating structural paths, collinearity between predictor constructs was examined using variance inflation factor values, with values above accepted thresholds indicating potential multicollinearity concerns (Hair *et al.*, 2013). Once collinearity had been addressed, path coefficients were estimated to evaluate the direction and strength of relationships between constructs. Statistical significance was assessed using a bootstrapping procedure that generates *t*- and *p*-values through repeated resampling, enabling robust estimation of direct and mediating effects.

Model explanatory power was evaluated using the coefficient of determination, which indicates the proportion of variance in endogenous constructs explained by the model (Chin, 2010). Effect sizes were then assessed to determine the relative contribution of each predictor to the explained variance, providing insight into the substantive impact of individual relationships beyond statistical significance. Finally, predictive relevance was examined using the blindfolding procedure to compute predictive accuracy measures that indicate how well the model predicts observed outcomes. Together, these analyses provide a comprehensive assessment of both the measurement and structural components of the proposed model and support rigorous evaluation of the study hypotheses.

11.3.1 Response rate

Data were collected using an online questionnaire administered by way of Google Forms over four weeks with individuals as the unit of analysis. To maximise reach within the limited timeframe, the survey link was disseminated through multiple WhatsApp groups and other social media platforms. In total, 555 survey invitations were distributed, with an intended response rate of approximately 60%, consistent with typical expectations for online survey administration, to ensure an adequate number of usable cases for multivariate analysis.

Table 11.1 reports that from the 555 invitations, 291 questionnaires were returned and 223 responses were deemed usable for analysis after data screening, yielding an overall response rate of 52%. The final usable sample exceeded common minimum requirements for structural equation modelling. Following the guideline of at least 5 responses per measurement item, a questionnaire with 34 items requires at least 170 cases. The achieved sample size of 223, therefore, provides sufficient observations to proceed with PLS-based analysis.

To ensure alignment with the target population, the questionnaire incorporated an initial screening question to assess respondent eligibility. Only participants who indicated that they met the inclusion criteria were permitted to proceed with the full survey. In contrast, those who did not meet the criteria were routed to the end of the questionnaire. This procedure ensured that only internal stakeholders working in Malaysian for-profit organisations were retained for subsequent analysis, thereby strengthening the dataset's relevance and validity.

Data completeness was assessed during the data preparation stage through a case processing summary as shown in Table 11.2. This evaluation was conducted to identify missing values

Table 11.1 Summary of survey distribution and response rate

Description	Frequency	Percentage: %
Total questionnaires distributed	555	100
Total questionnaires returned	291	52
Usable questionnaires for analysis	223	40
Unusable questionnaires for analysis	68	12

Note: percentages are calculated based on total questionnaires distributed (n = 555).

Table 11.2 Case processing summary for missing value

Variable	Included		Excluded		Total	
	n	Percentage: %	*n*	Percentage: %	*n*	Percentage: %
PRO	223	100.0	0	0.0	223	100.0
PLA	223	100.0	0	0.0	223	100.0
PPL	223	100.0	0	0.0	223	100.0
ISR	223	100.0	0	0.0	223	100.0
SPS	223	100.0	0	0.0	223	100.0
SPM	223	100.0	0	0.0	223	100.0

and determine whether data imputation or case exclusion would be necessary. The results indicated that all 223 usable responses were complete across all study variables, including profit, planet, people, internal stakeholder role, sustainability project success, and the composite sustainability project management construct. The absence of missing values supports the integrity of the dataset, allowing analyses to proceed without imputation or listwise deletion, thereby reducing the risk of bias associated with incomplete data.

11.3.2 Demographic profile

Profiling respondents' demographic characteristics is essential for contextualising the dataset and assessing the extent to which the sample reflects the intended study population. Accordingly, descriptive statistics were reported for key variables including age, education level, years of work experience, designation, industry sector and organisational type, consistent with established practices in survey-based research. These characteristics provide an overview of the internal stakeholders represented in the analysis and support the interpretation of potential variations in sustainability-related perceptions and practices across different organisational roles and contexts. The demographic summary is presented in Table 11.3.

Table 11.3 Demographic profile of respondents (n = 223 internal stakeholders of a for-profit organisation in Malaysia) (continued on next page)

Category	Selection/classification	*n*	Percentage: %
Age	18–24 years	4	1.8
	25–34 years	75	33.6
	35–44 years	107	48.0
	45 + years	37	16.6
Field of study	Business administration/management	81	36.3
	Engineering/industrial management/STEM	57	25.6
	Accounting/finance/economics/banking	29	13.0
	Healthcare	14	6.3
	Other (HR, IT, Law, env. science etc.)	42	18.8
Professional experience	Internship	5	2.2
	1–2 years	17	7.6
	3–5 years	18	8.1
	6+ years	183	82.1
Designation	Executive employee	120	53.8
	Project manager	56	25.1
	Director/CEO/owner	20	9.0
	Non-executive employee	17	7.6
	Project engineer	10	4.5

Table 11.3 Continued

Category	Selection/classification	*n*	Percentage: %
Project management experience	None	50	22.4
	1–5 years	87	39.0
	6–10 years	43	19.3
	10–20 years	34	15.2
	20+ years	9	4.0
Industry sector	Financial services	50	22.4
	Manufacturing	48	21.5
	IT and telecommunications	31	13.9
	Construction and real estate	23	10.3
	Healthcare and pharmaceuticals	19	8.5
	Energy and utilities	18	8.1
	Other (retail, logistics etc.)	34	15.3
Region (location)	Kuala Lumpur & Selangor	159	71.3
	Southern Region	30	13.5
	Northern Region	18	8.1
	Borneo	11	4.9
	Eastern Region	5	2.2
Ownership structure	Malaysian-owned	137	61.4
	European-owned	29	13.0
	United States-owned	20	9.0
	Joint venture	13	5.8
	Other (China, Singapore etc.)	24	10.8
Organisational size	Very large (>1000 employees)	75	33.6
	Medium-sized	62	27.8
	Large	42	18.8
	Small	33	14.8
	Micro	11	4.9

A total of 223 valid responses were retained for analysis. Respondent profiling focused on age, field of study, years of working experience, current designation and experience in project management to contextualise the sample and support the interpretation of subsequent findings.

The age distribution indicates a predominantly mid-career sample. Most respondents were aged 35 to 44 years (48.0%), followed by those aged 25 to 34 years (33.6%). Respondents aged 45 years and above accounted for 16.6%, while the 18 to 24 age group was minimally represented (1.8%). This pattern suggests that most participants are likely to hold roles with substantive organisational responsibility and exposure to sustainability and project-related decision-making.

Respondents represented a broad range of academic backgrounds. Business administration or management was the most common field (36.3%), followed by engineering, industrial management and other STEM areas (25.6%), and accounting, finance, economics or banking (13.0%). Smaller proportions were drawn from healthcare (6.3%), human resource management (4.5%) and information technology (3.6%). Several additional disciplines, including architecture, law, environmental science, project management, and construction and real estate, each contributed less than 2%. Overall, the distribution reflects the cross-disciplinary nature of sustainability-driven project work across Malaysian industries.

The sample also comprised respondents with substantial professional tenure. Individuals reporting more than six years of work experience constituted 82.1% of the sample, whereas those with internship to five years of experience represented 17.9%. This indicates that the dataset is largely informed by experienced internal stakeholders with practical organisational exposure.

By designation, executive-level employees constituted the largest group (53.8%), followed by project managers (25.1%). Directors, CEOs or business owners accounted for 9.0%, while non-executive employees accounted for 7.6%. The representation across hierarchical levels supports a more comprehensive view of sustainability practices, capturing both strategic and operational perspectives.

Project management experience was distributed across different levels of maturity. Respondents with one to five years of project management experience formed the largest segment (39.0%), followed by those with six to ten years (19.3%) and those with ten to twenty years (15.2%). A notable proportion (22.4%) reported no project management experience, consistent with the inclusion of employees who contribute to sustainability efforts without formal project management responsibilities.

Overall, the sample reflects a substantial proportion of experienced professionals from diverse educational backgrounds and organisational roles, particularly within middle and upper levels of responsibility. This composition suggests that respondents are positioned to provide informed perspectives on sustainability practices and project implementation within their organisations. In addition to respondent characteristics, organisational context is also important for interpreting sustainability practices; therefore, the next section examines key organisational attributes, including industry sector, location, ownership structure and organisational size.

Respondents were drawn from a broad range of industries, supporting a cross-sector perspective on internal stakeholder involvement in sustainability efforts. The largest representations came from financial services (22.4%) and manufacturing (21.5%). This was followed by information technology and telecommunications (13.9%) and construction and real estate (10.3%). Additional contributions were recorded from healthcare and pharmaceuticals (8.5%)

and energy and utilities (8.1%). Other sectors, including retail and consumer goods, education and research, logistics and transportation, and miscellaneous industries, accounted for less than 5% each. Overall, the diversity of sectoral participation enhances the breadth of insights into sustainability-driven practices across Malaysian for-profit organisations.

Most respondents were employed in organisations located in Kuala Lumpur and Selangor (71.3%), reflecting the central region's concentration of corporate and economic activity. Additional representation was obtained from the Southern (13.5%) and Northern (8.1%) regions. By comparison, fewer responses were recorded from the Eastern region (2.2%) and Borneo (4.9%). Although participation is concentrated in the central corridor, including organisations across multiple regions provides broader geographical coverage. It supports the applicability of the findings to Malaysian urban and semi-urban industrial settings.

Furthermore, most respondents were employed in Malaysian-owned organisations (61.4%). A notable proportion reported employment in European-owned firms (13.0%) and United States-owned firms (9.0%). Joint venture organisations represented 5.8% of the sample, while the remaining responses were distributed across other foreign ownership categories, including firms associated with China, India, Dubai and Singapore. The presence of both domestic and international organisations suggests potential differences in exposure to global sustainability expectations, reporting norms and governance practices, which may shape how sustainability is embedded in project management.

The sample also captured organisations of varying sizes. Very large firms with more than 1000 employees constituted the largest segment (33.6%), followed by medium-sized organisations (27.8%) and large organisations (18.8%). Small firms represented 14.8% of respondents' organisations, while micro enterprises accounted for 4.9%. The predominance of medium to very large firms suggests that many respondents operate within more formalised organisational settings, where project governance structures and sustainability-related policies are more likely to be established.

Overall, the sample reflects a broad spread across Malaysian industries, with most organisations located in Kuala Lumpur and Selangor and a strong representation of medium to very-large enterprises. The inclusion of both locally and foreign-owned firms further strengthens the dataset by capturing potential differences in sustainability orientation and exposure to international standards, thereby supporting the broader applicability of the findings across domestic and internationally affiliated organisations in Malaysia.

As part of the organisational profile, respondents were asked to indicate the extent to which sustainability is incorporated into their organisation's project management practices. The responses suggest moderate but uneven maturity in adoption. Nearly half of the respondents (45.3%, $n = 101$) reported that sustainability is a fundamental component across all projects, indicating a relatively institutionalised approach. A further 27.8% ($n = 62$) noted that sustainability is applied only in selected projects, commonly when driven by specific client expectations or regulatory requirements. Another 16.1% ($n = 36$) described sustainability integration as partial, reflecting informal or ad hoc practices. Smaller proportions indicated that sustainability is not currently practised but is planned (4.5%, $n = 10$), while

6.3% ($n = 14$) reported that sustainability is not a consideration within their current project management approach. Overall, the distribution indicates that while sustainability is prioritised as a core element in a substantial share of organisations, a sizeable segment remains compliance-driven or project-contingent, reflecting varied levels of organisational readiness and commitment.

To assess whether organisational structures support organisational priorities, respondents were also asked whether sustainability-related roles or departments were present. A dedicated sustainability department was reported by 43.9% of respondents ($n = 98$), suggesting formalisation and strategic emphasis. Sustainability oversight was located within other functional units, such as corporate social responsibility (CSR), compliance or operations, for 21.5% of respondents ($n = 48$). In addition, 17.0% ($n = 38$) indicated that responsibility is assigned to specific individuals or teams despite the absence of a formal department. Conversely, 13.9% ($n = 31$) reported having no sustainability department or assigned personnel, while 3.6% ($n = 8$) were unsure. Taken together, these results indicate that most organisations have some form of structural responsibility for sustainability, whether formal or embedded within existing functions, which is broadly consistent with the high proportion of respondents reporting at least partial sustainability practices.

Collectively, the findings show that differing maturity levels characterise sustainability adoption among Malaysian for-profit organisations. While many firms report sustainability as a project priority supported by organisational arrangements, others remain selective, externally driven or minimally engaged. This variation reinforces the relevance of examining internal stakeholder roles as a mechanism that may explain how sustainability intentions translate into consistent practice and project outcomes.

11.3.3 Minimum, maximum, mean and standard deviation

All constructs in this study were measured using a five-point Likert scale ranging from 1 (strongly disagree) to 5 (strongly agree). Accordingly, the minimum possible score for each construct is 1 and the maximum is 5. Descriptive statistics, including minimum, maximum, mean and standard deviation (SD), are reported in Table 11.4 to provide an initial overview of response distribution across the measured constructs.

Table 11.4 Minimum, maximum, mean and standard deviation

Variable	*n*	Minimum	Maximum	Mean	SD
PRO	223	1.00	5.00	3.9776	0.81165
PLA	223	1.00	5.00	3.8368	0.84783
PPL	223	1.00	5.00	4.0224	0.82486
ISR	223	1.00	5.00	3.7475	0.97684
SPS	223	1.00	5.00	4.0055	0.73795
SPM	223	1.00	5.00	3.9456	0.75334

The mean represents the average level of agreement for each construct and indicates central tendency. To support interpretation, mean scores can be classified into three ranges where values between 1.00 and 2.49 indicate low agreement, values between 2.50 and 3.49 indicate neutrality and values between 3.50 and 5.00 indicate high agreement. Standard deviation reflects the dispersion of responses around the mean and therefore indicates the degree of consensus among respondents. Lower values signal greater agreement, whereas higher values indicate greater variation in perceptions.

Overall, the descriptive results in Table 11.4 indicate that the mean values for all six constructs exceed 3.50, suggesting generally positive perceptions across the variables. In addition, the standard deviation values fall within a narrow range of 0.73–0.97, indicating that responses are closely clustered around their respective means and reflecting a relatively high level of consistency among participants. Collectively, the pattern of high mean scores and low variability suggests that internal stakeholders in Malaysian for-profit organisations generally perceive sustainability-driven project management practices, stakeholder roles and sustainability project outcomes favourably and with substantial agreement.

11.3.4 Common method variance (CMV)

Common method variance (CMV) refers to variance attributable to the measurement approach rather than to the underlying constructs of interest (Podsakoff *et al.*, 2012). It is a particular concern in survey research when predictor and outcome variables are collected from the same respondents using the same instrument, as this may artificially inflate observed relationships and introduce systematic measurement error. If unaddressed, CMV can weaken the validity of conclusions by biasing estimates of associations among constructs.

To assess the potential presence of CMV, this study applied Harman's single-factor test. This procedure involves conducting an exploratory factor analysis of all measurement items to determine whether a single factor emerges or whether a single general factor accounts for the majority of the total variance. The results of this assessment are reported in Table 11.5.

Table 11.5 Total variance explained (continued on next page)

Component	Initial eigenvalues			Extraction sums of squared loadings		
	Total	% of variance	Cumulative %	Total	% of variance	Cumulative %
1	19.139	56.292	56.292	19.139	56.292	56.292
2	2.667	7.845	64.137	2.667	7.845	64.137
3	1.354	3.982	68.119	1.354	3.982	68.119
4	1.034	3.041	71.160	1.034	3.041	71.160
5	0.997	2.933	74.093			

Table 11.5 Continued

Component	Initial eigenvalues			Extraction sums of squared loadings		
	Total	% of variance	Cumulative %	Total	% of variance	Cumulative %
6	0.867	2.551	76.644			
7	0.712	2.095	78.739			
8	0.590	1.737	80.476			
9	0.537	1.579	82.054			
10	0.516	1.518	83.573			
11	0.479	1.409	84.981			
12	0.434	1.276	86.257			
13	0.395	1.163	87.420			
14	0.374	1.099	88.518			
15	0.346	1.018	89.537			
16	0.330	0.972	90.508			
17	0.321	0.946	91.454			
18	0.295	0.867	92.321			
19	0.277	0.815	93.136			
20	0.251	0.739	93.875			
21	0.233	0.686	94.560			
22	0.213	0.627	95.188			
23	0.194	0.570	95.758			
24	0.193	0.567	96.325			
25	0.177	0.520	96.845			
26	0.161	0.475	97.320			
27	0.155	0.457	97.777			
28	0.139	0.410	98.187			
29	0.130	0.383	98.569			
30	0.127	0.372	98.942			
31	0.109	0.321	99.263			
32	0.098	0.288	99.551			
33	0.081	0.237	99.788			
34	0.072	0.212	100.000			

Note: extraction method principal component analysis.

Common method variance was assessed using Harman's single-factor test, implemented through principal component analysis on all 34 measurement items. Under this procedure, common method bias is considered problematic when a single factor emerges or when the first factor accounts for more than 50% of the total variance (Podsakoff *et al.*, 2012).

The results indicate that the first extracted factor explains 56.292% of the variance, exceeding the conventional 50% per cent threshold. However, multiple factors with eigenvalues above 1 were extracted, and four components collectively explain 71.16% of the variance, suggesting that variance is distributed across several underlying dimensions rather than being driven by a single general factor. Consistent with methodological discussions that caution against relying solely on the 50% criterion, the presence of multiple substantive components indicates that common method variance is unlikely to be the dominant source of covariance in the dataset and therefore is not expected to pose a substantial threat to the validity of the findings (Fuller *et al.*, 2016).

Nonetheless, given that the first factor accounts for a relatively large share of variance, additional procedural and statistical safeguards remain advisable in future work. These may include improving item format and response structure to reduce response patterning and employing more rigorous diagnostics such as latent method factor approaches, which provide stronger controls for detecting and accounting for method effects. The subsequent analysis, therefore, proceeds to measurement model assessment to confirm that item loadings align with their intended constructs and that no single factor disproportionately drives the measurement structure.

11.3.5 Assessment of measurement model

After data preparation, the measurement model was assessed to establish the reliability and validity of the constructs, with particular attention to convergent and discriminant validity. This evaluation was conducted using partial least squares structural equation modelling, with SmartPLS used to estimate the measurement model through PLS path modelling. Such an assessment is essential to confirm that the indicators adequately represent their intended latent constructs and that the construct structure aligns with the proposed conceptual relationships.

The conceptual framework was specified in SmartPLS to link the three sustainability project management dimensions – profit (PRO), people (PPL) and planet (PLA) – to the mediating construct – internal stakeholder role (ISR) – and the outcome construct – sustainability project success (SPS), as illustrated in Figure 11.2. The measurement model comprised five indicators for PRO (PRO1–PRO5), five for PPL (PPL1–PPL5), five for PLA (PLA1–PLA5), ten for ISR (ISR1–ISR10) and nine for SPS (SPS1–SPS9).

11.3.6 Convergent validity

Convergent validity refers to the extent to which indicators intended to measure the same construct are empirically associated, indicating that the measures share a high proportion of common variance. Establishing convergent validity is essential to ensure that each construct is adequately captured by its observed items and that subsequent structural analysis is meaningful and reliable. In this study, convergent validity was assessed using three standard criteria, namely indicator loadings, composite reliability and average variance extracted. The assessment was performed in SmartPLS, which estimates the measurement model using partial least squares path modelling.

Figure 11.2 Conceptual framework in SmartPLS

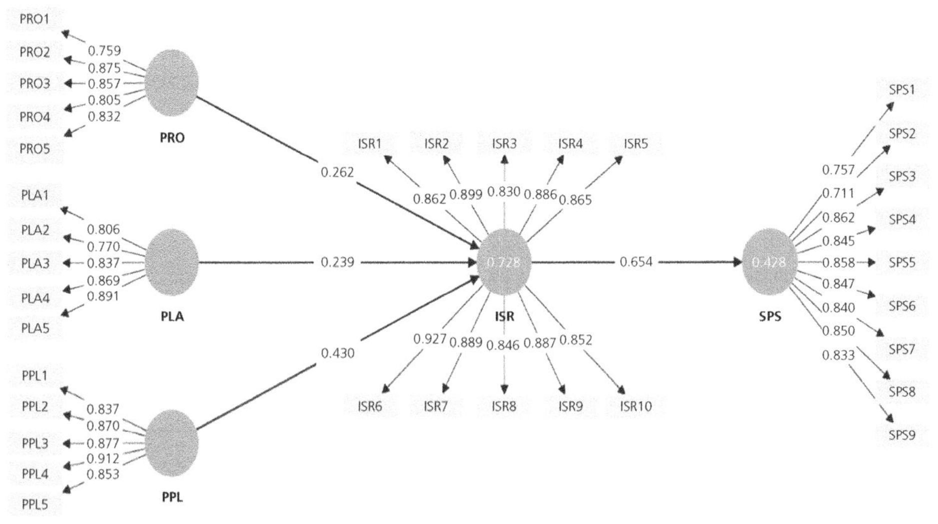

Indicator loadings assess the extent to which each item contributes to its intended construct, thereby supporting the adequacy of the measurement model. Loadings above 0.50 are commonly considered acceptable for establishing that an indicator contributes meaningfully to the latent variable (Ciavolino *et al.*, 2022). In this study, all items exceeded the minimum threshold, indicating that the indicators were sufficiently representative of their respective constructs. The outer loadings for the independent constructs profit, people and planet, the mediating construct internal stakeholder role and the dependent construct sustainability project success are reported in Tables 11.6–11.8.

Table 11.6 Factor loading of items to assess the independent variables (continued on next page)

Item ← variable	Outer loading
PLA1 ← PLA	0.806
PLA2 ← PLA	0.770
PLA3 ← PLA	0.837
PLA4 ← PLA	0.869
PLA5 ← PLA	0.891
PPL1 ← PPL	0.837
PPL2 ← PPL	0.870
PPL3 ← PPL	0.877
PPL4 ← PPL	0.912

Table 11.6 Continued

Item ← variable	Outer loading
PPL5 ← PPL	0.853
PRO1 ← PRO	0.759
PRO2 ← PRO	0.875
PRO3 ← PRO	0.857
PRO4 ← PRO	0.805
PRO5 ← PRO	0.832

Table 11.7 Factor loading of items to assess the mediating variable

Item ← variable	Outer loading
ISR1 ← ISR	0.862
ISR2 ← ISR	0.899
ISR3 ← ISR	0.830
ISR4 ← ISR	0.886
ISR5 ← ISR	0.865
ISR6 ← ISR	0.927
ISR7 ← ISR	0.889
ISR8 ← ISR	0.846
ISR9 ← ISR	0.887
ISR10 ← ISR	0.852

Table 11.8 Factor loading of items to assess the dependent variable

Item ← variable	Outer loading
SPS1 ← SPS	0.757
SPS2 ← SPS	0.711
SPS3 ← SPS	0.862
SPS4 ← SPS	0.845
SPS5 ← SPS	0.858
SPS6 ← SPS	0.847
SPS7 ← SPS	0.840
SPS8 ← SPS	0.850
SPS9 ← SPS	0.833

Internal consistency reliability was evaluated using Cronbach's alpha (α) and composite reliability. Reliability values above 0.70 indicate adequate internal consistency and suggest that items within each construct measure the same underlying concept. In PLS SEM, composite reliability can be reported as ρ_a and ρ_c, where ρ_a is often considered a more precise estimate because it accounts for the empirical construct structure. In contrast, ρ_c provides a traditional composite reliability estimate. As shown in Table 11.9, Cronbach's alpha and both composite reliability measures for all constructs exceed 0.70, demonstrating satisfactory reliability.

Convergent validity was further assessed using average variance extracted (AVE), which reflects the proportion of variance in the indicators explained by the construct. An AVE value above 0.50 indicates that a construct explains at least half of the variance in its indicators, supporting convergent validity. All AVE values reported in Table 11.9 surpass this threshold.

Overall, the results indicate that the measurement model demonstrates adequate indicator reliability, internal consistency and convergent validity. All items and constructs meet established criteria, supporting the instrument's robustness and suitability for subsequent structural analysis.

11.3.7 Discriminant validity

Discriminant validity evaluates whether each construct is empirically distinct from the others, thereby reducing the risk of construct redundancy and supporting the interpretability of structural relationships. In this study, discriminant validity was assessed using the Fornell–Larcker criterion and the Heterotrait–Monotrait ratio.

The Fornell–Larcker criterion requires that the square root of the AVE for each construct exceeds its correlations with other constructs in the model. As shown in Table 11.10, the diagonal values representing the square roots of AVE are higher than the corresponding inter-construct correlations for each construct, indicating that each latent variable shares more variance with its own indicators than with other constructs. This provides evidence that the constructs are sufficiently distinct.

Table 11.9 Composite reliability and average variance extracted

	Cronbach's alpha, α	Composite reliability (ρ_a)	Composite reliability (ρ_c)	Average variance extracted (AVE)
ISR	0.966	0.967	0.970	0.765
PLA	0.891	0.894	0.920	0.699
PPL	0.920	0.924	0.940	0.757
PRO	0.884	0.892	0.915	0.683
SPS	0.941	0.948	0.950	0.679

Table 11.10 Fornell–Larcker criterion

	ISR	PLA	PPL	PRO	SPS
ISR	**0.875**				
PLA	0.749	**0.836**			
PPL	0.813	0.760	**0.870**		
PRO	0.760	0.699	0.768	**0.826**	
SPS	0.654	0.631	0.740	0.707	**0.824**

Discriminant validity was further examined using the Heterotrait–Monotrait ratio, which compares correlations across constructs with correlations within constructs. Values below 1.00 indicate that constructs are not excessively similar and therefore are likely to be distinct. As reported in Table 11.11, all Heterotrait–Monotrait values are below 1.00, providing additional support for discriminant validity.

Taken together, the Fornell–Larcker results and Heterotrait–Monotrait values indicate that the measurement model demonstrates adequate discriminant validity. Combined with the convergent validity results, the measurement model is considered reliable and valid for subsequent structural model assessment and hypothesis testing.

Overall, as reported in Table 11.12, the variance inflation factor (VIF) results indicate that multicollinearity is not a major concern, as most indicator values fall within acceptable limits. However, two ISR indicators, ISR2 and ISR6, exceed the recommended threshold of 5, suggesting potential redundancy and overlap in their coverage. While this does not undermine the validity of the measurement model, future research may refine the ISR scale by revising item wording, removing one of the overlapping indicators where theoretically justified, or combining them to strengthen construct clarity and distinctiveness.

11.3.8 Path coefficient and hypotheses test results

This subsection presents the structural model results by reporting the estimated path coefficients that represent the strength and direction of hypothesised relationships between constructs. Path coefficients range from minus one to plus one, with values closer to plus

Table 11.11 Heterotrait–Monotrait ratio

	ISR	PLA	PPL	PRO	SPS
ISR					
PLA	0.803				
PPL	0.856	0.838			
PRO	0.814	0.784	0.851		
SPS	0.672	0.681	0.789	0.772	

Table 11.12 Outer VIF model

IV	VIF	MV	VIF	DV	VIF
PLA1	1.915	ISR1	5.007	SPS1	2.883
PLA2	1.805	ISR2	6.176	SPS2	2.305
PLA3	2.338	ISR3	2.965	SPS3	3.247
PLA4	3.694	ISR4	3.914	SPS4	3.194
PLA5	4.409	ISR5	4.101	SPS5	3.692
PPL1	2.359	ISR6	6.542	SPS6	3.108
PPL2	2.741	ISR7	4.291	SPS7	2.842
PPL3	2.969	ISR8	3.828	SPS8	3.589
PPL4	3.692	ISR9	4.109	SPS9	3.280
PPL5	2.503	ISR10	3.711		
PRO1	2.027				
PRO2	2.921				
PRO3	2.302				
PRO4	2.031				
PRO5	2.311				

Note: IV, independent variable; MV, mediating variable; DV dependent variable.

one indicating stronger positive relationships and values closer to minus one indicating stronger negative relationships. Statistical significance was assessed using bootstrapping, a nonparametric resampling procedure commonly applied in PLS SEM to obtain standard errors, confidence intervals and inferential statistics for hypothesis testing (Hair *et al.*, 2013). Consistent with recommended practice, 5000 bootstrap resamples were generated to provide stable estimates (Hayes, 2009). Hypotheses were evaluated using a one-tailed criterion, where support is indicated by a t-value greater than 1.645 and a p-value below 0.05. Confidence intervals were also examined, with intervals that do not include zero indicating robust statistical support.

The direct effects, as shown in Table 11.13, indicate that all hypothesised paths are positive and statistically significant. Internal stakeholder role has a strong effect on sustainability project success, with a path coefficient of 0.654 and high statistical significance, confirming that internal stakeholder engagement is the most influential driver of sustainable project outcomes. The three sustainability project management dimensions also significantly predict internal stakeholder role. Among these, the people dimension shows the largest effect on ISR with a coefficient of 0.430, followed by profit at 0.262 and planet at 0.239. The pattern suggests that social and human-centred sustainability practices are most strongly associated with internal stakeholder mobilisation, while financial and environmental practices also contribute meaningfully to strengthening internal roles.

Table 11.13 Direct relationship hypotheses

Hypothesis	Path coefficient, β	t-value	p-value	Result	Confidence interval (5%, 95%)
ISR → SPS	0.654	13.055	0.000	Supported	(0.573, 0.738)
PLA → ISR	0.239	2.755	0.003	Supported	(0.102, 0.389)
PPL → ISR	0.430	4.764	0.000	Supported	(0.269, 0.569)
PRO → ISR	0.262	3.761	0.000	Supported	(0.154, 0.380)

The mediation results, as shown in Table 11.14, further support the proposed mechanism. All three indirect paths from the sustainability project management dimensions to sustainability project success by way of the internal stakeholder role are positive and statistically significant. The people dimension exhibits the strongest indirect effect on project success with a coefficient of 0.281, indicating that people-related sustainability practices translate most effectively into successful outcomes when internal stakeholders are engaged. The profit and planet dimensions also demonstrate significant indirect effects on success, with coefficients of 0.171 and 0.156, respectively. For all indirect relationships, the confidence intervals do not include zero, reinforcing the reliability of the mediation estimates. Collectively, these findings provide empirical support for the structural model and highlight internal stakeholder role as a key pathway through which sustainability practices are translated into sustainability project success, with the people dimension emerging as the most influential antecedent in this process.

11.3.9 Coefficient of determination (R^2)

To complement the structural model results, the coefficient of determination was examined for each endogenous construct. In PLS SEM, R^2 indicates the proportion of variance in an endogenous latent variable explained by its predictor constructs, thereby reflecting the model's explanatory capability (Chin, 2010). Higher R^2 values indicate stronger explanatory power, while lower values suggest that additional factors outside the model may account for the outcome. In exploratory research, modest explanatory power is common because multiple contextual variables typically influence behavioural and organisational outcomes.

Table 11.14 Indirect relationship hypotheses

Hypothesis	Path coefficient, β	t-value	p-value	Result	Confidence interval (5%, 95%)
PLA → SPS	0.156	2.529	0.006	Supported	(0.063, 0.266)
PPL → SPS	0.281	4.781	0.000	Supported	(0.178, 0.373)
PRO → SPS	0.171	3.562	0.000	Supported	(0.098, 0.255)

Table 11.15 Coefficient of determination (R^2)

	R^2
ISR	0.728
SPS	0.428

As shown in Table 11.15, the R^2 value for ISR is 0.728, indicating that PLA, PPL and PRO collectively explain 72.8% of the variance in internal stakeholder role. This represents strong explanatory power and suggests that sustainability-oriented project management practices are substantial predictors of internal stakeholder engagement in sustainability efforts. The R^2 value for SPS is 0.428, indicating that ISR, together with the three sustainability practice dimensions, explains 42.8% of the variance in sustainability project success. Although lower than the value for ISR, this level of explanation is acceptable within sustainability and social science research, where project success is shaped by many additional organisational and environmental factors beyond the constructs specified in the model.

11.3.10 Effect size (f^2)

To further assess the structural model, effect sizes were examined using f^2, which indicates the practical contribution of each predictor construct to the endogenous construct beyond statistical significance. Effect size complements path coefficients and R^2 by showing how much an endogenous construct's explained variance changes when a specific predictor is removed from the model. The commonly used interpretation is where values below 0.02 indicate a small effect, values between 0.15 and 0.35 indicate a medium effect, and values above 0.35 indicate a large effect (Cohen, 2013).

As shown in Table 11.16, ISR has a very large effect on SPS with an f^2 value of 0.747, indicating that the internal stakeholder role is a dominant contributor to sustainability project success. This reinforces the structural path findings and highlights ISR as the most influential driver in the model. The people dimension shows a medium effect on ISR with an f^2 value of 0.213, suggesting that social sustainability practices meaningfully strengthen internal stakeholder engagement in sustainability efforts. In contrast, planet and profit show smaller effects on ISR, with f^2 values of 0.082 and 0.096, respectively, indicating that environmental and

Table 11.16 Effect size (f^2)

	ISR	PLA	PPL	PRO	SPS
ISR					0.747
PLA	0.082				
PPL	0.213				
PRO	0.096				
SPS					

economic sustainability practices contribute to internal stakeholder roles but with comparatively weaker practical impact.

Overall, the effect size results further underscore the centrality of internal stakeholders in achieving sustainability project outcomes and emphasise the relative importance of people-oriented sustainability practices in activating internal stakeholder roles.

11.3.11 Predictive relevance (Q^2)

Predictive relevance was assessed using Q^2 to evaluate the model's out-of-sample predictive capability within the PLS SEM framework. Q^2 was computed using the blindfolding procedure, which omits portions of the data and then predicts them to assess whether the model accurately reproduces the values of endogenous constructs. A Q^2 value greater than zero indicates that the exogenous constructs provide meaningful predictive relevance for the endogenous construct (Henseler, 2017).

As shown in Table 11.17, ISR recorded a Q^2 value of 0.714, indicating strong predictive relevance and suggesting that the planet, people and profit dimensions effectively predict internal stakeholder roles. Sustainability project success produced a Q^2 value of 0.539, also indicating strong predictive relevance and demonstrating that ISR, together with the three sustainability practice dimensions, provides substantial predictive capability for project success. Overall, these findings strengthen the evidence for model robustness by confirming that the structural relationships offer practical predictive value for understanding internal stakeholder engagement and sustainability project outcomes.

11.3.12 Assessment of structural model

Collinearity assessment is an essential step in structural model evaluation because high correlations among predictor constructs can distort path estimates and reduce the model's ability to isolate unique effects. Even when discriminant validity is established, collinearity may persist when multiple predictors capture overlapping variance, potentially biasing causal inferences (Kock and Lynn, 2012). In this study, collinearity was assessed using variance inflation factor (VIF) values; VIFs of 5.0 or higher indicate a potentially serious multicollinearity concern (Hair *et al.*, 2013).

The inner VIF values reported in Table 11.18 assess collinearity among the predictor constructs in the structural model. The VIF values for PLA, PPL and PRO are 2.563, 3.200 and 2.641, respectively, all well below the recommended threshold. This indicates that the three sustainability project management dimensions are not excessively correlated and can be

Table 11.17 Predictive relevance (Q^2)

	Q^2
ISR	0.714
SPS	0.539

Table 11.18 Inner VIF model

	ISR	PLA	PPL	PRO	SPS
ISR					1.000
PLA	2.563				
PPL	3.200				
PRO	2.641				
SPS					

interpreted as conceptually distinct predictors of the endogenous constructs. The VIF value of 1.000 for ISR similarly indicates no collinearity concern in the specified structural relationship, supporting the stability of the estimated paths.

Collinearity was also examined at the indicator level using outer VIFs, which are particularly relevant for identifying redundancy among measurement items. For the PRO, PLA and PPL indicators, all VIF values are below 5, indicating no problematic multicollinearity. For the dependent construct SPS, outer VIF values range from 2.305 to 3.692, also within acceptable limits. In contrast, two ISR indicators exceed the threshold – ISR2 (VIF = 6.176) and ISR6 (VIF = 6.542) – while the remaining ISR indicators fall largely within acceptable ranges. These elevated VIF values suggest that ISR2 and ISR6 may capture overlapping content within the ISR construct, implying potential redundancy.

Although the presence of two high VIF values does not necessarily invalidate the measurement model, it suggests that refining the ISR measurement could improve construct clarity. Future research may address this issue by revising item wording to reduce conceptual overlap, removing one of the redundant indicators where theoretically appropriate or forming a composite indicator if both items capture essential aspects of ISR and are retained for content validity. Overall, the VIF results indicate no severe multicollinearity issues at the construct level, while highlighting limited indicator-level redundancy within the ISR construct, which warrants attention in further instrument development.

11.4. Practical insights

The study provides clear practical implications for improving sustainability outcomes in Malaysian for-profit organisations by shifting attention from the external adoption of sustainability frameworks to the internal conditions that enable those frameworks to translate into project results. The most consequential insight is that the internal stakeholder role serves as the key mechanism through which sustainability project management practices lead to project success. The strong direct effect of the internal stakeholder role on sustainability project success indicates that project outcomes depend less on the presence of sustainability policies and more on the extent to which employees, project managers and senior leaders actively enact sustainability in decision-making, resource allocation and day-to-day execution. Practically, organisations should therefore prioritise building internal ownership by strengthening participation, accountability and shared responsibility for sustainability across hierarchical levels, rather than treating sustainability primarily as a compliance response to regulators, clients or investors.

A second insight is the relative dominance of the people dimension in activating internal stakeholder roles and in improving the success of sustainability projects. Compared with profit and planet, people-related sustainability practices exert the strongest influence, implying that investments in employee wellbeing, ethical standards, learning and capability development, cross-functional collaboration and meaningful engagement are more effective levers for sustained implementation. This suggests that sustainability strategies should be embedded within human capital systems, including training, internal communication, recognition mechanisms and performance management, so that sustainability becomes integrated into role expectations and career incentives. In organisational terms, sustainability maturity is strengthened when leadership commitment is operationalised through clear policies, visible sponsorship and governance structures that assign responsibility for sustainability, whether through a dedicated department or through designated roles within existing functions. This also implies that economic and environmental initiatives are more likely to generate measurable project benefits when they are socially anchored, meaning they are communicated and implemented in ways that connect sustainability practices to employees' values, responsibilities and perceived relevance to organisational and societal outcomes.

Finally, the demographic patterns reported here indicate heterogeneous sustainability maturity among Malaysian firms, with many organisations adopting sustainability selectively rather than systematically. This variability reinforces the need for practical interventions that institutionalise sustainability as a consistent project management principle across projects rather than as a one-off requirement. For Malaysian industries, the overall implication is that meaningful progress depends on cultivating intrinsic motivation and internal accountability structures that complement external pressures, enabling sustainability-driven project management to function as an enduring organisational capability rather than a superficial reporting exercise.

11.5. Conclusion and recommendations

The data analyses confirm that all hypothesised relationships in the structural model are supported, demonstrating significant direct and indirect linkages between sustainability practices and sustainability project success. All three sustainability dimensions, namely profit, people and planet, exhibit statistically significant effects on internal stakeholder role, with the people dimension emerging as the most influential. This finding underscores the importance of social factors such as employee participation, stakeholder engagement and community responsibility in shaping stakeholder behaviour and influence, which is consistent with prior research emphasising empowerment and engagement as central to organisational sustainability efforts (Murray and Holmes, 2021; Sethumadhavan *et al.*, 2025). Although the profit and planet dimensions also contribute positively, their comparatively smaller effects suggest that economic and environmental considerations alone may be insufficient to generate strong internal commitment in the absence of social alignment.

The internal stakeholder role exerts the strongest direct influence on success in sustainability projects, underscoring its pivotal role in translating sustainability practices into tangible project outcomes. This suggests that internal stakeholders enhance sustainability success by shaping decision-making processes, enforcing standards and fostering accountability within organisations, aligning with earlier studies that emphasise stakeholder governance in sustainable project delivery (Eskerod and Huemann, 2024; Zhang *et al.*, 2023). In addition, all three sustainability dimensions indirectly influence project success through internal stakeholder role, with the people dimension again demonstrating the strongest effect. While economic and environmental initiatives remain important, the findings indicate that project outcomes are most effectively enhanced when sustainability efforts prioritise human-centred practices that promote motivation, acceptance and participation.

The mediating role of internal stakeholders confirms their function as a critical transmission mechanism linking sustainability practices to project success, supporting stakeholder theory by illustrating how strategy is operationalised through stakeholder involvement. The model demonstrates strong explanatory and predictive capabilities, as evidenced by R^2, effect size and predictive relevance measures, indicating robustness and practical applicability in the Malaysian context. Overall, the findings align closely with prior sustainability and project management research that highlights the centrality of people-oriented approaches (Malik *et al.*, 2024; Zhang, 2024), while the relatively weaker influence of the planet dimension may reflect differing levels of environmental maturity and emphasis within Malaysian industries, where social and economic considerations remain more prominent drivers of sustainability adoption.

11.5.1 Implications and strategic considerations

This study provides important implications for scholarship, practice and policy regarding sustainability-driven project success within Malaysian for-profit organisations. Academically, the findings extend the sustainable project management literature by empirically demonstrating that the internal stakeholder role functions as a key mechanism linking sustainability practices to sustainability project success. By integrating the triple bottom line dimensions with stakeholder theory, the study strengthens the theoretical argument that internal stakeholder engagement is not merely supportive. Rather, it constitutes an essential condition for sustainability performance in project environments. The results further refine existing conceptualisations by showing that the people dimension is the strongest catalyst for both stakeholder engagement and project success, exceeding the relative influence of the planet and profit dimensions. This suggests that future models of sustainability-oriented project management, particularly in emerging economy contexts, should more explicitly incorporate social and relational drivers and treat them as central explanatory elements rather than peripheral considerations.

Practically, the findings highlight that sustainability outcomes are unlikely to be achieved through external compliance pressures alone unless organisations cultivate internal capability, commitment and accountability. The strong influence of internal stakeholder role on project success implies that sustainability strategies must be translated into clear responsibilities and behaviours across organisational levels, including leadership, project management and operational execution. Firms should embed sustainability expectations within job roles, strengthen staff participation through communication and training, and ensure that employee wellbeing, ethical culture and stakeholder engagement are integrated into project planning and delivery routines. The evidence also indicates value in formalising sustainability governance through dedicated functions or cross-functional sustainability teams that sustain oversight and continuous improvement. To institutionalise sustainability, organisations may incorporate sustainability-related indicators into performance evaluation systems and leadership assessment frameworks, ensuring that accountability is embedded rather than discretionary. Given the comparatively weaker activation of the planet dimension, environmental targets may be more effectively adopted when operationalised through clear organisational benefits, such as energy savings, innovation potential and risk reduction, and supported by structured training and cross-functional implementation mechanisms. Over time, these internal measures can support a transition from compliance-based sustainability to value-creation models that strengthen competitiveness in an increasingly ESG-oriented economy.

From a policy perspective, the study suggests that institutional efforts to accelerate sustainability adoption should move beyond prescriptive reporting requirements and focus on mechanisms that strengthen internal stakeholder engagement. Regulatory agencies and industry bodies may encourage firms to embed sustainability indicators into project monitoring systems, while also supporting capacity-building programmes that develop sustainability leadership, project governance competence and cross-sector collaboration. Policies that incentivise internal participation, knowledge transfer and public–private partnerships may be particularly effective in improving the social foundations of sustainability adoption, thereby increasing the likelihood that sustainability initiatives are implemented consistently and translated into sustainable project success across Malaysian industries.

11.5.2 Limitations

Despite its contributions to understanding the relationships between sustainability practices, internal stakeholder roles and project success in Malaysian for-profit organisations, this study has several limitations. First, the research adopted a cross-sectional design with data collected at a single point in time in early 2025. As a result, the findings provide evidence of associations between constructs but do not allow definitive causal inference. Longitudinal designs would be better suited to examine how internal stakeholder engagement and sustainability-related routines evolve and how these dynamics influence project success over time.

Second, the study relied on self-reported perceptual data collected through an online questionnaire. Although common method variance was assessed and addressed using Harman's single-factor test, the use of a single data source may still introduce method-related bias. Respondents may also have provided socially desirable answers or interpreted survey items differently, which could affect measurement accuracy and the robustness of the conclusions.

Third, the sample was drawn from Malaysian for-profit organisations, with stronger representation of medium- and large-sized firms concentrated in major urban and industrial regions such as Kuala Lumpur and Selangor. This distribution may limit generalisability to micro and small enterprises and to organisations operating in rural contexts, where sustainability priorities, resource constraints and governance structures may differ.

Finally, while the study operationalised sustainability practices through the triple bottom line dimensions, other potentially influential factors that may shape project outcomes were not incorporated. Variables such as organisational culture, leadership style and technological capability could provide additional explanatory power and help refine the model in future research.

11.5.3 Recommendations and future research

Future research can strengthen the understanding of sustainability-driven project success by addressing design, measurement and contextual gaps identified in the present study. Longitudinal research conducted over extended periods would be particularly valuable, as it would allow researchers to examine how internal stakeholder engagement develops over time and to make stronger causal inferences about the long-term effects of stakeholder roles on project success, especially in dynamic project environments.

Methodological refinement is also recommended to reduce potential common method variance. Studies could incorporate multi-source evidence by combining survey responses with objective project documentation, performance indicators, sustainability reports and external stakeholder feedback. Such triangulation would enhance the credibility and validity of findings by reducing reliance on single respondent perceptions.

Further extension of the conceptual model is encouraged to include additional explanatory variables such as leadership style, organisational culture, innovation capability and regulatory pressure. These factors may serve as moderators or additional mediators, shaping the strength and direction of relationships between sustainability practices, internal stakeholder roles and project outcomes. Integrating these variables would provide a more comprehensive account of sustainability performance in project settings.

Complementary qualitative or mixed-methods approaches could deepen insight into motivations, behaviours and implementation realities that survey instruments do not fully capture. Interviews, focus groups and case studies can reveal how sustainability priorities are interpreted, negotiated and enacted by different stakeholder groups across organisational contexts, thereby enriching quantitative conclusions with contextual evidence.

Finally, future studies should pay greater attention to the influence of Malaysian cultural and institutional conditions on stakeholder activation. Cultural patterns associated with collectivism, hierarchical norms and deference to authority may affect employees' willingness to initiate or voice sustainability initiatives, particularly when leadership support is perceived as limited. Institutional drivers such as Bursa Malaysia requirements, ESG reporting expectations and government incentives may also shape sustainability engagement differently across firms. Comparative research across organisations with varying levels of regulatory exposure, ownership structures and size categories, including contrasts between multinational corporations and local small and medium-sized enterprises (SMEs), or between private- and public-sector settings, would improve external validity and support more context-sensitive explanations of how sustainability-driven project success is achieved in Malaysia.

REFERENCES

Alnsour MA (2022) Using modified grounded theory for conducting systematic research study on sustainable project management field. *MethodsX* **9**: 101897.

Blak Bernat G, Qualharini EL, Castro MS, Barcaui AB and Soares RR (2023) Sustainability in project management and project success with virtual teams: a quantitative analysis considering stakeholder engagement and knowledge management. *Sustainability* **15(12)**: 9834.

Chakrabarti PD (2024) Integrating climatic disaster risk reduction and climate change adaptation into project management cycle. In *Disaster Risk and Management Under Climate Change*. Springer, Singapore, pp. 539–556.

Chin WW (2010) How to write up and report PLS analyses. In *Handbook of Partial Least Squares: Concepts, Methods and Applications*. Springer, Berlin, Germany, pp. 655–690.

Chow TC, Zailani S, Rahman MK *et al.* (2021) Impact of sustainable project management on project plan and project success of the manufacturing firm: structural model assessment. *PLOS ONE* **16(11)**: e0259819.

Ciavolino E, Aria M, Cheah JH and Roldán JL (2022) A tale of PLS structural equation modelling: Episode I – a bibliometrix citation analysis. *Social Indicators Research* **164(3)**: 1323–1348.

Cohen J (2013) *Statistical Power Analysis for the Behavioral Sciences*, 2nd edn. Routledge, New York, NY, USA.

Elaigwu M, Abdulmalik SO and Talab HR (2024) Corporate integrity, external assurance and sustainability reporting quality: evidence from the Malaysian public listed companies. *Asia-Pacific Journal of Business Administration* **16(2)**: 410–440.

Eskerod P and Huemann M (2024) Project stakeholder orientation principles for managing projects sustainably. In *Research Handbook on Sustainable Project Management* (Silvius G and Huemann M (eds)). Edward Elgar, Cheltenham, UK, pp. 240–251.

Fernandes G, Ward S and Araújo M (2015) Improving and embedding project management practice in organisations – a qualitative study. *International Journal of Project Management* **33(5)**: 1052–1067.

Freeman RE, Menghwar PS and Grushka-Cockayne Y (2025) A stakeholder theory perspective for project management. *International Journal of Project Management* **43(7)**: 102764.

Fuller CM, Simmering MJ, Atinc G, Atinc Y and Babin BJ (2016) Common methods variance detection in business research. *Journal of Business Research* **69(8)**: 3192–3198.

Ghanbaripour AN, Tumpa RJ, Skitmore M and Hon C (2025) Project delivery success and sustainability: comparative strategies in the post-pandemic era. *Management Decision*. [In press/Early cite]. https://doi.org/10.1108/MD-03-2024-0445.

Grunwald G and Fischer K (2022) Corporate sustainability crisis: theoretical framework and stakeholder-oriented typology. *Business and Society Review* **127(1)**: 23–48.

Hair JF, Black WC, Babin BJ and Anderson RE (2013) *Multivariate Data Analysis: Pearson New International Edition*, 7th edn. Pearson, Harlow, UK.

Hamada H, Abu Raje HM, Alrashed AA, Shakah G and Chikho AH (2025) Decision-making and policy support for take-back systems: enhancing sustainable performance in Malaysia's construction industry. In *2025 International Conference on Decision Aid Sciences and Applications (DASA)*. IEEE, pp. 2350–2354.

Hayes AF (2009) Beyond Baron and Kenny: statistical mediation analysis in the new millennium. *Communication Monographs* **76(4)**: 408–420.

Henseler J (2017) Partial least squares path modeling. In *Advanced Methods for Modeling Markets*. Springer, Cham, Switzerland, pp. 361–381.

In SY, Rook D and Monk A (2019) Integrating alternative data (also known as ESG data) in investment decision making. *Global Economic Review* **48(3)**: 237–260.

Jum'a L, Zimon D, Ikram M and Madzik P (2022) Towards a sustainability paradigm; the nexus between lean green practices, sustainability-oriented innovation and Triple Bottom Line. *International Journal of Production Economics* **245**: 108393.

Keshavarzian S and Silvius G (2022) The perceived relationship between sustainability in project management and project success. *The Journal of Modern Project Management* **9(3)**: 66–85.

Khan FA (2024) Sustainable business management: balancing profit and purpose. *Journal of Management Science Research Review* **2(1)**: 24–32.

Kock N and Lynn GS (2012) Lateral collinearity and misleading results in variance-based SEM: an illustration and recommendations. *Journal of the Association for Information Systems* **13(7)**: 2.

Li X, Cheng L, Jiao H and Li H (2026) Exploring the impact of information technology integration capability on project management performance in Chinese construction industry: a moderated mediation analysis. *Engineering, Construction and Architectural Management* **33(2)**: 1235–1257.

Malik MY, Wang L and Zhu F (2024) Understanding variations of governmentality and governance structures at the project level in project-based organizations. *International Journal of Managing Projects in Business* **17(4–5)**: 644–685.

Molina A and Rajagopal (2023) People, planet, and profit: crossing the triple bottom line. In *Challenge-Based Learning, Research, and Innovation: Leveraging Industry, Government, and Society*. Springer, Cham, Switzerland, pp. 35–65.

Moshood TD, Rotimi JO and Shahzad W (2025) Enhancing sustainability considerations in construction industry projects. *Environment, Development and Sustainability* **27(12)**: 29287–29313.

Murray WC and Holmes MR (2021) Impacts of employee empowerment and organizational commitment on workforce sustainability. *Sustainability* **13(6)**: 3163.

Park SR and Jang JY (2021) The impact of ESG management on investment decision: institutional investors' perceptions of country-specific ESG criteria. *International Journal of Financial Studies* **9(3)**: 48.

Podsakoff PM, MacKenzie SB and Podsakoff NP (2012) Sources of method bias in social science research and recommendations on how to control it. *Annual Review of Psychology* **63(1)**: 539–569.

Qiu Y and Chen H (2023) A systematic review of the knowledge domain of institutional theory in construction project management. *Engineering, Construction and Architectural Management* **30(6)**: 2523–2544.

Sankaran S, Müller R and Drouin N (2024) Sustainable project management and its governance in the context of the United Nations sustainable development goals. In *Research Handbook on Sustainable Project Management*, Edward Elgar, Cheltenham, UK, pp. 83–96.

Sethumadhavan RPPS, Sajimon PP, Selvakumar P *et al.* (2025) Environmental sustainability and employee participation. In *Green Leadership Strategies for Increased Employee Engagement and Performance* (Şeker C, Örücü E and Kaya A (eds)). IGI Global, Beijing, China, pp. 275–298.

Shokouhi M and Bachari MS (2025) An overview of the aspects of sustainability in project management. *Progress in Engineering Science* **2(1)**: 100048.

Soares I, Fernandes G and Santos JM (2024) Sustainability in project management practices. *Sustainability* **16(10)**: 4275.

Sood K, Pathak P and Gupta S (2025) How do the determinants of investment decisions get prioritized? Peeking into the minds of investors. *Kybernetes* **54(4)**: 2175–2203.

Zhang L (2024) How to measure the impact of landsenses ecology on sustainable development? A review of people-oriented emerging approaches. *Environmental Sciences Europe* **36(1)**: 190.

Zhang Y, Phanniphong K and Li P (2023) Corporate greening initiatives and management governance: analysing their impact on project performance and stakeholder engagement. *The Journal of Modern Project Management* **11(2)**: 132–153.

Part IV

Communication and culture

Advancing Leadership and Collaboration in Project Management: Culture, Knowledge and Team Dynamics

Mohammed Reyasudin Basir Khan
ISBN 978-1-83742-803-8
https://doi.org/10.1108/978-1-83742-802-120261015
Emerald Publishing Limited: All rights reserved

Chapter 12
The pivotal role of communication and leadership in enhancing project manager effectiveness within the logistics sector

Tanaraj Muniandy and Mohammed Reyasudin Basir Khan

Tun Razak Graduate School, Universiti Tun Abdul Razak (UNIRAZAK), Kuala Lumpur, Malaysia

12.1. Introduction

Malaysia's strategic geographical location, together with the rapid expansion of global trade and e-commerce, has positioned the logistics industry as a key contributor to national economic growth (Bakar and Jaafar, 2016). As logistics operations continue to expand in scale and complexity, project-based initiatives have become increasingly central to managing modern supply chain networks, regulatory compliance requirements and advanced technological integration. Within this environment, project managers play a pivotal role in ensuring the successful execution of logistics projects, particularly in balancing operational efficiency, cost control and timely delivery.

Contemporary research highlights that project manager effectiveness is shaped by an interaction of personal competencies, organisational factors and technical capabilities (Szczepańska-Woszczyna and Gatnar, 2022). Leadership and communication skills are consistently identified as critical enablers of project success, especially in culturally diverse and dynamic contexts such as Malaysia. Prior studies indicate that effective project managers must demonstrate adaptability, emotional intelligence, and strategic thinking to manage diverse teams and stakeholder expectations (Jayaweera *et al.*, 2025; Sposito *et al.*, 2024). In addition, technical competencies in digital technologies, such as the Internet of Things, cloud computing and digital supply chain platforms, are increasingly essential for real-time monitoring, efficient resource allocation and informed decision-making.

Organisational support further strengthens project manager effectiveness through top management commitment, collaborative cultures and adequate training opportunities

(Sepúlveda-Rivillas *et al.*, 2022). Firms that promote open communication, cross-functional collaboration and continuous learning are better able to execute complex logistics projects effectively. At the same time, external pressures arising from fluctuating global supply chains, regulatory changes and sustainability demands have intensified the need for resilient, proactive project management. Understanding the combined influence of leadership, communication skills, organisational culture, and training and development is therefore essential to improving project outcomes and sustaining Malaysia's competitiveness as a regional logistics hub.

Despite growing recognition of project manager effectiveness as a determinant of project success, its practical implementation within the Malaysian logistics industry remains uneven (Wahab *et al.*, 2024). Leadership practices, particularly transformational leadership, are widely acknowledged as effective in fostering collaboration and innovation, yet they are not consistently applied across the sector. Inadequate leadership approaches have contributed to suboptimal team performance, project delays and cost overruns.

Communication challenges further constrain project success (Swart *et al.*, 2022). Ineffective communication channels, unclear project directives and poorly defined scopes frequently lead to misunderstandings and inefficiencies. Communication skills mediate the relationship between leadership and project outcomes, requiring project managers to demonstrate both interpersonal and organisational communication competence. However, such skills are often underdeveloped in high-pressure logistics environments.

Organisational culture also significantly influences project execution (AlSaied and Alkhoraif, 2024). Supportive and flexible cultures enhance team cohesion and adaptability, while rigid or misaligned cultures tend to inhibit innovation and foster resistance to change. Although organisational culture is widely recognised as a critical factor, Malaysian logistics firms continue to struggle to align cultural practices with project management requirements.

Training and development represent another area of concern (Ramasamy Velar and Kee, 2025). The effectiveness of project managers is closely linked to the relevance and quality of training opportunities available to them. Inadequate or poorly aligned training programmes contribute to skill gaps, flawed decision-making and reduced project performance. Given the rapid evolution of logistics operations, there is a clear need to assess how training and development initiatives influence project manager effectiveness within the Malaysian context. The study presented in this chapter addresses these gaps by examining the relationships between leadership, communication skills, organisational culture, training and development, and project manager effectiveness in the Malaysian logistics industry.

This study is grounded in stakeholder theory, which provides a useful lens for understanding project manager effectiveness in complex organisational environments (Freeman *et al.*, 2025). Stakeholder theory emphasises the importance of identifying, prioritising and engaging stakeholders to achieve project success. Effective stakeholder management requires strong leadership, clear communication and alignment between organisational practices and stakeholder expectations.

Based on the reviewed literature and stakeholder theory, this study proposes a conceptual framework linking leadership, communication skills, organisational culture, and training and development to project manager effectiveness. Leadership is expected to positively influence project effectiveness by motivating team members, aligning with the vision and demonstrating adaptability. Communication skills are anticipated to enhance coordination, stakeholder engagement and decision-making. Organisational culture is proposed to shape the environment in which project managers operate, affecting collaboration and innovation. Training and development are expected to strengthen managerial competencies and project performance.

The conceptual framework for this study is shown in Figure 12.1. The study formulated the following four hypotheses to examine the factors influencing project manager effectiveness in the Malaysian logistics industry.

- H1: there is a significant relationship between leadership and project manager effectiveness.
- H2: there is a significant relationship between communication skills and project manager effectiveness.
- H3: there is a significant relationship between organisational culture and project manager effectiveness.
- H4: there is a significant relationship between training and development and project manager effectiveness.

The primary objective of this study was to assess project manager effectiveness in the Malaysian logistics industry by examining the relationships between leadership, communication skills, organisational culture, training and development and project manager effectiveness, and to determine whether leadership, communication skills, organisational culture and training and development are significantly related to project manager effectiveness in the Malaysian logistics industry.

Figure 12.1 Conceptual framework

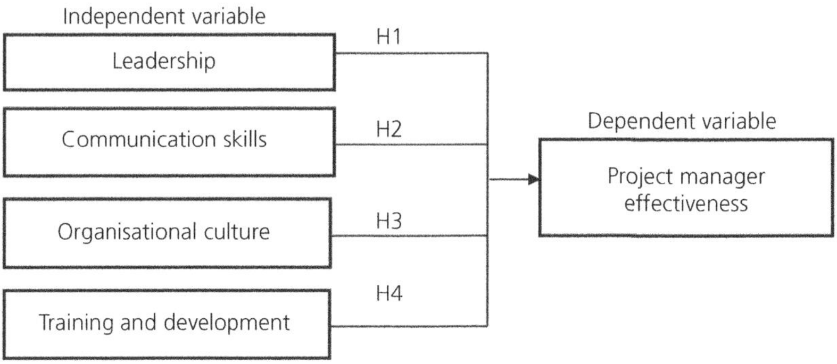

12.2. Methodology

12.2.1 Research design

This study adopted a quantitative research design. Quantitative research explains phenomena by collecting numerical data that can be analysed using statistical techniques, enabling systematic examination of relationships between variables. The study employed a survey approach, using structured questionnaires as the primary data collection instrument. Surveys are widely used to collect quantitative information through standardised procedures and are suitable for drawing conclusions about a broader population. Survey designs are often regarded as efficient because they can reach geographically dispersed participants, support anonymity and provide data that can be analysed systematically, while remaining relatively cost-effective.

Surveys may be conducted using different temporal structures. Cross-sectional surveys collect data once at a single point in time to describe current conditions and examine associations among variables. By contrast, longitudinal surveys involve repeated data collection over an extended period to observe change. Explanatory or correlational surveys focus on analysing relationships between two or more variables, particularly in contexts where experimental designs are not feasible. In line with the objectives of this study, a cross-sectional survey design was employed because it enabled data collection within a defined timeframe and provided a practical and efficient means of assessing relationships between the study variables.

12.2.2 Population and sampling procedures

The target population for this study comprised employees in Malaysia's logistics sector, including transportation, warehousing, inventory management and supply chain functions. The logistics and transportation sector employed approximately 350 000 individuals in 2022, with employment levels recovering following a decline during the COVID-19 pandemic (Chien, 2023). The sampling frame was constructed from official sources such as the Department of Statistics Malaysia, the Malaysian Logistics and Transport Council, and the Federation of Malaysian Manufacturers, supplemented by employment data from major logistics firms and labour market platforms. Efforts were made to ensure representation across operational, supervisory and managerial roles.

Sampling methods are broadly classified into probability and non-probability approaches. Probability sampling provides each population element with a known and equal chance of selection, thereby reducing sampling bias and enhancing generalisability. In contrast, non-probability sampling is more commonly used in qualitative and exploratory studies. This study used simple random sampling, which ensures that all elements in the population have an equal likelihood of inclusion, provided that a comprehensive sampling frame is available.

Determining an appropriate sample size is essential to minimise sampling error and support valid statistical analysis. While larger samples generally improve precision, excessively large samples offer limited additional benefits and may strain available resources. Conversely, small sample sizes reduce statistical power and limit meaningful inference. A sample size of approximately 384 respondents is considered adequate to estimate population characteristics with 95% confidence and a 5% margin of error.

The study includes respondents from all organisational levels within the logistics industry, as project manager effectiveness influences employees across functional and hierarchical boundaries. Senior roles, such as vice presidents and assistant vice presidents, were included despite their lower representation because of their strategic influence on project outcomes through decision-making, policy formulation and resource allocation. Their inclusion enriched the dataset by capturing strategic-level perspectives and allowed for comparison across organisational hierarchies, thereby enhancing the robustness and explanatory value of the findings.

12.2.3 Data collection method

The survey instrument for this study was developed using Google Forms and distributed electronically to employees in the Malaysian logistics industry. Questionnaires were disseminated through established networks and platforms commonly used by logistics professionals, including email and professional or industry-related social media channels such as LinkedIn and relevant Facebook groups. To support adequate participation, the distribution volume was increased to account for typical response rates in online surveys. Given that the response rate for electronic surveys is approximately 30%, the number of questionnaires was multiplied by 5 to achieve the required sample size. Accordingly, approximately 1920 questionnaires were distributed.

12.2.4 Operationalisation and measurement

The questionnaire items were adapted from established studies on project manager effectiveness, leadership, communication skills, organisational culture, and training and development to ensure relevance to the research objectives and the Malaysian logistics context (Abolghasemi *et al.*, 2026; Đajić *et al.*, 2024; Pomaza-Ponomarenko *et al.*, 2023; Radhakrishnan *et al.*, 2022). The instrument comprises 20 items, with five items for each independent variable and five items for the dependent variable. Responses were recorded using a five-point Likert scale ranging from 1 (strongly disagree) to 5 (strongly agree), with 3 representing a neutral position. Likert scales are widely used because they are simple to administer, easy for respondents to understand and suitable for producing reliable measurements of attitudes and perceptions. Including a neutral midpoint helps reduce forced-choice responses and may mitigate response bias when respondents are uncertain or lack a clear opinion.

Leadership was conceptualised as the behavioural processes through which leaders guide, motivate and influence their teams to achieve goals. Leadership styles differ in the extent of control and participation, ranging from centralised decision-making to shared authority and employee autonomy. These styles shape team motivation and performance, with participative leadership often associated with stronger engagement and creativity, particularly in skilled work settings. In this study, leadership items focus on leadership behaviours relevant to project contexts, particularly transformational and transactional dimensions, drawing on project leadership measures developed from the multifactor leadership questionnaire and adapted for project environments (Bajcar and Babiak, 2022; Romero *et al.*, 2023).

Communication skills are the ability to transmit and interpret information effectively through verbal, written and nonverbal channels. Effective communication requires clarity of expression, active listening, accurate interpretation and feedback to ensure shared understanding, particularly in team-based and stakeholder-driven environments. Communication competence also includes conflict management, negotiation and persuasion skills that support coordination and performance in organisational settings. The communication items used in this study were adapted from measures that link project success to the project manager's ability to manage communication plans, align stakeholders and maintain team cohesion (Elkbuli *et al.*, 2025; Muhammad *et al.*, 2014).

Organisational culture refers to shared values, beliefs and norms that influence how members behave and interpret organisational practices. Schein's model frames culture through observable artefacts, espoused values and underlying assumptions that shape organisational stability and adaptability (Khaddour, 2024). Complementing this view, the competing values framework classifies organisational culture into clan, adhocracy, market and hierarchy orientations, reflecting different emphases on collaboration, innovation, competition and control. Organisational culture is commonly assessed using instruments such as the organisational culture assessment instrument, which has been applied in studies examining cultural influences on project management outcomes (AlSaied and Alkhoraif, 2024; Waseem *et al.*, 2025).

Training and development were assessed as organisational practices that enhance employees' competencies and support performance improvement. The Kirkpatrick Four-Level Model provides a widely used framework for evaluating training outcomes through reactions, learning, behavioural transfer and results (Rucks *et al.*, 2024). Training effectiveness is also influenced by alignment with organisational objectives and employees' attitudes toward learning, as positive perceptions increase the likelihood that acquired skills will be applied. The transfer-of-training model further explains how training inputs and workplace conditions shape the extent to which learning is effectively applied in practice (Baldwin and Ford, 1988). These perspectives inform the measurement of training and development in this study.

Project manager effectiveness, as the dependent variable, is defined as the extent to which project objectives are achieved within time, cost and quality constraints while meeting stakeholder expectations. Effectiveness reflects a combination of technical competence and behavioural capabilities, including leadership, decision-making, risk management, conflict handling and adaptability to changing project requirements. The measurement of project manager effectiveness in this study draws on prior competence frameworks and project success perspectives, including those associated with the Project Management Institute and performance models that consider both delivery outcomes and stakeholder satisfaction (Ciric Lalic *et al.*, 2022; Ika and Pinto, 2022).

12.2.5 Data analysis technique

Data analysis was conducted using IBM SPSS Statistics. Descriptive statistics, including frequency distributions, mean and standard deviation, were used to summarise respondent characteristics and provide an initial overview of the data. Inferential analyses were then applied to examine relationships between the study variables, specifically using Pearson correlation and multiple regression procedures (Hair *et al.*, 2010).

Pearson correlation analysis was performed as a preliminary step before regression to assess the direction and strength of linear associations between the independent variables and the dependent variable. The Pearson correlation coefficient ranges from −1.0 to 1.0, where values closer to −1.0 indicate a stronger negative relationship and values closer to 1.0 indicate a stronger positive relationship. Coefficients near zero indicate weak associations. Correlation strength is commonly interpreted using threshold ranges: values above 0.70 indicate strong relationships, values between 0.40 and 0.69 indicate moderate relationships, and values between 0.10 and 0.39 indicate weak relationships.

Multiple regression analysis was used to test the hypotheses and evaluate the extent to which the independent variables predict project manager effectiveness. Multiple linear regression estimates the unique contribution of each predictor while controlling for the effects of other predictors in the model, and it also indicates the proportion of variance in the dependent variable explained collectively by the predictors (Hair *et al.*, 2010). Because predictors may be intercorrelated, preliminary correlation results were reviewed to ensure the suitability of the regression model. In addition, a stepwise regression procedure was applied to evaluate mediation effects.

12.3. Results and analysis

12.3.1 Response rate

In this study, data were collected through online platforms due to their cost efficiency and ability to reach a broad audience. Malaysia recorded approximately 26.69 million internet users in January 2020, representing an internet penetration rate of about 83%, which supports the suitability of online data collection methods. Online surveys enable faster data collection, reduce human error and offer researchers advantages such as increased sample size, greater participant diversity, greater accessibility, and lower time and cost requirements.

The survey instrument was developed using Google Forms and distributed to employees in the logistics industry through simple random sampling. Respondents were approached via social media platforms such as Facebook and LinkedIn, as well as through email. To ensure an adequate sample size, 1920 questionnaires were distributed electronically.

Questionnaire-based studies are commonly subject to non-response bias and common method variance. Non-response bias occurs when there are differences between respondents and non-respondents, potentially limiting the generalisability of the findings. This bias is often associated with low response rates and may distort conclusions if the sample does not adequately represent the population. To mitigate non-response bias, this study employed the extrapolation method, which compares early and late responses to estimate potential bias in survey results.

Common method variance arises when variation in responses is attributable to the measurement method rather than to the constructs being measured. Failure to address this issue may lead to inflated or deflated relationships between variables, affecting the validity of the findings (Yıldız, 2023). In this study, common method bias was assessed using Harman's single-factor test using exploratory factor analysis, in which all variables are loaded onto a single unrotated factor (Harman, 1976).

Data screening was conducted to ensure data accuracy and to meet the assumptions required for multivariate analysis. This process helps prevent distortions in correlations and enhances the reliability and interpretability of research findings. Missing values may occur when respondents fail to answer certain items or when technical issues, such as system errors, occur. These missing values can reduce statistical power and complicate data interpretation. In this study, missing data were addressed using mean substitution or by ensuring questionnaire completeness before analysis.

Outlier detection is another essential component of data screening, as extreme values can disproportionately influence statistical results and compromise validity (Hair *et al.*, 2010). Outliers may be identified using visual methods such as box plots and histograms, as well as statistical techniques such as z-scores and Mahalanobis distance. Identified outliers were carefully evaluated and addressed to ensure robust and reliable analytical outcomes.

12.3.2 Demographic profile

The demographic profile of the 161 respondents in the Malaysian logistics industry is presented in Table 12.1, covering age, educational attainment, tenure and organisational position.

The respondents represent a relatively young and energetic segment of the workforce, with a significant 85.7% falling between the ages of 26 and 35. The largest age group is the 31–35 bracket (43.5%), while those over 40 years old comprise the smallest portion of the sample at just 3.7%. This concentration among young-to-mid-career professionals suggests a workforce that likely possesses the adaptability and technological familiarity required to manage modern, complex logistics operations in Malaysia.

The participants' educational backgrounds indicate a highly qualified and professionalised workforce, as 100% of respondents have attained tertiary-level education. While the Bachelor's degree is the most common qualification (59.0%), over 30% of the sample holds postgraduate degrees, including Master's (20.5%) and Doctorate degrees (9.9%). This high level of academic achievement is consistent with the increasing demand for technical competencies in digital supply chain platforms, cloud computing and strategic decision-making within the industry.

Regarding organisational stability and experience, the majority of respondents demonstrate a solid commitment to their current firms. Over 83% of participants have held their current roles for 4–7 years, with 6–7 years of tenure the most common (42.2%). Only 3.1% of respondents are new to their organisations, with less than a year of service. This suggests that the data reflects the insights of individuals well-versed in their company's internal culture, project management requirements and long-term operational goals.

The sample is primarily composed of mid-to-senior-level professionals, with senior executives (42.2%) and management level staff (40.4%) representing the vast majority of respondents. Although top-tier leadership roles, such as vice presidents and assistant vice presidents, account for only 6.2% of the total, their inclusion was strategic to ensure the study captured high-level perspectives on resource allocation and policy formulation.

Table 12.1 Demographic profile of respondents

Category	Sub-category	Frequency, *n*	Percentage: %
Age	20–25 years	9	5.6
	26–30 years	68	42.2
	31–35 years	70	43.5
	36–40 years	8	5.0
	Over 40 years	6	3.7
Education level	Advanced diploma	17	10.6
	Bachelor's degree	95	59.0
	Master's degree	33	20.5
	Doctorate degree (PhD)	16	9.9
Tenure in current job	Less than 1 year	5	3.1
	2–3 years	12	7.5
	4–5 years	66	41.0
	6–7 years	68	42.2
	More than 7 years	10	6.2
Position in firm	Junior executive	6	3.7
	Senior executive	68	42.2
	Management level	65	40.4
	Assistant manager	12	7.5
	Assistant vice president	5	3.1
	Vice president	5	3.1

This distribution across hierarchical boundaries enables a comprehensive analysis of how leadership and communication influence project effectiveness from both operational and strategic perspectives.

Finally, the respondents were geographically dispersed across various Malaysian states and federal territories, ensuring that the findings were not concentrated in a single region. Pahang recorded the highest representation (11.2%), followed by Penang (7.5%), while Labuan contributed the smallest share (1.2%). Other regions, such as Melaka, Perak, Perlis and Sarawak, showed relatively even participation ranging from 4.3% to 6.8%, providing a broad national context for the study's assessment of the Malaysian logistics sector.

12.3.3 Mean and standard deviation of variables

Table 12.2 reports the mean and standard deviation (SD) for all items measuring project manager effectiveness. Item means ranged from 3.86 to 3.98, with an overall mean of 3.942.

Table 12.2 Mean of project manager effectiveness

No.	Item	Mean	SD
1.	My project manager effectively communicates project goals and objectives to the team	3.95	1.071
2.	My project manager successfully manages project timelines and ensures deadlines are met	3.86	1.239
3.	My project manager effectively allocates resources and ensures optimal utilisation	3.96	1.140
4.	My project manager resolves issues and challenges that arise during the project in a timely manner	3.96	1.134
5.	My project manager fosters a collaborative work environment within the project team	3.98	1.129

Standard deviations ranged from 1.071 to 1.239. Based on the adopted classification, the overall mean indicates a high level of project manager effectiveness in the logistics industry, with mean values of 4.21 and above considered very high, 3.41 to 4.20 considered high and 3.41 and below considered moderate.

Table 12.3 presents the mean and standard deviation for the leadership items measured on a five-point scale. Item means ranged from 3.94 to 4.03, with an overall mean of 3.978. Standard deviations ranged from 1.064 to 1.158. Overall, the findings indicate a relatively high level of leadership involvement in the logistics industry.

Table 12.4 reports the mean and standard deviation for communication skills items measured on a five-point scale. Item means ranged from 3.86 to 4.02, with an overall mean of

Table 12.3 Mean of leadership

No.	Item	Mean	SD
1.	My project manager encourages team members to contribute ideas and solutions	3.95	1.071
2.	My project manager provides clear direction and sets specific goals for the team	3.94	1.158
3.	My project manager empowers team members by delegating authority and responsibility	4.01	1.064
4.	My project manager recognises and rewards good performance	4.03	1.086
5.	My project manager is approachable and listens to team members' concerns	3.96	1.109

Table 12.4 Mean of communication skills

No.	Item	Mean	SD
1.	My project manager communicates project goals and expectations clearly	3.90	1.168
2.	My project manager ensures all team members are well-informed about project updates	3.86	1.217
3.	My project manager provides constructive feedback to team members	3.96	1.169
4.	My project manager demonstrates confidence and clarity during presentations or meetings	4.02	1.043
5.	My project manager actively listens to team members' concerns and suggestions	3.93	1.173

3.934. Standard deviations ranged from 1.043 to 1.217. Overall, communication skills in the logistics industry were relatively high.

Table 12.5 presents the mean and standard deviation for organisational culture items measured on a five-point scale. Item means ranged from 3.87 to 4.04, with an overall mean of 3.954. Standard deviations ranged from 0.987 to 1.173. Overall, organisational culture in the logistics industry was at a relatively high level.

Table 12.6 reports the mean and standard deviation for training and development items measured on a five-point scale. Item means ranged from 3.84 to 3.99, with an overall mean of 3.946. Standard deviations ranged from 1.072 to 1.143. Overall, training and development in the logistics industry were at a relatively high level.

Table 12.5 Mean of organisational culture

No.	Item	Mean	SD
1.	The organisation promotes open communication and transparency across all levels	3.94	1.131
2.	The organisation encourages collaboration and teamwork among employees	4.04	0.987
3.	The organisation fosters a culture of respect and inclusion	3.87	1.173
4.	The organisation recognises and rewards employees for their achievements	3.94	1.171
5.	The organisation supports continuous learning and professional growth for employees	3.98	1.098

Table 12.6 Mean of training and development

No.	Item	Mean	SD
1.	The company provides equal training opportunities for all project managers	3.98	1.072
2.	The training provided is relevant to my current project's responsibilities	3.84	1.143
3.	The content of the training is up-to-date and aligned with the industry's best practices	3.94	1.119
4.	The training I received has improved my ability to manage logistics projects effectively	3.98	1.118
5.	The organisation has a structured career development plan for project managers	3.99	1.126

12.3.4 Normality analysis

Normality tests were conducted to assess whether the study data were approximately normally distributed. Normality was evaluated using skewness and kurtosis. Kurtosis reflects the concentration of scores around the centre relative to the tails, where a normal distribution has an excess kurtosis of 0. Skewness indicates the degree of asymmetry in a distribution, with a normal distribution having a skewness value of 0. Positive values suggest a positively skewed distribution, whereas negative values indicate a negatively skewed distribution. As shown in Table 12.7, the data were approximately normally distributed and negatively skewed.

12.3.5 Multicollinearity

Multicollinearity refers to high intercorrelations between independent variables, which may compromise the stability of regression estimates. It is commonly assessed using tolerance and the variance inflation factor (VIF), which indicate the extent to which an independent variable is explained by other predictors. Hair *et al.* (2010) suggest that tolerance values should exceed 0.10 and VIF values should be below 10. As shown in Table 12.8, all tolerance values

Table 12.7 Skewness and kurtosis

Variable	Mean	SD	Skewness	Kurtosis
Leadership	3.9776	0.79278	−1.095	−0.162
Communication skills	3.9354	0.83654	−0.968	−0.689
Organisational culture	3.9540	0.83044	−1.103	−0.218
Training and development	3.9466	0.81724	−0.984	−0.566

Table 12.8 Multicollinearity test of variables

Variable	Tolerance	VIF
Leadership	0.202	4.944
Communication skills	0.174	5.754
Organisational culture	0.218	4.586
Training and development	0.183	5.460

Note: dependent variable, project manager effectiveness.

were above 0.10 and all VIF values were below 10, indicating that multicollinearity was not a concern in this study.

Linearity was assessed using a regression scatterplot of standardised residuals against standardised predicted values of the dependent variable. The residuals were randomly dispersed around the horizontal axis at zero, indicating that the assumptions of linearity and homoscedasticity were generally satisfied. Although several unusually large and small residuals were observed, suggesting the possible presence of outliers or influential cases, the overall plot showed no systematic pattern. Figure 12.2 presents the standardised residual plot for the regression model.

Homoscedasticity refers to a condition in which the variance of residuals remains constant across all levels of the predicted values. The residuals appear evenly dispersed, indicating that the assumption is satisfied. Accordingly, the regression results are considered reliable, as the estimated standard errors are appropriate and the resulting statistical inferences remain valid.

Figure 12.2 Regression standardised residual

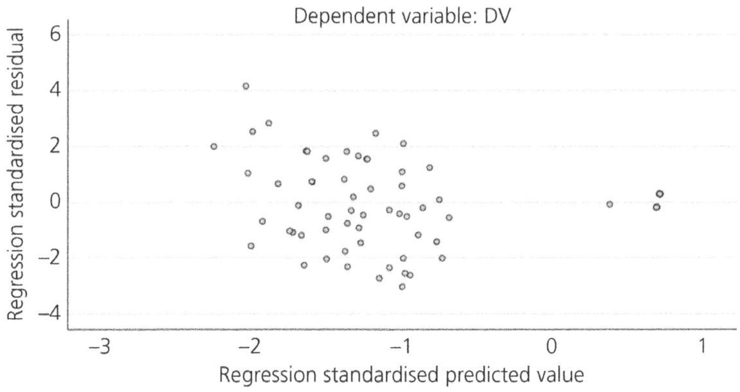

12.3.6 Factor analysis

Factor analysis is a multivariate technique used to identify underlying dimensions and determine the most suitable items for each construct. Prior to analysis, sampling adequacy was assessed using the Kaiser-Meyer-Olkin measure and Bartlett's test of sphericity. Bartlett's test should be significant at $p < 0.05$, while the Kaiser-Meyer-Olkin value should be at least 0.60. In this study, the five constructs, namely project manager effectiveness, leadership, communication skills, organisational culture, and training and development, were examined using principal component analysis. Items with factor loadings of 0.40 and above were retained for the final analysis. As shown in Table 12.9, the Kaiser-Meyer-Olkin value was 0.696, and Bartlett's test of sphericity was significant ($p < 0.001$), indicating that the data were appropriate for factor analysis.

Table 12.10 summarises the factor analysis results for project manager effectiveness. The average variance extracted (AVE) was 0.3293, which is below the recommended 0.50 threshold, indicating limited convergent validity. In contrast, the composite reliability was 0.705, exceeding the 0.70 criterion and supporting acceptable internal consistency. Overall, the scale is sufficiently reliable, but the low AVE suggests that item refinement is required to better capture the construct.

Table 12.9 Kaiser-Meyer-Olkin and Bartlett's test for project manager effectiveness

Kaiser-Meyer-Olkin measure of sampling adequacy		0.696
Bartlett's test of sphericity	Approx. χ^2	299.422
	df	10
	Sig.	<0.001

Table 12.10 Factor analysis for project manager effectiveness

No.	Item	Factor loading
1.	My project manager effectively communicates project goals and objectives to the team	0.599
2.	My project manager successfully manages project timelines and ensures deadlines are met	0.459
3.	My project manager effectively allocates resources and ensures optimal utilisation	0.455
4.	My project manager resolves issues and challenges that arise during the project in a timely manner	0.703
5.	My project manager fosters a collaborative work environment within the project team	0.613
	Percentage of average variance extracted (AVE)	32.93%
	Composite reliability	0.705

Factor analysis was conducted on leadership using a five-item scale. Prior to principal component analysis, data suitability was assessed and items with factor loadings below 0.40 were removed. Sampling adequacy was examined using the Kaiser-Meyer-Olkin measure. As shown in Table 12.11, the Kaiser-Meyer-Olkin value was 0.641 and Bartlett's test of sphericity was significant ($p < 0.001$), indicating that the data were appropriate for factor analysis.

Table 12.12 summarises the factor analysis results for leadership. The recalculated average variance extracted was 0.5584, indicating that the construct accounts for more than half of the variance in its indicators and demonstrating adequate convergent validity. Composite reliability was 0.863, exceeding the recommended threshold of 0.70 and supporting strong internal consistency. Overall, these findings suggest that the leadership scale is both reliable and valid for use in this study context.

Factor analysis was performed on communication skills using a five-item scale. Before conducting principal component analysis, the data were assessed for suitability and items with factor loadings below 0.40 were removed. Sampling adequacy was evaluated using the Kaiser-Meyer-Olkin measure. As shown in Table 12.13, the Kaiser-Meyer-Olkin value was 0.72 and Bartlett's test of sphericity was significant ($p < 0.001$), indicating that the data were appropriate for factor analysis.

Table 12.11 Kaiser-Meyer-Olkin and Bartlett's test for leadership

Kaiser-Meyer-Olkin measure of sampling adequacy		0.641
Bartlett's test of sphericity	Approx. χ^2	260.590
	df	10
	Sig.	<0.001

Table 12.12 Factor analysis for leadership

No.	Item	Factor loading
1.	My project manager encourages team members to contribute ideas and solutions	0.759
2.	My project manager provides clear direction and sets specific goals for the team	0.759
3.	My project manager empowers team members by delegating authority and responsibility	0.747
4.	My project manager recognises and rewards good performance	0.816
5.	My project manager is approachable and listens to team members' concerns	0.645
	Percentage of average variance extracted (AVE)	55.84%
	Composite reliability	0.863

Table 12.13 KMO and Barlett's test for communication skills

Kaiser-Meyer-Olkin measure of sampling adequacy		0.752
Bartlett's test of sphericity	Approx. χ^2	236.030
	df	10
	Sig.	<0.001

Table 12.14 Factor analysis for communication skills

No.	Item	Factor loading
1.	My project manager communicates project goals and expectations clearly	0.744
2.	My project manager ensures all team members are well-informed about project updates	0.732
3.	My project manager provides constructive feedback to team members	0.687
4.	My project manager demonstrates confidence and clarity during presentations or meetings	0.755
5.	My project manager actively listens to team members' concerns and suggestions	0.744
	Percentage of average variance extracted (AVE)	53.7%
	Composite reliability	0.853

Table 12.14 summarises the factor analysis results for communication skills. The average variance extracted was 0.537, exceeding the recommended threshold of 0.50 and indicating acceptable convergent validity. Composite reliability was 0.853, reflecting strong internal consistency among the items. Overall, the results support the scale's suitability for measuring communication skills in this study context.

Organisational culture was examined using a five-item scale and factor analysis. Before conducting principal component analysis, the data were assessed for suitability and items with factor loadings below 0.40 were removed. Sampling adequacy was evaluated using the Kaiser-Meyer-Olkin measure. As shown in Table 12.15, the Kaiser-Meyer-Olkin value was 0.713, and Bartlett's test of sphericity was significant ($p < 0.001$), indicating that the data were appropriate for factor analysis.

Table 12.16 summarises the factor analysis results for organisational culture. The five items yielded an average variance extracted of 62.22% and a composite reliability of 0.891, both

Table 12.15 Kaiser-Meyer-Olkin and Bartlett's test for organisational culture

Kaiser-Meyer-Olkin measure of sampling adequacy		0.713
Bartlett's test of sphericity	Approx. χ^2	317.668
	df	10
	Sig.	<0.001

Table 12.16 Factor analysis for organisational culture

No.	Item	Factor loading
1.	The organisation promotes open communication and transparency across all levels	0.723
2.	The organisation encourages collaboration and teamwork among employees	0.830
3.	The organisation fosters a culture of respect and inclusion	0.792
4.	The organisation recognises and rewards employees for their achievements	0.806
5.	The organisation supports continuous learning and professional growth for employees	0.789
	Percentage of average variance extracted (AVE)	62.22%
	Composite reliability	0.891

exceeding the recommended thresholds of 0.50 and 0.70, respectively, indicating strong convergent validity and internal consistency. However, the reported AVE and composite reliability values appear inconsistent with the factor loadings, suggesting a calculation error or discrepancies in the reported results.

Training and development were examined using a five-item scale and factor analysis. Prior to conducting principal component analysis, data suitability was assessed, and items with factor loadings below 0.40 were removed. Sampling adequacy was evaluated using the Kaiser-Meyer-Olkin measure. As shown in Table 12.17, the Kaiser-Meyer-Olkin value was 0.712 and Bartlett's test of sphericity was significant ($p < 0.001$), indicating that the data were appropriate for factor analysis.

Table 12.18 summarises the factor analysis results for training and development. Based on the factor loadings of the five items, the average variance extracted was 58.5%, and the composite reliability was 0.875. These values exceed the recommended thresholds, indicating satisfactory convergent validity and strong internal consistency. Overall, the indicators provide an adequate measurement of the construct.

Table 12.17 Kaiser-Meyer-Olkin and Bartlett's test for training and development

Kaiser-Meyer-Olkin measure of sampling adequacy		0.712
Bartlett's test of sphericity	Approx. χ^2	272.374
	df	10
	Sig.	<0.001

Table 12.18 Factor analysis for training and development

No.	Item	Factor loading
1.	The company provides equal training opportunities for all project managers	0.831
2.	The training provided is relevant to my current project's responsibilities	0.751
3.	The content of the training is up-to-date and aligned with the industry's best practices	0.710
4.	The training I received has improved my ability to manage logistics projects effectively	0.827
5.	The organisation has a structured career development plan for project managers	0.694
	Percentage of average variance extracted (AVE)	58.5%
	Composite reliability	0.875

12.3.7 Correlation analysis

The Pearson correlation coefficient measures the strength and direction of linear relationships between variables, with values ranging from −1 to 1. A value of 1 indicates a perfect positive relationship, −1 indicates a perfect negative relationship and 0 indicates no linear relationship. Statistical significance is determined by the *p*-value, where values below 0.05 indicate a significant correlation. Table 12.19 presents the Pearson correlation coefficients and corresponding significance levels for each pair of variables.

Correlation analysis indicated strong, positive and statistically significant relationships between project manager effectiveness and the four explanatory variables. Project manager effectiveness was strongly correlated with leadership ($r = 0.863, p < 0.001$), communication skills ($r = 0.849, p < 0.001$), organisational culture ($r = 0.803, p < 0.001$) and training and development ($r = 0.823, p < 0.001$). Overall, the results suggest that higher project manager effectiveness is associated with stronger leadership, communication, organisational culture, and training and development, supporting the interrelated nature of these factors within the logistics industry.

Table 12.19 Pearson correlation coefficients

		Project manager effectiveness	Leadership	Communication skills	Organisational culture	Training and development
Project manager effectiveness	Pearson correlation	1	0.863[a]	0.849[a]	0.803[a]	0.823[a]
	Sig. (2-tailed)		<0.001	<0.001	<0.001	<0.001
Leadership	Pearson correlation	0.863[a]	1	0.849[a]	0.851[a]	0.835[a]
	Sig. (2-tailed)	<0.001		<0.001	<0.001	<0.001
Communication skills	Pearson correlation	0.849[a]	0.849[a]	1	0.832[a]	0.882[a]
	Sig. (2-tailed)	<0.001	<0.001		<0.001	<0.001
Organisational culture	Pearson correlation	0.803[a]	0.851[a]	0.832[a]	1	0.832[a]
	Sig. (2-tailed)	<0.001	<0.001	<0.001		<0.001
Training and development	Pearson correlation	0.823[a]	0.835[a]	0.882[a]	0.832[a]	1
	Sig. (2-tailed)	<0.001	<0.001	<0.001	<0.001	

[a]Correlation is significant at the 0.01 level (2-tailed).

12.3.8 Multiple regression analysis

Multiple regression analysis was used to examine the effects of multiple independent variables on a dependent variable. The standardised β coefficient indicates the direction and magnitude of change in the dependent variable associated with a one-unit change in an independent variable. The R^2 value reflects the proportion of variance in the dependent variable explained by the predictors, while the F-statistic assesses the overall model fit. In this study, multiple regression was employed to test the four hypotheses by assessing the relationships between leadership, communication skills, organisational culture, training and development, and project manager effectiveness. The analysis also identifies the strongest predictor of project manager effectiveness. The regression results are presented in Tables 12.20, 12.21 and 12.22.

Table 12.20 Regression model summary

R	R^2	Adjusted R^2	Std. error of the estimate
0.894[a]	0.799	0.794	0.38907

[a]Predictors (constant): leadership, communication skills, organisational culture, and training and development.
[b]Dependent variable: project manager effectiveness.

Table 12.21 ANOVA (significance of regression model)

Model	Sum of squares	*df*	Mean square	F	Sig.
Regression	93.717	4	23.429	154.775	<0.001[b]
Residual	23.615	156	0.151		
Total	117.331	160			

[a]Predictors (constant): leadership, communication skills, organisational culture, and training and development.
[b]Dependent variable: project manager effectiveness.

Table 12.22 Multiple regression results for leadership, communication skills, organisational culture, training and development, and project manager effectiveness

Model	Standardised coefficient, β	T	Sig.
(Constant)			
Leadership	0.430	5.380	<0.001
Communication skills	0.306	3.553	<0.001
Organisational culture	0.069	0.899	0.370
Training and development	0.137	1.630	0.105

[a]Dependent variable: project manager effectiveness.

The results indicate that leadership and communication skills positively and significantly predicted the dependent variable. Leadership showed the strongest effect ($\beta = 0.430$, $p < 0.001$), followed by communication skills ($\beta = 0.306$, $p < 0.001$). In contrast, organisational culture ($p = 0.370$) and training and development ($p = 0.105$) were not statistically significant predictors, as their p-values exceeded 0.05. Overall, the findings suggest that strengthening leadership and communication skills is most relevant to improving outcomes.

12.3.9 Hypothesis testing

Table 12.23 summarises the hypotheses evaluated in this study. Two of the four hypotheses studied were supported, while the other two were not. The multiple regression results indicate that leadership had a significant positive relationship with project manager effectiveness, with a standardised β coefficient of 0.430 and a p-value of < 0.001. This finding suggests that stronger leadership is associated with greater project manager effectiveness. The result is consistent with prior studies highlighting the roles of transformational and servant leadership in motivating employees, fostering innovation and aligning organisational goals within the Malaysian logistics sector (Xian, 2022). Previous research also demonstrates that leadership plays a critical role in supporting sustainability initiatives, enhancing team climate, and clarifying project goals, thereby contributing to improved project outcomes. Overall, the findings reinforce the importance of leadership as a key driver of project manager effectiveness in the logistics industry.

Communication skills were also found to be positively and significantly related to project manager effectiveness, with a standardised β coefficient of 0.306 and $p < 0.001$. This indicates that improved communication skills are associated with greater effectiveness among project managers. The result aligns with prior studies in the Malaysian logistics context, which emphasise that effective communication systems enhance decision-making, project monitoring and operational performance (Muhammad *et al.*, 2014). Use of integrated communication platforms, real-time data sharing and mobile technologies has been shown to improve coordination and responsiveness, leading to better project outcomes (Swart *et al.*, 2022). These findings are further supported by evidence linking strong communication and interpersonal skills to improved risk management, employee satisfaction and overall project success (Elkbuli *et al.*, 2025).

Table 12.23 Summary of results for the hypotheses tested

No.	Hypothesis	Result
H1	There is a significant relationship between leadership and project manager's effectiveness	Supported
H2	There is a significant relationship between communication skills and project manager's effectiveness	Supported
H3	There is a significant relationship between organisational culture and project manager effectiveness	Not supported
H4	There is a significant relationship between training and development and project manager effectiveness	Not supported

In contrast, organisational culture did not demonstrate a significant relationship with project manager effectiveness, as indicated by a low standardised β coefficient of 0.069 and a non-significant p-value. Although this finding differs from some earlier studies that associate organisational culture with communication quality and project performance (Abdullahi and Luketero, 2018; Muneer *et al.*, 2022), recent research suggests that the influence of organisational culture may be indirect or context dependent (Stare, 2011; Watanabe *et al.*, 2024). Factors such as resistance to cultural change, national cultural norms and mediating variables, such as employee engagement, may reduce the direct impact of organisational culture on individual managerial effectiveness. Therefore, the non-significant result does not imply the absence of influence; rather, it reflects the complex, indirect nature of organisational culture in project settings.

Similarly, training and development did not emerge as a significant predictor of project manager effectiveness, with a standardised β coefficient of 0.137 and a non-significant p-value. This finding contrasts with studies that highlight the benefits of training (Ramasamy Velar and Kee, 2025; Swart *et al.*, 2022) but is consistent with research suggesting that poorly designed or misaligned training programmes may fail to translate into improved managerial performance (Nixon *et al.*, 2012). Previous studies also note that limitations in training evaluation methods and the absence of standardised performance metrics make it difficult to capture the direct impact of training on project manager effectiveness (Kealey *et al.*, 2005; Ramasamy Velar and Kee, 2025). Furthermore, supporting factors such as leadership quality, communication effectiveness and organisational climate may moderate the relationship between training and performance, reducing its observable direct effect.

Overall, the findings suggest that leadership and communication skills play a more direct and influential role in enhancing project manager effectiveness in the logistics industry, whereas organisational culture and training and development may exert their influence indirectly, through interactions with other contextual factors.

12.4. Practical insights

Based on the empirical findings of this study, several practical insights are provided for organisations in the Malaysian logistics industry to enhance their project manager performance.

Since leadership emerged as the most powerful predictor of project manager effectiveness, logistics firms should prioritise it as a core business strategy. Organisations should move beyond basic management and instead institutionalise development programmes that specifically cultivate transformational and servant leadership capabilities. These initiatives, such as formal mentoring, succession planning and structured leadership workshops, should empower managers to motivate diverse teams, clarify project goals and align operational tasks with broader sustainability values. By fostering leaders who can inspire a shared vision, companies can better navigate the high-pressure and complex nature of modern supply chain projects.

Communication skills were identified as a significant driver of success, necessitating a dual approach that addresses both technical and interpersonal capabilities. From a technical standpoint, logistics firms should leverage advanced information technologies, such as real-time fleet management systems, mobile communication tools and integrated data-sharing platforms, to ensure transparency and rapid responsiveness. Interpersonally, project managers require targeted training in conflict management, negotiation and professional writing to ensure that project directives are clear and that stakeholders remain aligned. Implementing clear communication protocols and regular feedback mechanisms will further reduce misunderstandings and support the timely delivery of projects.

While organisational culture and training and development did not show a direct predictive impact in the regression analysis, their strong positive correlations with effectiveness indicate they are essential supporting factors. Organisations must foster a collaborative, transparent culture to reduce resistance to change and improve team performance. Furthermore, to ensure training and development are not demotivating or ineffective, programmes must be closely aligned with actual job requirements and current industry best practices. Rather than generic training, firms should prioritise measurable skill development in both technical and behavioural competencies, supported by evaluation mechanisms to ensure these skills are effectively transferred to the workplace.

Ultimately, logistics firms seeking a competitive advantage should adopt an integrated strategy that aligns leadership development, communication infrastructure and cultural management. Relying on a single factor is insufficient; instead, a comprehensive approach ensures that the organisational environment and training systems are specifically designed to reinforce core leadership and communication competencies. This holistic alignment is likely to improve project performance, increase employee satisfaction and strengthen the firm's credibility and competitiveness in the global logistics market.

12.5. Conclusion and recommendations

The study presented in this chapter has provided a robust assessment of the drivers of project manager effectiveness in the Malaysian logistics sector, a field that is increasingly complex due to global trade expansion and technological integration. The empirical results, derived from a sample of 161 industry professionals, demonstrate that the proposed model has exceptional predictive power, explaining 79.9% of the variance in project manager effectiveness ($R^2 = 0.799$). The statistical significance of the model ($F = 154.775, p < 0.001$) confirms that the combination of leadership, communication, culture and training provides a highly reliable framework for understanding managerial success.

The findings pinpoint leadership as the most influential determinant of effectiveness, with the highest standardised β coefficient ($\beta = 0.430, p < 0.001$). This suggests that the ability to inspire a shared vision and motivate teams is the single most critical asset for a logistics project manager. Furthermore, the study highlights the centrality of the communication theme; communication skills emerged as the second significant predictor ($\beta = 0.306, p < 0.001$), emphasising that technical expertise alone is insufficient without the ability to align stakeholders and maintain clear information flows in high-pressure environments.

Interestingly, while organisational culture and training and development did not emerge as significant direct predictors in the regression analysis, they showed strong, positive and statistically significant correlations with effectiveness ($r = 0.803$ and $r = 0.823$, respectively). This nuance indicates that while they may not 'drive' effectiveness in a direct cause-and-effect manner, they serve as essential environmental supports. Without a collaborative culture or relevant training, even the most skilled leaders may find their efforts constrained.

12.5.1 Implications and strategic considerations

From a theoretical perspective, this study advances stakeholder theory by demonstrating how leadership and communication act as the primary mechanisms for managing complex stakeholder expectations in project-based environments. It specifically enriches transformational and servant leadership theories by showing that managers who prioritise individual development and sustainability-oriented values achieve more effective project delivery. Furthermore, the study adds a layer of complexity to existing literature on organisational behaviour by suggesting that the impact of culture is indirect and context-dependent, potentially mediated by factors such as employee engagement or national norms.

12.5.2 Recommendations for future research

This study has limitations that suggest avenues for future investigation. First, the use of a cross-sectional design means the data only captured a 'snapshot' in time; future researchers should adopt longitudinal designs to observe how leadership and communication influence effectiveness over the entire lifecycle of long-term projects. Second, the study focused specifically on the logistics sector; broadening the context to include other high-tech or service industries would enable valuable comparative analysis.

Additionally, future work could explore mediating variables, such as emotional intelligence or digital literacy, to further explain the 20% of variance not covered by the current model. Finally, expanding the sampling frame to include a higher proportion of executive and senior management perspectives would provide a more top-down view of how strategic resource allocation directly filters down to project-level effectiveness. Such efforts will continue to refine the understanding of what makes a project manager truly effective in an ever-evolving global economy.

REFERENCES

Abdullahi RI and Luketero SW (2018) Influence of organisational culture on project performance in Waso Trust Land Project Organisation Isiolo County, Kenya. *International Academic Journal of Information Sciences and Project Management* **3(2)**: 472–497.

Abolghasemi M, Al-Mhdawi MKS, Rahimian F and Qazi A (2026) Exploring the nexus of stakeholder management, project performance and stakeholder satisfaction in Malaysian residential building projects: a PLS-SEM approach. *Engineering, Construction and Architectural Management* **33(3)**: 2089–2117.

AlSaied MK and Alkhoraif AA (2024) The role of organizational learning and innovative organizational culture for ambidextrous innovation. *The Learning Organization* **31(2)**: 205–226.

Bajcar B and Babiak J (2022) Transformational and transactional leadership in the Polish organizational context: validation of the full and short forms of the multifactor leadership questionnaire. *Frontiers in Psychology* **13**: 908594.

Bakar MAA and Jaafar HS (2016) Malaysian logistics performance: a manufacturer's perspective. *Procedia-Social and Behavioral Sciences* **224**: 571–578.

Baldwin TT and Ford JK (1988) Transfer of training: a review and directions for future research. *Personnel Psychology* **41(1)**: 63–105.

Chien F (2023) The role of corporate governance and environmental and social responsibilities on the achievement of sustainable development goals in Malaysian logistic companies. *Economic Research-Ekonomska Istraživanja* **36(1)**: 1610–1630.

Ciric Lalic D, Lalic B, Delić M, Gracanin D and Stefanovic D (2022) How project management approach impact project success? From traditional to agile. *International Journal of Managing Projects in Business* **15(3)**: 494–521.

Đajić MJ, Lalic DC, Vujičić MD *et al.* (2024) Development and validation of the project manager skills scale (PMSS): an empirical approach. *Heliyon* **10(3)**: e25055.

Elkbuli M, Azmy N and Lee CK (2025) Enhancing project manager communication soft skills and risk management practices in Libyan oil and gas construction projects: the role of experience. *Journal of Engineering, Design and Technology* **23(3)**: 1017–1038.

Freeman RE, Menghwar PS and Grushka-Cockayne Y (2025) A stakeholder theory perspective for project management. *International Journal of Project Management* **43(7)**:102764.

Hair Jr JF, Black WC, Babin BJ and Anderson RE (2010) *Multivariate Data Analysis*, 7th edn. Prentice Hall, Upper Saddle River, NJ, USA.

Harman HH (1976) *Modern Factor Analysis*, 3rd edn. University of Chicago Press, Chicago, IL, USA.

Ika LA and Pinto JK (2022) The 're-meaning' of project success: updating and recalibrating for a modern project management. *International Journal of Project Management* **40(7)**: 835–848.

Jayaweera M, Perera BAKS and Liyanawatta TN (2025) Use of emotional intelligence to enhance the leadership skills of project managers in construction: a qualitative Delphi study. *Journal of Legal Affairs and Dispute Resolution in Engineering and Construction* **17(3)**: 04525011.

Kealey DJ, Protheroe DR, MacDonald D and Vulpe T (2005) Re-examining the role of training in contributing to international project success: a literature review and an outline of a new model training program. *International Journal of Intercultural Relations* **29(3)**: 289–316.

Khaddour WN (2024) Critical realism and organizational culture: framing Schein's model of studying organizational culture by using critical realism philosophy. *Open Journal of Philosophy* **15(1)**: 77–97.

Muhammad M, Saahar S, Hasan H, Fiah AFM and Nor AM (2014) Effective communication systems for Malaysian logistics industry. *Procedia – Social and Behavioral Sciences* **130**: 204–215.

Muneer M, Khan N, Awais Hussain M *et al.* (2022) A quantitative study of the impact of organizational culture, communication management, and clarity in project scope on constructions' project success with moderating role of project manager's competencies to enhance constructions management practices. *Buildings* **12(11)**: 1856.

Nixon P, Harrington M and Parker D (2012) Leadership performance is significant to project success or failure: a critical analysis. *International Journal of Productivity and Performance Management* **61(2)**: 204–216.

Pomaza-Ponomarenko A, Kryvova S, Hordieiev A, Hanzyuk A and Halunko O (2023) Innovative risk management: identification, assessment and management of risks in the context of innovative project management. *Economic Affairs (New Delhi)* **68(4)**: 2263–2275.

Radhakrishnan A, Zaveri J, David D and Davis JS (2022) The impact of project team characteristics and client collaboration on project agility and project success: an empirical study. *European Management Journal* **40(5)**: 758–777.

Ramasamy Velar V and Kee DMH (2025) Enhancing project success through transformational leadership, top management support and training and development and knowledge sharing. *Asian Education and Development Studies* **14(4)**: 832–855.

Romero RAM, Vaca CAM, Bejarano MAG, Hermenegildo AM and De la Torre Tejada JA (2023) Leadership styles in Ecuadorian companies: application of the Multifactor Leadership Questionnaire (MLQ). *Business: Theory and Practice* **24(2)**: 544–556.

Rucks L, Wingate L, López M *et al.* (2024) Leveraging the Kirkpatrick four-level model to evaluate evaluation capacity building work. *New Directions for Evaluation* **2024(183)**: 81–94.

Sepúlveda-Rivillas CI, Alegre J and Oltra V (2022) Impact of knowledge-based organizational support on organizational performance through project management. *Journal of Knowledge Management* **26(4)**: 993–1013.

Sposito L, Scafuto IC, Serra FR and Ferreira MP (2024) Influence of the project managers' expertise and experience in the success of projects: the moderating effect of emotional intelligence. *International Journal of Managing Projects in Business* **17(1)**: 1–26.

Stare A (2011) The impact of the organisational structure and project organisational culture on project performance in Slovenian enterprises. *Management: Journal of Contemporary Management Issues* **16(2)**: 1–22.

Swart K, Bond-Barnard T and Chugh R (2022) Challenges and critical success factors of digital communication, collaboration and knowledge sharing in project management virtual teams: a review. *International Journal of Information Systems and Project Management* **10(4)**: 84–103.

Szczepańska-Woszczyna K and Gatnar S (2022) Key competences of research and development project managers in high technology sector. *Forum Scientiae Oeconomia* **10(3)**: 107–130.

Wahab SN, Tan A and Roche O (2024) Identifying supply chain manager leadership skills and competencies gaps in Malaysia. *Higher Education, Skills and Work-Based Learning* **14(5)**: 921–937.

Watanabe WC, Shafiq M, Nawaz MJ, Saleem I and Nazeer S (2024) The impact of emotional intelligence on project success: mediating role of team cohesiveness and moderating role of organizational culture. *International Journal of Engineering Business Management* **16**: 18479790241232508.

Waseem M, Iqbal S and Khan K (2025) Effect of humble leadership on project success: the mediating role of team engagement and the moderating role of organizational culture. *Journal of Facilities Management* **23(1)**: 98–121.

Xian GS (2022) The relationship between transformational leadership and job satisfaction in a logistic company (small and medium enterprises in Johor Bahru). *Journal of Arts and Social Sciences* **5(2)**: 15–27.

Yıldız O (2023) PLS-SEM bias: traditional vs consistent. *Quality & Quantity* **57(Suppl 4)**: 537–552.

Advancing Leadership and Collaboration in Project Management: Culture, Knowledge and Team Dynamics

emerald PUBLISHING ice Publishing

Mohammed Reyasudin Basir Khan
ISBN 978-1-83742-803-8
https://doi.org/10.1108/978-1-83742-802-120261016
Emerald Publishing Limited: All rights reserved

Chapter 13
Navigating industrial instability: the interplay of advanced tools, stakeholder communication and governance in Middle Eastern oil and gas risk management

Hussain Ibrahim Saleh Saad and Mohammed Reyasudin Basir Khan

Tun Razak Graduate School, Universiti Tun Abdul Razak (UNIRAZAK), Kuala Lumpur, Malaysia

13.1. Introduction

Risk management plays a central role in ensuring organisational stability and long-term sustainability, particularly in complex, high-risk industries such as infrastructure development and energy (Secundo *et al.*, 2024). Effective risk management is essential for navigating uncertainty and maintaining business continuity in dynamic sectors (Tchankova, 2002). This importance is especially pronounced in the oil and gas industry, where operational complexity, capital intensity and exposure to external shocks elevate risk management from a supporting function to a strategic imperative.

The oil and gas sector is exposed to a wide range of risks, including technological change, financial volatility, environmental hazards and the challenge of managing diverse stakeholder interests. In politically volatile environments, these risks are further intensified. Libya's oil and gas industry represents a particularly compelling case due to the profound political and institutional disruptions that followed the 2011 revolution.

The Libyan context presents a uniquely complex risk landscape shaped by unstable governance structures, contested political authority and persistent security challenges. In Libya, the convergence of political instability, regulatory inconsistency and internal conflict necessitates risk management approaches that are both adaptive and resilient (Ghanem, 2023). These conditions highlight the need for comprehensive strategies that can respond to rapid change while safeguarding operational continuity and national economic contributions.

The study presented in this chapter was motivated by the need to examine how Libya's post-2011 political environment has reshaped risk management practices in its oil and gas sector.

It identifies the key challenges organisations face and assesses the strategies employed to mitigate risks in an environment marked by ongoing instability (Dwa, 2025). Understanding these dynamics is critical not only for organisational performance but also for Libya's broader economic stability, which remains heavily dependent on hydrocarbon revenues.

Political instability has emerged as a defining challenge for risk management in the Libyan oil and gas industry. Security threats, including terrorism, further complicate the risk environment. Against this backdrop, this study investigates the relationship between political instability and the effectiveness of risk management strategies in Libya's oil and gas sector. It examines how political factors influence organisational decision-making and explores the extent to which risk management practices can adapt to Libya's evolving sociopolitical landscape. The study also considers the challenges enterprises face in maintaining legitimacy and operational effectiveness amid political turmoil (Darendeli and Hill, 2016).

Political instability has direct implications for oil production, market stability and investment confidence. The disruptions in governance and security have tangible effects on oil output and global energy markets. These disruptions underscore the necessity for organisations to develop flexible risk management frameworks capable of responding to sudden political shifts. The broader economic consequences of political unrest, as discussed, further reinforce the importance of effective risk management, as instability in the oil and gas sector has ripple effects across the national economy.

The primary objective of this research was to analyse how various factors influence risk management strategies in Libyan oil and gas companies. Specifically, it examined the role of advanced risk management tools, the impact of political stability or instability, the influence of stakeholder engagement and the integration of social and environmental considerations. These objectives were addressed through research questions that examine the relationships between risk management techniques, technological tools, political conditions, stakeholder communication involvement and diverse environmental and social factors.

The scope of the study is confined to Libya's oil and gas industry, with particular emphasis on the post-revolution period. It generates insights that are relevant to both academic research and industry practice by enhancing understanding of risk management in politically unstable contexts. The significance of the study lies in its potential to improve organisational safety, operational resilience and risk mitigation practices. By identifying key internal and external risk factors and evaluating existing control mechanisms, the research advances knowledge of risk management in high-risk industrial environments.

Risk management techniques are defined in this study as the methods used to identify, assess and mitigate risks arising from both internal and external sources. Risk management tools are systems and practices that enhance the efficiency and effectiveness of these techniques. Political factors encompass the regulatory environment, governance structures and levels of political volatility that affect organisational operations. Stakeholder communication involvement involves the participation of internal and external actors in the risk management process, while social and environmental factors relate to sustainability, ethical considerations and community impacts.

The oil and gas industry plays a pivotal role in shaping global economic dynamics. The strategic importance of these resources places significant demands on organisations to manage risks effectively, particularly given the technical expertise required to exploit them.

Risk management theory is grounded in the principle that potential threats can be systematically identified and addressed. Risk management is a structured process that enables decision-makers to align organisational objectives with mitigation strategies (Kerzner, 2017; Pomaza-Ponomarenko, 2023). Within the oil and gas industry, several theoretical approaches inform risk management practice. Event chain methodology emphasises the interconnected nature of risks and their cumulative impact on project timelines. Portfolio theory provides a framework for balancing risk and return through diversification. Decision tree analysis supports complex investment decisions by evaluating alternative outcomes under uncertainty (Fabozzi *et al.*, 2008).

Empirical studies provide further insight into risk management practices in the oil and gas sector (Bjerga and Aven, 2016; Jagoda and Wojcik, 2019). These theoretical and empirical insights inform a conceptual framework in which risk management techniques are influenced by political factors, stakeholder communication involvement, technological tools and diverse environmental and social factors. These variables interact dynamically, shaping organisational responses to risk.

Based on the theoretical and empirical literature reviewed, the following hypotheses were formulated to guide the study.

- H1: there is a significant relationship between risk management techniques and cutting-edge risk management tools.
- H2: there is a correlation between the degree of political stability and risk management techniques in oil and gas companies.
- H3: there is a role for stakeholder engagement in risk management techniques in oil and gas projects.
- H4: there is an impact of social and environmental issues on the effectiveness of risk management techniques in the oil and gas industry.

Figure 13.1 Conceptual framework

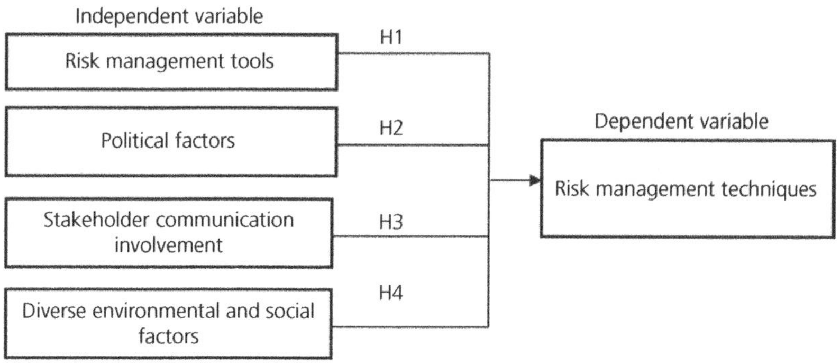

Risk itself is defined as the possibility of events that hinder the achievement of objectives. Risk identification is widely regarded as the most critical stage of the risk management process, as it determines how effectively potential threats can be anticipated and addressed. Risks are typically classified according to their probability and impact, with high-impact, low-probability events presenting significant challenges due to their potentially catastrophic consequences (Kerzner, 2017).

Risk identification relies heavily on experience, historical data and expert knowledge. Risks may be categorised as business or insurable risks, internal or external, and technical or non-technical. The estimation of a risk event is commonly expressed as the product of probability and consequence. Effective risk identification results in a comprehensive risk register that supports informed decision-making (George, 2020).

Integrating risk management into project management is essential for improving project performance. Risks affect all project dimensions, including scope, quality, schedule, cost, safety, procurement, human resources, communication and integration. Risk exposure varies across the project lifecycle, with the highest uncertainty in the conceptual phase and the highest stakes in execution. A systematic risk management approach enables organisations to reduce uncertainty early, thereby improving outcomes during implementation.

The risk management process typically includes risk identification, assessment, response, monitoring and documentation. Oil and gas projects face both internal risks, such as financial and managerial challenges, and external risks, including political instability, regulatory uncertainty, environmental concerns and market volatility. Emerging risks arise from the interaction of multiple risk factors and require continuous monitoring. Compliance with international environmental standards is increasingly essential for mitigating environmental and social risks and maintaining organisational legitimacy.

Effective project management in the oil and gas industry requires balancing stakeholder expectations with operational constraints. Project success is measured not only by achieving predefined objectives but also by managing competing demands in complex environments (Kerzner, 2017). Risk management is therefore an integral component of project management, contributing to organisational resilience and long-term sustainability.

13.2. Methodology

13.2.1 Research design

The research strategy adopted for this study is a descriptive quantitative design, consistent with approaches frequently employed in comparable investigations. This design was intended to identify and explain the key characteristics of the phenomenon under study, namely the risk management practices applied in Libya's oil and gas industry. By measuring defined variables such as behaviours, attitudes and organisational practices, the approach enables a structured and systematic representation of prevailing conditions within the sector. In this way, it supports a comprehensive assessment of the current situation through quantifiable evidence.

13.2.2 Population and sampling procedures

The target demographic for this study comprises oil and gas workers. Because this population is so large, a stratified random sampling approach was used. To ensure a more representative sample, this strategy entailed splitting the population into homogeneous subgroups and randomly selecting participants from each segment (Trochim, 2006).

13.2.3 Data collection method

The main tool used in this study to collect data was the questionnaire. A five-point Likert-scale questionnaire was administered to the selected participants. Participants' responses to statements on risk management in the oil and gas industry were evaluated on a scale from 'strongly disagree' to 'strongly agree'.

13.2.4 Operationalisation and measurement

The independent variables in this study comprise the range of risk management approaches applied within the oil and gas industry. These variables were operationalised through multiple questionnaire items, with respondents indicating their level of agreement based on their professional experience and observations of organisational practice (Hair *et al.*, 2010).

The dependent variable reflects the outcomes associated with risk management techniques, with particular emphasis on operational efficiency. Respondents evaluated this construct using the designated Likert scale, drawing on their perceptions of measurable improvements or declines in organisational activities and operational procedures (Cooper and Schindler, 2011).

13.2.5 Data analysis techniques

The dataset obtained from the structured questionnaires required summarisation and simplification to enable clear interpretation, making descriptive statistical analysis essential. To provide an initial overview of the data, this study used measures of central tendency, including mean, median and mode, as well as measures of dispersion, such as standard deviation and variance. These techniques enabled a concise, coherent representation of respondents' views and experiences in the oil and gas industry. They are widely regarded as fundamental for establishing an empirical profile of the sample.

In addition to descriptive analysis, the study employed inferential statistical techniques to generate conclusions that extend beyond the sample and are relevant to the wider population. Regression analysis was used to examine the extent to which risk management techniques predict or influence operational efficiency. In contrast, correlation analysis assessed the strength and direction of relationships between the study variables.

13.2.6 Sample size

A stratified random sampling approach was used to recruit participants from Libya's oil and gas sector. A total of 101 usable responses were obtained and included in the analysis. The degrees of freedom reported across the correlation and regression analyses ($df = 99$) confirm an effective analytical sample of 101 participants.

13.3. Results and analysis

13.3.1 Demographic profile

The demographic profile of the participants provides a foundational context for understanding the socioeconomic and political structure of the workforce within Libya's Oil Crescent.

This strategically important region is characterised by a workforce with high levels of professional experience and technical qualification, as detailed in Table 13.1.

The gender distribution is highly skewed, with 99% of the sample being male. This disparity reflects the 'workforce realities' in the Libyan oil and gas sector. While such an imbalance

Table 13.1 Demographic profile of respondents

Demographic category	Sub-category	Frequency	Percentage: %
Gender	Male	100	99.0
	Female	1	1.0
Age group	18–29 years	6	5.9
	30–39 years	22	21.8
	40–49 years	43	42.6
	50–59 years	22	21.8
	Over 59 years	8	7.9
Education level	Secondary school or equivalent	3	3.0
	High school or equivalent	5	5.0
	High diploma	20	19.8
	Bachelor's (BS/BA) level	46	45.5
	Master's degree	23	22.8
	Doctoral level	4	4.0
Work experience	Less than 5 years	2	2.0
	5–15 years	24	23.8
	16–25 years	46	45.5
	More than 25 years	29	28.7
Current position	Engineer	19	18.8
	Supervisor	14	13.9
	Coordinator	12	11.9
	Head of department	11	10.9
	Other	10	9.9
	Technician	9	8.9
	Project engineer	8	7.9
	Group supervisor	7	6.9
	Department manager	5	5.0
	Superintendent	3	3.0
	Project manager	2	2.0
	General manager	1	1.0

necessitates caution regarding generalisability, statistical analysis indicates that gender is not associated with the effectiveness of risk management variables in this context.

The age and experience data indicate a mature and seasoned workforce. The largest segment of participants (42.6%) is in the 40–49 age bracket, and 74.2% of the sample has more than 16 years of professional experience. This creates a core group with strong technical competence and extensive operational knowledge to navigate the sector's inherent complexity. Interestingly, while experience shows a statistically significant positive relationship with the use of risk management practices, its overall influence is relatively weak compared to the impact of advanced technological tools.

Regarding academic background, the sample is predominantly well-educated, with 72.3% of participants holding at least a Bachelor's degree. However, a key empirical finding is that formal education levels do not significantly correlate with risk management effectiveness. This suggests that in the volatile environment of Libya's Oil Crescent, practical competence, targeted training and the adoption of specialised tools are more critical drivers of success than academic qualifications alone.

The occupational distribution reflects a technical and supervisory focus, following a 'conventional hierarchical pyramid' where senior decision-making roles are limited. Engineers and supervisors constitute the largest groups, emphasising the importance of frontline oversight. A notable observation is the negative association between higher organisational positions and direct tool usage. This implies that senior management may be somewhat detached from the operational risk data and monitoring systems used at the ground level. Collectively, these demographic insights highlight a workforce that is technically capable but must increasingly prioritise integrated technological and political intelligence to maintain operations amid ongoing instability.

13.3.2 Validity test

Data suitability for factor analysis was assessed using the Kaiser-Meyer-Olkin (KMO) measure and Bartlett's test of sphericity. The KMO value of 0.921, as shown in Table 13.2, indicates excellent sampling adequacy and strong intercorrelations between variables, supporting the appropriateness of factor analysis. Bartlett's test was significant ($p < 0.001$), rejecting the null hypothesis that the correlation matrix is an identity matrix and confirming that the variables are sufficiently correlated. Together, these results demonstrate that the dataset contains substantial shared variance and is well-suited for extracting underlying factor structures.

Table 13.2 Validity tests

Kaiser-Meyer-Olkin and Bartlett's test		
Kaiser-Meyer-Olkin measure of sampling adequacy		0.921
Bartlett's test of sphericity	Approx. χ^2	2195.022
	df	300
	Sig.	<0.001

13.3.3 Reliability test

Reliability analysis is a fundamental component of the quantitative methodology employed to verify that the survey instrument consistently measures the intended constructs. To assess the internal consistency of the scales used for both independent and dependent variables, Cronbach's alpha (α) coefficients were calculated. The standard interpretation of these coefficients holds that values above 0.90 are excellent, while those between 0.80 and 0.89 are good. Table 13.3 consolidates the reliability statistics for the constructs examined in the study.

The reliability statistics in Table 13.3 indicate that the measurement instruments exhibit very high internal consistency and reliability. For the dependent variable risk management techniques, the Cronbach's alpha of 0.936 across five items suggests that respondents interpreted and answered the questions about organisational risk practices in a highly coherent manner. This level of reliability ensures that the construct accurately captures the intended operational outcomes and efficiency markers defined in the research framework.

The independent variables, collectively encompassing risk management tools, political factors, stakeholder communication involvement and diverse environmental and social factors, were measured using a 20-item scale. This scale yielded an excellent Cronbach's alpha of 0.948. Such a high coefficient indicates that the 20 items are strongly interrelated and effectively capture the multidimensional nature of the factors influencing risk management amid political instability.

While these elevated alpha values strengthen the credibility of the findings, they also warrant a nuanced interpretation. In academic literature, extremely high reliability coefficients (above 0.90), particularly for longer scales, may sometimes indicate potential item redundancy. This suggests that some questions in the 20-item independent variable scale may capture overlapping information, and future applications of the instrument might benefit from a review to ensure maximum efficiency without sacrificing content coverage.

Table 13.3 Cronbach's alpha for dependent variables

Variable category	Construct measured	Number of items	Cronbach's alpha, α	Standardised Cronbach's alpha, α	Internal consistency rating
Dependent variable, Y	Risk management techniques	5	0.936	0.937	Excellent
Independent variables, X	Risk management tools (X1), political factors (X2), stakeholder communication involvement (X3), diverse environmental and social factors (X4)	20	0.948	0.949	Excellent

Ultimately, demonstrating robust reliability for both scales provides a vital empirical foundation for subsequent inferential analyses. Because the measures are proven to be stable and consistent, greater confidence can be placed in the Pearson correlation and multiple regression results used to test the study's hypotheses. These reliability results confirm that the data collected from the 101 participants in Libya's Oil Crescent is suitable for explaining and predicting the complex determinants of risk management effectiveness in volatile industrial environments.

13.3.4 Descriptive analysis

As reported in Table 13.4, risk management techniques recorded a moderately high mean of 3.55, with a median of 3.80 and a mode of 4.00, indicating that respondents generally reported frequent use or strong recognition of these practices. Responses showed moderate dispersion (standard deviation (SD) 1.06) across the full scale range from 1 to 5, with a modest negative skew suggesting a greater concentration of higher ratings.

Table 13.4 Descriptive statistics of dependent and independent variables

Descriptive statistics						
		Risk management techniques	Risk management tools	Political factors	Stakeholder communication involvement	Diverse environmental and social factors
n	Valid	101	101	101	101	101
	Missing	0	0	0	0	0
Mean		3.5505	3.3089	3.7347	3.3901	3.2218
Median		3.8000	3.4000	4.0000	3.4000	3.4000
Mode		4.00	4.00	4.00	3.20[a]	3.60[a]
SD		1.06448	0.97469	1.03966	0.87881	0.93975
Variance		1.133	0.950	1.081	0.772	0.883
Skewness		−0.700	−0.281	−1.147	−0.519	−0.437
Std. error of skewness		0.240	0.240	0.240	0.240	0.240
Kurtosis		−0.409	−0.593	0.446	0.089	−0.340
Std. error of kurtosis		0.476	0.476	0.476	0.476	0.476
Range		4.00	4.00	4.00	4.00	4.00
Minimum		1.00	1.00	1.00	1.00	1.00
Maximum		5.00	5.00	5.00	5.00	5.00

[a]Multiple modes exist. The smallest value is shown.

Risk management tools produced a mean of 3.31, supported by a median of 3.40 and a mode of 4.00. Variation was slightly lower (SD 0.97), indicating more concentrated agreement, while the mild negative skew reflects a tendency toward higher evaluations.

Political factors received the highest overall rating, with a mean of 3.73 and both the median and mode at 4.00. The distribution shows moderate variability (SD 1.04) but a stronger negative skew, indicating that many respondents rated political considerations as highly influential.

Stakeholder communication involvement yielded a mean of 3.39, with central tendency values suggesting moderate to moderately high agreement. Dispersion was relatively low (SD 0.88) and the distribution was close to normal.

Diverse environmental and social factors had the lowest mean at 3.22, although the median and mode were slightly higher, indicating moderate recognition overall. Responses showed moderate variability (SD 0.94) and a slight tendency toward higher ratings.

Overall, the variables trend toward the upper end of the scale, suggesting broad recognition of the multiple dimensions shaping risk management practice.

13.3.5 Correlation analysis

This study applied partial correlation analysis to examine relationships between the main variables while controlling for demographic factors, specifically position, experience and education. By accounting for shared variance with these controls, the analysis isolates the unique association between key risk management constructs. Pearson correlation and multiple regression were then used to evaluate the relationships between political, communication, technological and socioenvironmental factors and risk management effectiveness.

Table 13.5 indicates significant positive associations between political factors, stakeholder communication involvement, risk management techniques, risk management tools and diverse environmental and social factors. These patterns suggest that effective risk management extends beyond technical mechanisms and is closely linked to political awareness, stakeholder engagement and socioenvironmental integration. Experience shows a statistically significant but weak relationship with effectiveness, implying a limited influence compared to other factors. Regression analysis supports these findings, with high explanatory power ($R^2 = 0.759$) and significant contributions from political factors and risk management tools. Employee position and education show limited direct relationships with effectiveness, but their inclusion as controls strengthens the interpretation of the core relationships.

Correlation results also show strong links between risk management techniques and tools (0.835). Political factors correlate positively with techniques ($r = 0.694$) and tools ($r = 0.606$). Stakeholder communication involvement correlates strongly with techniques ($r = 0.708$) and tools ($r = 0.749$). Diverse environmental and social factors show strong correlations with techniques ($r = 0.781$) and tools ($r = 0.836$). Current position shows weak, non-significant negative associations, suggesting that risk management effectiveness is influenced more by systemic and contextual factors than by job level alone.

The correlation matrix, reported in Table 13.6, indicates a strong positive association between risk management techniques and tools ($r = 0.835$), suggesting that more advanced techniques

Table 13.5 Partial correlation between all variables controlling for current position (continued on next page)

Control variables			Y	X1	X2	X3	X4	X5: current position
-none-[a]	y	Correlation	1.000	0.835	0.694	0.708	0.781	–0.105
		Significance (2-tailed)		0.000	0.000	0.000	0.000	0.298
		df	0	99	99	99	99	99
	X1	Correlation	0.835	1.000	0.606	0.749	0.836	–0.258
		Significance (2-tailed)	0.000		0.000	0.000	0.000	0.009
		df	99	0	99	99	99	99
	X2	Correlation	0.694	0.606	1.000	0.593	0.666	–0.170
		Significance (2-tailed)	0.000	0.000		0.000	0.000	0.090
		df	99	99	0	99	99	99
	X3	Correlation	0.708	0.749	0.593	1.000	0.772	–0.211
		Significance (2-tailed)	0.000	0.000	0.000		0.000	0.034
		df	99	99	99	0	99	99
	X4	Correlation	0.781	0.836	0.666	0.772	1.000	–0.210
		Significance (2-tailed)	0.000	0.000	0.000	0.000		0.035
		df	99	99	99	99	0	99
	Current position	Correlation	–0.105	–0.258	–0.170	–0.211	–0.210	1.000
		Significance (2-tailed)	0.298	0.009	0.090	0.034	0.035	
		df	99	99	99	99	99	0

Table 13.5 Continued

Control variables			Y	X1	X2	X3	X4	X5: current position
Current position	Y	Correlation	1.000	0.841	0.690	0.705	0.780	
		Significance (2-tailed)		0.000	0.000	0.000	0.000	
		df	0	98	98	98	98	
	X1	Correlation	0.841	1.000	0.591	0.736	0.827	
		Significance (2-tailed)	0.000		0.000	0.000	0.000	
		df	98	0	98	98	98	
	X2	Correlation	0.690	0.591	1.000	0.578	0.655	
		Significance (2-tailed)	0.000	0.000		0.000	0.000	
		df	98	98	0	98	98	
	X3	Correlation	0.705	0.736	0.578	1.000	0.761	
		Significance (2-tailed)	0.000	0.000	0.000		0.000	
		df	98	98	98	0	98	
	X4	Correlation	0.780	0.827	0.655	0.761	1.000	
		Significance (2-tailed)	0.000	0.000	0.000	0.000		
		df	98	98	98	98	0	

[a]Cells contain zero-order (Pearson) correlations.

Note: A significance value of 0.000 as reported by SPSS indicates $p < 0.001$ (i.e. $p < 0.0005$ before rounding). This applies to all tables in this chapter.

Table 13.6 Partial correlation between all variables controlling for experience (continued on next page)

Correlation								
Control variables			Y	X1	X2	X3	X4	X5: experience
-none-[a]	Y	Correlation	1.000	0.835	0.694	0.708	0.781	0.222
		Significance (2-tailed)		0.000	0.000	0.000	0.000	0.026
		df	0	99	99	99	99	99
	X1	Correlation	0.835	1.000	0.606	0.749	0.836	0.246
		Significance (2-tailed)	0.000		0.000	0.000	0.000	0.013
		df	99	0	99	99	99	99
	X2	Correlation	0.694	0.606	1.000	0.593	0.666	0.089
		Significance (2-tailed)	0.000	0.000		0.000	0.000	0.374
		df	99	99	0	99	99	99
	X3	Correlation	0.708	0.749	0.593	1.000	0.772	0.198
		Significance (2-tailed)	0.000	0.000	0.000		0.000	0.047
		df	99	99	99	0	99	99
	X4	Correlation	0.781	0.836	0.666	0.772	1.000	0.223
		Significance (2-tailed)	0.000	0.000	0.000	0.000		0.025
		df	99	99	99	99	0	99
	X5	Correlation	0.222	0.246	0.089	0.198	0.223	1.000
		Significance (2-tailed)	0.026	0.013	0.374	0.047	0.025	
		df	99	99	99	99	99	0

Table 13.6 Continued

Correlation								
Control variables			Y	X1	X2	X3	X4	X5: experience
Experience level in years	Y	Correlation	1.000	0.826	0.694	0.694	0.769	
		Significance (2-tailed)		0.000	0.000	0.000	0.000	
		df	0	98	98	98	98	
	X1	Correlation	0.826	1.000	0.605	0.737	0.827	
		Significance (2-tailed)	0.000		0.000	0.000	0.000	
		df	98	0	98	98	98	
	X2	Correlation	0.694	0.605	1.000	0.589	0.666	
		Significance (2-tailed)	0.000	0.000		0.000	0.000	
		df	98	98	0	98	98	
	X3	Correlation	0.694	0.737	0.589	1.000	0.762	
		Significance (2-tailed)	0.000	0.000	0.000		0.000	
		df	98	98	98	0	98	
	X4	Correlation	0.769	0.827	0.666	0.762	1.000	
		Significance (2-tailed)	0.000	0.000	0.000	0.000		
		df	98	98	98	98	0	

[a]Cells contain zero-order (Pearson) correlations.

tend to be used alongside more comprehensive tools. Political factors show moderate positive relationships with both techniques (r = 0.694) and tools (r = 0.606), indicating that the political environment is meaningfully related to the implementation of risk management. Stakeholder communication involvement is also strongly correlated with techniques (r = 0.708) and tools (r = 0.749), reinforcing the importance of engagement and involvement in effective risk management. Diverse environmental and social factors demonstrate strong positive correlations with techniques (r = 0.781) and tools (r = 0.836), highlighting their influence on risk management practice.

Experience level in years shows weak positive correlations with the other variables, indicating only a limited association between greater experience and the application of risk management practices. Most correlations are statistically significant (p <0.05), except the relationship between political factors and experience level (r = 0.089, p = 0.374), which is not significant. Overall, these findings suggest that contextual and operational factors are more strongly linked to risk management effectiveness than experience alone, while recognising that correlation does not establish causality.

The correlation matrix in Table 13.7 shows strong, statistically significant positive relationships between risk management techniques and key practice dimensions, including risk management tools (r = 0.835), political factors (r = 0.694), stakeholder communication involvement (r = 0.708) and diverse environmental and social factors (r = 0.781), indicating that these elements tend to increase together. In contrast, the level of education shows negligible and non-significant correlations with the risk management variables, including techniques (r = –0.097, p = 0.335) and tools (r = 0.009, p = 0.931), suggesting no meaningful linear association in this sample. Most correlations have df = 99, consistent with a sample size of 100. Overall, the findings indicate that risk management effectiveness is driven more by interconnected operational and contextual factors than by education level alone.

The correlation matrix indicates strong, statistically significant positive associations between risk management techniques and related dimensions, including risk management tools (r = 0.835), political factors (r = 0.694), stakeholder communication involvement (r = 0.708) and diverse environmental and social factors (r = 0.781), suggesting these components tend to co-occur in practice.

Gender shows very weak, non-significant correlations with all risk management variables (r = 0.026–0.061, p > 0.05), indicating no meaningful linear relationship in this sample. Most correlation reports use df = 99, consistent with a sample size of 100. Overall, the results suggest that risk management effectiveness is more closely linked to interconnected operational and contextual factors than to gender, although correlational evidence does not establish causality.

The correlation matrix in Table 13.8 shows a strong positive relationship between risk management techniques and risk management tools (r = 0.835, p <0.001), meaning greater use or competency in techniques is associated with greater use of tools. Experience level (years) has weak but statistically significant positive correlations with both risk management techniques (r = 0.222, p = 0.026) and risk management tools (r = 0.246, p = 0.013). With df = 99, the sample size is 100. Overall, techniques and tools are closely linked, while experience is associated with slightly higher use of both, but the effect is small compared to the relationship between techniques and tools.

Table 13.7 Partial correlation between all variables controlling for education level (continued on next page)

Correlation								
Control variables			Y	X1	X2	X3	X4	X5: level of education
-none-[a]	Y	Correlation	1.000	0.835	0.694	0.708	0.781	–0.097
		Significance (2-tailed)		0.000	0.000	0.000	0.000	0.335
		df	0	99	99	99	99	99
	X1	Correlation	0.835	1.000	0.606	0.749	0.836	0.009
		Significance (2-tailed)	0.000		0.000	0.000	0.000	0.931
		df	99	0	99	99	99	99
	X2	Correlation	0.694	0.606	1.000	0.593	0.666	0.012
		Significance (2-tailed)	0.000	0.000		0.000	0.000	0.905
		df	99	99	0	99	99	99
	X3	Correlation	0.708	0.749	0.593	1.000	0.772	–0.098
		Significance (2-tailed)	0.000	0.000	0.000		0.000	0.327
		df	99	99	99	0	99	99
	X4	Correlation	0.781	0.836	0.666	0.772	1.000	–0.023
		Significance (2-tailed)	0.000	0.000	0.000	0.000		0.819
		df	99	99	99	99	0	99
	X5	Correlation	–0.097	0.009	0.012	–0.098	–0.023	1.000
		Significance (2-tailed)	0.335	0.931	0.905	0.327	0.819	
		df	99	99	99	99	99	0

X5	Y	Correlation	1.000	0.840	0.698	0.705	0.782
		Significance (2-tailed)		0.000	0.000	0.000	0.000
		df	0	98	98	98	98
	X1	Correlation	0.840	1.000	0.606	0.754	0.836
		Significance (2-tailed)	0.000		0.000	0.000	0.000
		df	98	0	98	98	98
	X2	Correlation	0.698	0.606	1.000	0.597	0.667
		Significance (2-tailed)	0.000	0.000		0.000	0.000
		df	98	98	0	98	98
	X3	Correlation	0.705	0.754	0.597	1.000	0.774
		Significance (2-tailed)	0.000	0.000	0.000		0.000
		df	98	98	98	0	98
	X4	Correlation	0.782	0.836	0.667	0.774	1.000
		Significance (2-tailed)	0.000	0.000	0.000	0.000	
		df	98	98	98	98	0

[a]Cells contain zero-order (Pearson) correlations.

Table 13.8 Partial correlation between risk management techniques (Y) and risk management tools (X1) controlling for experience level in years

Control variables			Y	X1	X5: experience level in years
-none-[a]	Y	Correlation	1.000	0.835	0.222
		Significance (2-tailed)		0.000	0.026
		df	0	99	99
	X1	Correlation	0.835	1.000	0.246
		Significance (2-tailed)	0.000		0.013
		df	99	0	99
	Experience level in years	Correlation	0.222	0.246	1.000
		Significance (2-tailed)	0.026	0.013	
		df	99	99	0
Experience level in years	Y	Correlation	1.000	0.826	
		Significance (2-tailed)		0.000	
		df	0	98	
	X1	Correlation	0.826	1.000	
		Significance (2-tailed)	0.000		
		df	98	0	

[a]Cells contain zero-order (Pearson) correlations.

The results in Table 13.9 show a strong, significant correlation between risk management techniques and risk management tools ($r = 0.835, p < 0.001$), indicating that increased use or proficiency in techniques is associated with greater use of supporting tools. Level of education is not significantly related to either techniques ($r = -0.097, p = 0.335$) or tools ($r = 0.009$, $p = 0.931$), suggesting that formal education does not meaningfully influence their use in this sample. Controlling for education level does not alter the association between techniques and tools, as the partial correlation remains strong and significant ($r = 0.840, p < 0.001$). This supports the interpretation that the relationship between the techniques and tools is largely independent of educational background and more likely reflects practical competence or targeted training.

After controlling for experience, education and current position, the association between risk management techniques and risk management tools remains strong and significant ($r = 0.835$, $p < 0.001$; controlled $r = 0.836, p < 0.001$), indicating a stable link between technique use and tool adoption. Experience shows weak but significant positive correlations with both techniques ($r = 0.222, p = 0.026$) and tools ($r = 0.246, p = 0.013$), suggesting a modest tendency for more experienced practitioners to use more advanced practices.

Table 13.9 Partial correlation between risk management techniques (Y) and risk management tools (X1) controlling for level of education

Correlation					
Control variables			Y	X1	X5: level of education
-none-[a]	Y	Correlation	1.000	0.835	–0.097
		Significance (2-tailed)		0.000	0.335
		df	0	99	99
	X1	Correlation	0.835	1.000	0.009
		Significance (2-tailed)	0.000		0.931
		df	99	0	99
	Level of education	Correlation	–0.097	0.009	1.000
		Significance (2-tailed)	0.335	0.931	
		df	99	99	0
Level of education	Y	Correlation	1.000	0.840	
		Significance (2-tailed)		0.000	
		df	0	98	
	X1	Correlation	0.840	1.000	
		Significance (2-tailed)	0.000		
		df	98	0	

[a]Cells contain zero-order (Pearson) correlations.

Education is not meaningfully related to either techniques ($r = -0.097$, $p = 0.335$) or tools ($r = 0.009$, $p = 0.931$), indicating limited influence of formal educational attainment in this sample. Current position is negatively associated with techniques ($r = -0.105$, $p = 0.298$) and more clearly with tools ($r = -0.258$, $p = 0.009$), implying that higher roles may be less involved in operational tool use. Political factors, stakeholder communication involvement and diverse environmental and social factors are strongly interrelated ($r = 0.593$ to 0.781, all $p < 0.001$) and remain significant under controls, highlighting the interconnected external context shaping organisational risk management. Overall, technique and tool use are closely aligned and largely independent of background characteristics, while contextual factors remain important influences.

Pearson correlation analysis was used to assess the strength and direction of linear associations between the study variables. The correlation coefficient (r) ranges from -1 to $+1$, with values closer to ± 1 indicating stronger relationships. The analysis examined four independent variables, namely risk management tools, political factors, stakeholder communication involvement and diverse environmental and social factors. As shown in Table 13.10, risk management techniques demonstrated significant positive correlations with each variable.

Table 13.10 Pearson correlation between the variables

Correlations						
		Y: Risk management techniques	X1: Risk management tools	X2: Political factors	X3: Stakeholder communication involvement	X4: Diverse environmental and social factors
Y	Pearson correlation	1	0.835[a]	0.694[a]	0.708[a]	0.781[a]
	Sig. (2-tailed)		0.000	0.000	0.000	0.000
	N	101	101	101	101	101
X1	Pearson correlation	0.835[a]	1	0.606[a]	0.749[a]	0.836[a]
	Sig. (2-tailed)	0.000		0.000	0.000	0.000
	N	101	101	101	101	101
X2	Pearson correlation	0.694[a]	0.606[a]	1	0.593[a]	0.666[a]
	Sig. (2-tailed)	0.000	0.000		0.000	0.000
	N	101	101	101	101	101
X3	Pearson correlation	0.708[a]	0.749[a]	0.593[a]	1	0.772[a]
	Sig. (2-tailed)	0.000	0.000	0.000		0.000
	N	101	101	101	101	101
X4	Pearson correlation	0.781[a]	0.836[a]	0.666[a]	0.772[a]	1
	Sig. (2-tailed)	0.000	0.000	0.000	0.000	
	N	101	101	101	101	101

[a]Correlation is significant at the 0.01 level (2-tailed).

A strong association was observed between risk management techniques and risk management tools ($r = 0.835$, $p < 0.001$), indicating that greater use or proficiency in techniques corresponds with higher tool utilisation. Risk management techniques also correlated positively with political factors ($r = 0.694$, $p < 0.001$), suggesting that more advanced approaches tend to incorporate political considerations. In addition, stakeholder communication involvement was strongly related to risk management techniques ($r = 0.708$, $p < 0.001$), reinforcing the importance of stakeholder engagement in effective risk management. Risk management techniques show a strong positive correlation with diverse environmental and social factors ($r = 0.781$, $p < 0.001$), indicating that greater use of techniques is associated with greater attention to environmental and social considerations.

Risk management tools are also significantly related to key contextual factors. Tool use correlates moderately with political factors ($r = 0.606$, $p < 0.001$) and strongly with stakeholder communication involvement ($r = 0.749$, $p < 0.001$). The strongest association is between tools and diverse environmental and social factors ($r = 0.836$, $p < 0.001$), suggesting that tool utilisation is closely aligned with managing risks across environmental and social domains.

Stakeholder communication involvement is strongly related to diverse environmental and social factors ($r = 0.772$, $p < 0.001$). Political factors show moderate positive correlations with both stakeholder communication involvement ($r = 0.593$, $p < 0.001$) and diverse environmental and social factors ($r = 0.666$, $p < 0.001$), supporting the relevance of political context to stakeholder and sustainability-related risk considerations. All reported relationships are significant at the 0.01 level, indicating a low probability that the observed correlations are due to chance, with larger coefficients reflecting stronger associations.

13.3.6 Multiple regression

Multiple regression was used to examine how well risk management tools, political factors, stakeholder communication involvement, and diverse environmental and social factors predict risk management techniques.

The correlation matrix, as shown in Table 13.11, indicates that the dependent variable Y is strongly and positively correlated with X1 and exhibits strong positive associations with X2, X3 and X4, suggesting that higher values of these predictors tend to correspond to higher values of Y. The independent variables are also positively intercorrelated, indicating that they generally covary, although not perfectly. All correlations are significant at the 0.01 level in a sample of 101, supporting the stability of these associations and providing an empirical basis for subsequent multiple regression modelling.

Multiple regression was conducted to assess the combined effects of the predictors on risk management techniques. The model summary shown in Table 13.12 indicates strong model fit and predictive performance. Model 1 yielded a multiple correlation of $R = 0.871$, indicating a strong positive relationship between the dependent variable and the predictors. The model explains a substantial proportion of variance, with $R^2 = 0.759$, indicating that the independent variables account for 75.9% of the variability in risk management techniques. The adjusted R^2 remains high at 0.749, suggesting that the explanatory power is retained after accounting for the number of predictors. The standard error of the estimate is 0.53303, reflecting a relatively

Table 13.11 Correlation analysis between the variables

Correlation						
		Y	X1	X2	X3	X4
Pearson correlation	Y	1.000	0.835	0.694	0.708	0.781
	X1	0.835	1.000	0.606	0.749	0.836
	X2	0.694	0.606	1.000	0.593	0.666
	X3	0.708	0.749	0.593	1.000	0.772
	X4	0.781	0.836	0.666	0.772	1.000
Sig. (1-tailed)	Y		0.000	0.000	0.000	0.000
	X1	0.000		0.000	0.000	0.000
	X2	0.000	0.000		0.000	0.000
	X3	0.000	0.000	0.000		0.000
	X4	0.000	0.000	0.000	0.000	
n	Y	101	101	101	101	101
	X1	101	101	101	101	101
	X2	101	101	101	101	101
	X3	101	101	101	101	101
	X4	101	101	101	101	101

Table 13.12 Model summary of multiple regression

Model summary[a]			
R	R^2	Adjusted R^2	Std. error of the estimate
0.871[b]	0.759	0.749	0.53303

[a]Dependent variable: risk management techniques.
[b]Predictors: (constant), diverse environmental social factors, political factors, stakeholder communication involvement, risk management tools.

small average deviation between observed and predicted values. Overall, these statistics support the conclusion that the model is robust and provides strong predictive validity for risk management techniques.

The analysis of variance (ANOVA) results shown in Table 13.13 indicate that the regression model provides a statistically significant prediction of risk management techniques. The regression variance exceeds the residual variance, demonstrating strong explanatory power. With four predictors, the model is significant overall ($F = 75.703, p < 0.001$), confirming that the independent variables jointly explain variation in risk management techniques.

Table 13.13 Analysis of variance

ANOVA[a]					
Model	Sum of squares	*df*	Mean square	*F*	Sig.
Regression	86.037	4	21.509	75.703	<0.001[b]
Residual	27.276	96	0.284		
Total	113.313	100			

[a]Dependent variable: risk management techniques.
[b]Predictors: (constant), diverse environmental and social factors, political, stakeholder communication involvement, risk management tools.

X1 is the strongest predictor in the regression model, as indicated in Table 13.14, showing a substantial positive effect on the dependent variable ($B = 0.588$, $\beta = 0.538$). X2 also contributes significantly, with a positive association ($B = 0.260$, $\beta = 0.254$). In contrast, X3 ($B = 0.087$, $p = 0.393$) and X4 ($B = 0.121$, $p = 0.309$) are not statistically significant predictors. Multicollinearity appears moderate, with X1 showing a variance inflation factor (VIF) of 3.664, indicating some shared variance with other predictors. Accordingly, the effects of X1 and X2 should be interpreted with awareness of overlap among the independent variables.

13.3.7 Hypothesis testing

Validation of the research hypotheses was conducted through a dual-analytical approach, utilising both Pearson correlation analysis to determine the strength of associations and multiple regression analysis to assess the predictive power of the independent variables. The regression model demonstrated robust explanatory power, with an R^2 of 0.759, indicating that the independent variables collectively account for 75.9% of the variance in risk management techniques. Table 13.15 summarises the hypothesis testing results.

Table 13.14 Multiple regression

Coefficient[a]							
Model	Unstandardised coefficients		Standardised coefficients	*t*	Sig.	Collinearity statistics	
	B	Std. error	β			Tolerance	VIF
(Constant)	–0.048	0.230		–0.208	0.835		
X1	0.588	0.105	0.538	5.600	0.000	0.273	3.664
X2	0.260	0.070	0.254	3.716	0.000	0.538	1.860
X3	0.087	0.101	0.072	0.859	0.393	0.361	2.767
X4	0.121	0.118	0.106	1.022	0.309	0.231	4.327

Table 13.15 Summary of hypothesis testing results

Hypothesis	Relationship	Correlation, r	Regression, β	p-value	Status
H1	Risk management tools → risk management techniques	0.835[a]	0.538	<0.001	Supported
H2	Political factors → risk management techniques	0.694[a]	0.254	<0.001	Partially supported
H3	Stakeholder communication involvement → risk management techniques	0.708[a]	0.072	0.393	Not supported (by correlation)
H4	Diverse environmental and social factors → risk management techniques	0.781[a]	0.106	0.309	Not supported (by correlation)

The statistical evidence presented in Table 13.15 confirms that risk management effectiveness in the Middle Eastern oil and gas sector is multidimensional, although primarily driven by technological and political intelligence.

The primary driver, technological integration (H1): the analysis identifies advanced risk management tools (X1) as the most influential determinant of success ($B = 0.588$, $\beta = 0.538$, $p < 0.001$). The strong correlation ($r = 0.835$) underscores a critical industrial culture where technical proficiency and specialised software directly enhance the ability to identify and mitigate risks. This finding suggests that organisations that prioritise digital risk monitoring systems are better equipped to navigate the operational complexities of volatile environments.

Political intelligence and contextual awareness (H2): political factors (X2) also emerged as significant predictors of risk effectiveness ($B = 0.260$, $p < 0.001$). While the relationship is positive, the status is deemed 'partially supported' because the direction and nature of political stability effects require further nuanced exploration. Nevertheless, the data confirms that in post-revolutionary landscapes, integrating political risk monitoring into routine project planning is essential for maintaining operational continuity.

The communication and social dimension (H3 and H4): interestingly, while stakeholder communication involvement (X3) and diverse environmental and social factors (X4) show strong, significant positive correlations with risk management techniques ($r = 0.708$ and $r = 0.781$, respectively), they did not reach statistical significance in the final regression model ($p = 0.393$ and $p = 0.309$, respectively). This suggests that while communication and environmental and social sustainability are vital components of the organisational framework, their impact on risk management effectiveness is likely mediated or overshadowed by the more dominant influence of technical tools and political navigation. In practice, this implies that stakeholder engagement is a necessary baseline for risk identification but must be paired with advanced analytical tools to deliver measurable efficiency gains.

Implications for professional culture: the results highlight a professional culture that values practical competence over formal education. The negligible correlation between education level and risk effectiveness ($r = -0.097$) indicates that hands-on experience and tool-specific training are far more valuable in high-stakes environments than academic credentials alone. Ultimately, the study advocates an integrated risk management framework that coordinates the technical, political and communication dimensions to safeguard assets amid persistent instability.

13.4. Practical insights

The strongest empirical result shows that advanced risk management tools are the most influential driver of effective risk management techniques. Organisations should prioritise investing in specialised software, analytical tools and digital risk monitoring systems, as these directly enhance risk identification, mitigation and adaptability in volatile environments.

Political factors are significantly associated with risk management practices. Rather than viewing political instability as an external constraint, oil and gas firms should integrate political risk monitoring into routine project planning, scenario analysis and contingency strategies, especially in conflict-prone regions.

Strong correlations between stakeholder communication involvement and risk management effectiveness indicate that structured and continuous stakeholder engagement is not optional. Companies should formalise communication channels with governments, local communities, contractors and security actors to improve early detection of risks and conflict mitigation.

Environmental and social factors are strongly related to both risk management tools and techniques. Practical risk management should therefore include environmental impact assessments, community relations strategies and compliance with international standards as core risk controls, not as peripheral sustainability activities.

Education level shows no significant relationship with risk management effectiveness, while tools and experience matter more. This suggests that organisations should emphasise hands-on training, simulations and project-based learning rather than relying solely on formal academic qualifications.

Higher organisational positions are negatively associated with direct tool usage, indicating that senior managers may be detached from operational risk practices. Firms should encourage leadership involvement in practical risk processes and ensure that decision-makers remain informed by ground-level risk data.

The regression model demonstrates that risk management effectiveness is best explained by the combined influence of tools, political awareness, stakeholder communication involvement and diverse environmental and social factors. Practically, this means silo-based risk management should be replaced with integrated frameworks that coordinate technical, political, social and environmental risk responses.

In politically unstable regions such as Libya's Oil Crescent, the findings highlight the need for flexible, adaptive risk-management systems. Organisations that proactively align tools, communication and contextual awareness are better positioned to sustain operations and protect assets under uncertainty.

In summary, the study provides a clear practical message: effective risk management in the oil and gas sector, especially under political instability, depends less on demographics or formal education and more on advanced tools, political intelligence, stakeholder collaboration and integrated socioenvironmental awareness.

13.5. Conclusion and recommendations

The investigation into risk management within the Middle Eastern oil and gas sector, specifically focusing on the post-2011 transition in the Oil Crescent, demonstrates that navigating systemic instability requires a shift from traditional administrative oversight to an integrated, technology-driven strategic framework. The empirical evidence indicates that while the region's seasoned workforce provides a stable technical foundation, operational resilience is primarily determined by the synergy of advanced analytical tools and political intelligence.

Hypothesis testing, conducted through a combination of Pearson correlation and multiple regression analysis ($R^2 = 0.759$), yielded the following refined outcomes.

- H1 was fully supported, identifying advanced risk management tools (X1) as the most influential driver of effectiveness ($B = 0.588$, $\beta = 0.538$, $p < 0.001$). This underscores a professional culture that prioritises digital monitoring and specialised software to manage complex industrial risks.
- H2 was partially supported. While political factors (X2) were confirmed as a significant predictor in the regression model ($B = 0.260$, $p < 0.001$), the nuanced nature of political volatility suggests that its impact on operational decisions requires continuous, adaptive assessment.
- H3 and H4 were supported by way of correlation analysis but not as independent predictors. Although stakeholder communication involvement (X3) and diverse environmental and social factors (X4) demonstrated strong positive correlations with risk management techniques ($r = 0.708$ and $r = 0.781$, respectively), they failed to reach statistical significance in the final multivariate regression model ($p = 0.393$ and $p = 0.309$). This suggests that while communication and sustainability are vital baseline components of industrial practice, their unique predictive influence is likely mediated by the adoption of advanced technical tools and political navigation strategies.

13.5.1 Implications and strategic considerations

The study contributes to the literature by demonstrating that in conflict-prone environments, risk management effectiveness is multidimensional and primarily driven by practical competence rather than formal academic attainment. The negligible correlation between education levels and risk effectiveness ($r = -0.097$) indicates that the industry values hands-on experience and tool-specific proficiency over traditional credentials.

For industry practitioners, it is recommended to move away from siloed risk functions toward integrated frameworks that coordinate technical, political and communication responses.

Organisations should prioritise capital investment in digital risk monitoring while formalising communication channels with local stakeholders to facilitate early conflict detection.

13.5.2 Limitations

The study has several limitations to consider when interpreting the findings. First, the research is geographically limited to Libya's Oil Crescent, which restricts the generalisability of the results to other regions or industries with different political, economic or regulatory contexts. Second, the sample demographics are heavily skewed, particularly in terms of gender, with respondents almost exclusively male, reflecting workforce realities but limiting broader representation. Third, the study relies on self-reported questionnaire data, which may be subject to response bias, personal interpretation or social desirability effects. Finally, the cross-sectional research design captures perceptions at a single point in time and does not account for changes in political conditions or organisational practices over time, limiting causal inference.

13.5.3 Recommendations and future research

The study recommends that oil and gas companies adopt integrated risk management frameworks that systematically combine technical tools, political risk assessment, stakeholder engagement and environmental and social considerations. Organisations should also enhance continuous professional training focused on applied risk management rather than relying solely on formal education.

For policy makers, the study recommends developing or refining regulatory guidelines that encourage proactive risk management, political risk preparedness and environmental and social accountability.

Future research should expand the scope beyond Libya to include comparative studies across different countries or regions, enabling stronger generalisation. Longitudinal studies are also recommended to capture how risk management practices evolve over time under changing political conditions. Additionally, future research could incorporate qualitative methods, such as interviews or case studies, to gain deeper insights into organisational decision-making processes and emergent risks that are not captured by survey-based approaches.

REFERENCES

Bjerga T and Aven T (2016) Some perspectives on risk management: a security case study from the oil and gas industry. *Proceedings of the Institution of Mechanical Engineers, Part O: Journal of Risk and Reliability* **230(5)**: 512–520.

Cooper DR and Schindler PS (2011) *Business Research Methods*, 12th edn. McGraw-Hill/Irwin, New York, NY, USA.

Darendeli IS and Hill TL (2016) Uncovering the complex relationships between political risk and MNE firm legitimacy: insights from Libya. *Journal of International Business Studies* **47(1)**: 68–92.

Dwa MMM (2025) Risk management in Libyan construction projects during times of political unrest. *Science and Technology* **13(1)**.

Fabozzi FJ, Markowitz HM and Gupta F (2008) Portfolio selection. In *Handbook of Finance*. Wiley, Hoboken, NJ, USA, vol. 2, pp. 3–13.

George C (2020) The essence of risk identification in project risk management: an overview. *International Journal of Science and Research (IJSR)* **9(2)**: 1553–1557.

Ghanem S (2023) The oil industry and the Libyan economy: the past, the present and the likely future. In *The Economic Development of Libya*. Routledge, London, UK, pp. 58–72.

Hair Jr JF, Black WC, Babin BJ and Anderson RE (2010) *Multivariate Data Analysis*, 7th edn. Prentice Hall, Upper Saddle River, NJ, USA.

Jagoda K and Wojcik P (2019) Implementation of risk management and corporate sustainability in the Canadian oil and gas industry: an evolutionary perspective. *Accounting Research Journal* **32(3)**: 381–398.

Kerzner H (2017) *Project Management: A Systems Approach to Planning, Scheduling, and Controlling*, 12th edn. Wiley, Hoboken, NJ, USA.

Pomaza-Ponomarenko A, Kryvova S, Hordieiev A, Hanzyuk A and Halunko O (2023) Innovative risk management: identification, assessment and management of risks in the context of innovative project management. *Economic Affairs* **68(4)**: 2263–2275.

Secundo G, Mele G, Passiante G and Ligorio A (2024) How machine learning changes project risk management: a structured literature review and insights for organizational innovation. *European Journal of Innovation Management* **27(8)**: 2597–2622.

Tchankova L (2002) Risk identification – basic stage in risk management. *Environmental Management and Health* **13(3)**: 290–297.

Trochim WM (2006) *Research Methods Knowledge Base: Types of Reliability*. http://www.socialresearchmethods.net/kb/reltypes.php (accessed 16/05/2026).

Mohammed Reyasudin Basir Khan
ISBN 978-1-83742-803-8
https://doi.org/10.1108/978-1-83742-802-120261017
Emerald Publishing Limited: All rights reserved

Chapter 14
Global virtual engineering teams: evaluating the role of communication, cultural intelligence and operational success factors in distributed projects

Ahmad Fairuz Mohd Amin and Mohammed Reyasudin Basir Khan

Tun Razak Graduate School, Universiti Tun Abdul Razak (UNIRAZAK), Kuala Lumpur, Malaysia

14.1. Introduction

Businesses such as multinational corporations (MNCs) have expanded globally over recent decades to capture the benefits of globalisation. With the growth of internet-enabled work, complex sectors such as construction, oil and gas, marine engineering and infrastructure have increasingly executed major engineering design projects across geographically dispersed locations. This shift has been enabled by fast communication and collaboration technologies, including email, chat platforms, video conferencing and online information management systems. Improvements in digital infrastructure and the normalisation of remote collaboration in the post-COVID-19 era have further accelerated acceptance of virtual project execution as a standard operating model.

A central motivation for globalisation is access to cost-competitive resources, especially the engineering workforce. When engineering work is executed by dispersed teams that rely primarily on digital communication, these teams are commonly referred to as global virtual engineering teams (GVETs). Beyond cost advantages, globalisation enables organisations to access specialised talent without relocating personnel, thereby reducing travel, relocation and associated administrative burdens while enhancing organisational competency without incurring proportional costs (Alami, 2016). Expertise can also be transferred across regional operations, supporting capability development in multiple business locations.

Operating across multiple regions also encourages the development of local teams, which can enhance innovation and problem-solving through diverse perspectives (Swart *et al.*, 2022). Additionally, dispersed operations improve scalability. Instead of maintaining excess overhead during slow periods, organisations can scale resources up when new projects are awarded and scale down when demand declines. Another frequently cited benefit is continuous productivity. When teams work across time zones, tasks can progress through planned handovers, enabling near-round-the-clock progress. For example, a team in Kuala Lumpur can pass work to London, which then hands it over to California, supporting a continuous workflow. Cultural diversity can further support continuity when managed strategically, such as rebalancing workload during long regional holiday periods.

Global execution can also reduce software licensing costs. When expensive engineering software licenses can be shared across locations with limited working-hour overlap, MNCs may maximise utilisation through global licensing arrangements (Yousif *et al.*, 2022). A team operating across Southeast Asia, Europe and North America can achieve effective 24-hour utilisation with fewer licenses than separate per-location licensing, reducing costs while maintaining access.

Despite these benefits, GVETs face significant challenges. Effective engineering delivery requires coordination across geographic distance, time-based barriers and perceived distance that arises from limited visibility and weaker social connections. These physical and temporal constraints interact with social and emotional factors such as trust, motivation, and conflict, which can undermine performance if not managed deliberately (Morrison-Smith and Ruiz, 2020). Against this background, the study presented in this chapter investigates how task distribution, work practice alignment, cultural intelligence, trust and time zone management influence project performance in global engineering contexts. The aim is to understand how these variables affect key performance outcomes such as timely delivery, person-hours control, budget adherence, quality of deliverables and customer satisfaction, and to provide actionable recommendations to improve collaboration and efficiency in GVETs.

Organisations increasingly depend on GVETs to execute complex engineering projects and to move away from conventional co-located delivery to achieve cost efficiency, access high-competency resources, scale flexibly, leverage diverse expertise and enable continuous productivity (Reiff and Schlegel, 2022). However, these gains are often constrained by operational challenges rooted in the complexity of distributed work environments. These challenges can reduce project performance and create inefficiencies that offset the intended advantages. Several factors are frequently associated with GVET effectiveness, including task distribution, work practice alignment, cultural diversity and cultural intelligence, trust and time zone management. The following issues illustrate the practical gaps that motivate this research.

Misaligned work practices create inconsistencies across sites and introduce bottlenecks. In many GVET settings, workflows and processes vary by region, producing fragmented execution and rework (Alami, 2016). Differences in digital tools and systems can also hinder collaboration, creating a brittle digital ecosystem that consumes resources and reduces efficiency. When work practices remain misaligned, teams spend effort reconciling differences rather than producing deliverables, weakening synergy and compromising productivity and quality.

Ineffective task distribution is another persistent challenge. While task allocation is central to distributed delivery, there is limited agreement on practical frameworks for allocating engineering tasks across locations. In practice, ad hoc assignment decisions made with incomplete information can cause frequent reassignments, delays and scope expansion (Imtiaz and Ikram, 2020). Allocation must account for expertise, task dependencies and cost considerations. Yet, these factors are often not managed systematically, leading to workload imbalance in which some teams are overloaded and others are underutilised. Poor allocation can also create role ambiguity, duplication of work or scope gaps.

Cultural diversity can be both an advantage and a risk. Diverse teams can contribute broader perspectives and innovation, but unmanaged cultural differences can lead to miscommunication, misunderstandings and reduced cohesion (Yousef, 2024). Language barriers, hidden assumptions and differing work ethics can introduce friction and increase the likelihood of conflict. When cultural intelligence is limited, teams may struggle to build mutual respect and shared norms, which can indirectly affect performance even when technical capabilities are strong.

Trust is widely viewed as essential in virtual collaboration, yet it is fragile in dispersed settings. Limited face-to-face interaction, geographic separation and varying local priorities can hinder trust formation (Sagar *et al.*, 2023). Without trust, teams may become reluctant to share information, avoid decision-making and resist empowerment, leading to silos and inefficiencies. Trust is closely linked to communication openness and accountability, both of which can weaken when teams lack intentional trust-building strategies.

Time zone differences also create operational barriers. Limited overlap in working hours slows communication cycles and extends feedback loops, thereby delaying decisions and disrupting coordination (Taras, 2025). In global collaborations, these delays can create ripple effects that negatively impact schedules and stakeholder satisfaction. Effective scheduling tools, flexible work practices and clear communication protocols become necessary to reduce time zone friction and maintain momentum.

These challenges also highlight limitations in widely used project management references. While the project management body of knowledge (PMBOK: PMI, 2025) guides project management practice, its emphasis remains oriented toward traditional, co-located assumptions and does not address the full complexity of virtual, multi-location execution. Prior research has often examined these challenges in isolation and has focused heavily on global virtual teams in IT contexts (PMI, 2025). Research specifically addressing virtual engineering teams, particularly in the post-pandemic period, remains limited. This study addresses these gaps by examining how the five variables interact with project performance in global engineering design execution.

The main objective of the research was to evaluate how task distribution, work practice alignment, cultural intelligence, trust and time zone management affect project performance indicators in global engineering design projects. Project performance was assessed based on timely delivery, control of person-hours, budget adherence, quality of deliverables and customer satisfaction. Specifically, the study evaluated how task distribution influences performance, how aligning work practices improves outcomes, how cultural intelligence affects collaboration and performance, how trust contributes to success, and how time zone differences affect productivity and project outcomes. The research questions mirror these objectives, examining the extent to which each variable influences project performance within GVETs.

Academically, this study contributes to the literature on GVET management by addressing research gaps that have largely concentrated on IT and business services rather than engineering design projects. Engineering projects involve distinct workflows, technical dependencies and quality assurance demands that may change how virtual team factors influence performance. The study also contributes by examining the combined and comparative influence of multiple variables on project performance rather than treating them as isolated determinants. By extending insight into GVET operations in the post-COVID-19 context, this work supports the development of more context-specific theoretical frameworks and research approaches for virtual engineering project management.

Practically, the study guides organisations executing engineering design across locations. Findings can help organisations design task distribution more effectively based on location strengths and constraints, align procedures and tools to reduce rework, strengthen trust and cultural intelligence through policy and leadership practices, and implement time zone coordination measures to improve efficiency. These improvements can support better project outcomes while allowing organisations to sustain and scale global virtual execution.

Several theories underpin the proposed framework. Social exchange theory explains collaboration through a cost-benefit lens, in which reciprocal exchanges and perceived fairness shape relationships. In GVETs, trust is crucial because members must rely on each other's competence and commitment with limited physical interaction. Trust develops when interactions are perceived as transparent and reliable, enabling cooperation and cohesion. This supports the idea that trust-building mechanisms such as open communication and continuous feedback can enhance performance.

Cultural dimensions theory provides a lens for understanding how cultural differences influence communication, decision-making, and collaboration (Battistella *et al.*, 2024). Differences in power distance affect expectations about authority and participation, while individualism versus collectivism shapes approaches to teamwork and conflict. Uncertainty avoidance influences preferences for structure and tolerance for change. In global engineering projects, mismatches across these dimensions can lead to misunderstandings and delays, particularly in virtual settings where nonverbal cues are limited (Presbitero, 2020). This supports the role of cultural intelligence as a capability that can reduce friction, build mutual respect and improve team functioning.

The resource-based view (RBV) emphasises competitive advantage through effective utilisation of unique resources and capabilities (Barney, 1991). In GVETs, this highlights the importance of aligning tasks with location strengths and ensuring work practice alignment to maximise resource value. Misalignment can lead to inefficiency, increased cost and schedule delays. RBV also frames cultural diversity as a potential resource that can enhance innovation when managed effectively, underscoring the need to develop cultural intelligence and leadership practices to harness it.

Contingency theory argues that no single management approach is universally optimal (Patanakul, 2023). Instead, effective practices depend on contextual factors such as socio-cultural differences, regulatory environments, technological maturity and local constraints.

Drawing on the theoretical and empirical review, a conceptual framework was proposed in which project performance is the dependent variable measured by schedule adherence, person-hours adherence, budget adherence, quality and customer satisfaction. The independent variables are task distribution, work practice alignment, cultural intelligence, trust and time zone management. The framework, as shown in Figure 14.1, proposes that effective task allocation and aligned work practices reduce rework and improve coordination; cultural intelligence supports collaboration by reducing cultural friction; trust enables openness and accountability; and effective time zone management reduces delays and improves responsiveness, all of which contribute to improved project performance.

Based on this framework, five hypotheses are developed: task distribution designed to optimise competency and availability improves project performance; aligned work practices improve performance; cultural intelligence reduces conflict and improves performance; trust across locations significantly improves performance; and effective time zone management improves morale and performance.

In summary, the literature indicates that GVETs offer strong strategic advantages but face interdependent challenges that can undermine performance. Research has often focused on general virtual teams or IT contexts, leaving gaps in engineering-specific understanding, especially regarding the combined influence of key variables and the post-pandemic operating environment. The conceptual framework and hypotheses developed in this chapter provide a structured basis for empirical testing in subsequent chapters to advance both theory and practice for managing global virtual engineering project execution.

Figure 14.1 Conceptual framework

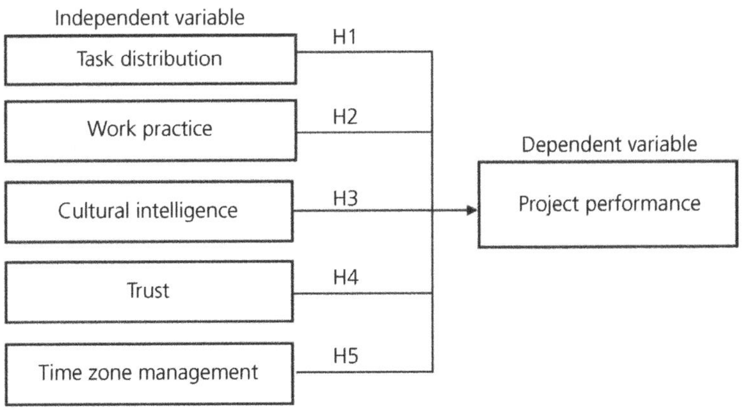

14.2. Methodology

14.2.1 Research design

This study adopts a quantitative, cross-sectional, non-experimental observational research design to examine relationships between the study variables within real-world GVET settings, without manipulating conditions. A quantitative approach is suitable because it enables the collection and analysis of measurable data and supports statistical testing of patterns, trends and associations, allowing findings to be generalised across similar populations. The cross-sectional design involves collecting data at a single point in time, providing a snapshot of the variables and enabling assessment of correlations between them within the study period. The non-experimental observational approach is appropriate because it investigates naturally occurring phenomena and preserves contextual authenticity, thereby supporting external validity for global virtual engineering project environments.

An experimental design was excluded because it requires manipulating and controlling variables, which would be impractical and inconsistent with examining real-world project conditions. A longitudinal design was also not feasible due to the study's 10-week timeframe, which limits repeated measurements over an extended period.

The statistical package for the social sciences (SPSS) was selected as the primary statistical tool because it supports comprehensive quantitative analysis, including descriptive statistics and diagnostic testing such as reliability and normality assessments. It also enables inferential procedures, including correlation and multiple regression, which are used to evaluate the strength and significance of relationships between variables and to test the study hypotheses (Field, 2024).

14.2.2 Population and sampling procedures

The study population comprises professionals actively involved in global engineering design projects, particularly within oil and gas, energy and utilities, renewables, manufacturing and infrastructure. Participants include owners, executives, senior managers, project managers and project team members, including engineers and designers. Inclusion of multiple role categories supports a broader assessment of factors influencing project performance across decision-making and delivery levels.

The estimated target population was 250 professionals. Sample size was determined using Cochran's formula, which is commonly applied in quantitative research to estimate the minimum sample required for generalisable results (Cochran, 1977). Based on a 95% confidence level and a 5% margin of error, the minimum required sample was 152 respondents, with allowance for non-response, consistent with common social science standards.

Respondents were recruited primarily through random sampling to reduce selection bias and enhance generalisability. The questionnaire was distributed by way of WhatsApp, LinkedIn and email. To expand reach within the target professional community, participants were invited to share the survey with relevant colleagues, applying a supplementary snowball approach.

The study emphasises Malaysia-based respondents to examine GVET challenges and practices involving Malaysian engineering participation, while also including respondents outside

Malaysia to capture international perspectives and strengthen comparative insight. Ethical safeguards include voluntary participation, anonymity, confidentiality and informed consent. No personally identifiable information was collected and the survey introduction clearly explained the study purpose and procedures before participation.

14.2.3 Data collection method

Data were collected through a structured online questionnaire administered using Microsoft Forms. Online surveys are appropriate for efficiently obtaining standardised responses from geographically dispersed participants and for supporting quantitative analysis. The survey link was distributed through LinkedIn, WhatsApp and email to maximise reach and participation across diverse roles and locations.

14.2.4 Operationalisation and measurement

The study applied established constructs for the independent variables, namely effective task distribution, aligned work practices, cultural intelligence, trust and time zone management, and for the dependent variable, namely project performance. Each construct was defined based on prior theory and empirical work and measured using multiple indicators adapted from validated studies. The indicators were designed to capture respondents' perceptions of how GVET practices influence project outcomes. Effective task distribution was measured through indicators assessing role clarity, alignment of tasks with expertise, avoidance of scope gaps, suitability of allocation to local resources and perceived effectiveness of task division (Alami, 2016; Imtiaz and Ikram, 2020). Aligned work practices were measured by process consistency across locations, ease of coordination, integration of workflow, standardisation of procedures and reduced rework attributable to aligned processes (de Souza Santos and Ralph, 2022; Kiely *et al.*, 2022; Nahar *et al.*, 2022). Cultural intelligence was measured by awareness of cultural differences, respect for norms, adaptability, cross-cultural communication and perceived reduction of conflict (Presbitero, 2020). Trust was measured through perceived reliability, communication transparency, openness in information sharing, mutual support and confidence in remote colleagues (Moosa *et al.*, 2023; Watanuki and Moraes, 2022). Time zone management was measured through perceived efficiency in handling time differences, structured handovers, responsiveness in asynchronous communication, scheduling flexibility and minimisation of delays due to time zone barriers (Alami, 2016; Meixner, 2018). Project performance was operationalised as timely delivery, efficient resource use (including person-hours), budget adherence, quality of deliverables, customer satisfaction and overall cost control (Alami, 2016). These operationalisations support construct validity and enable statistical testing of relationships aligned with the study objectives.

14.3. Results and analysis

14.3.1 Demographic profile

Table 14.1 consolidates the demographic profile of the 187 engineering professionals who participated in this study.

The demographic profile of the participants reveals a highly experienced and specialised workforce, which significantly strengthens the internal and external validity of the study's findings regarding GVETs.

Table 14.1 Demographic profile of respondents

Category	Demographic variable	Frequency	Percentage: %
Gender	Male	154	82
	Female	33	18
Age group	18–25	3	1.6
	26–35	25	13.4
	36–45	83	44.4
	46–55	49	26.2
	Over 55	27	14.4
Education	Doctorate	3	2
	Master's degree	59	32
	Bachelor's degree	116	62
	College diploma	7	4
	High school	2	1
Top roles	Dept. and senior managers	49	26
	Engineers and senior engineers	33	18
	Specialists and lead engineers	27	14
	Project managers	23	12
	Directors and business owners	15	8
Industry	Oil and gas	115	61
	Manufacturing and industrial	19	10
	Financial services	19	10
	Energy and utilities	17	9
Experience	>10 years	111	59
	4–6 years	32	17
	1–3 years	21	11
	7–10 years	17	9
	<1 year	6	4
Residence	Malaysia	169	90
	International (Norway, UK, India etc.)	18	10
Time zones	Two time zones	75	40
	Three time zones	58	31
	Four time zones	20	11
	Five or more time zones	34	18

A critical observation is the high level of professional maturity within the sample; approximately 85% of respondents are aged 36 or older, with the largest plurality (44.4%) falling into the 36–45 age bracket. Their extensive practical experience mirrors this maturity: 59% have worked on globally dispersed teams for more than a decade. This suggests that the data reflects 'informed views' from veterans who have navigated the evolution of virtual collaboration from early internet-enabled work to the post-pandemic 'standard operating model'. Furthermore, the sample is heavily weighted toward managerial and leadership roles (34% combined for managers, directors and owners), ensuring that the responses capture the strategic impacts of project underperformance and coordination challenges.

The sample is characterised by high educational attainment: 96% of respondents hold at least a Bachelor's degree and 34% hold postgraduate qualifications. From an academic perspective, this profile supports the study's focus on the strategic and technical aspects of project management, as higher education is positively correlated with the analytical capabilities required to manage complex variables like work practice alignment and task distribution. This aligns with the resource-based view (RBV), which views the team's human capital as a unique resource that must be effectively utilised to maintain competitive advantage.

The dominance of male respondents (82%) and the oil and gas sector (61%) reflects established industrial norms in heavy engineering and infrastructure construction. While the study includes a diverse range of sectors, including renewables and financial services, the concentration in oil and gas provides a specific lens into an industry that relies heavily on complex engineering design projects across geographically dispersed locations. This context is essential for understanding the high stakes of controlling person-hours and adhering to budgets, which are critical performance indicators in such capital-intensive fields.

Although the study centres on Malaysian engineering participation (90%), its inclusion of international respondents from Europe and the Middle East adds a necessary global comparative dimension. The operational reality of these professionals is inherently complex: 60% manage projects across at least three time zones, with 18% managing five or more. This high degree of temporal dispersion justifies the study's finding that time zone management is the strongest predictor of project performance, as these professionals face constant logistical and coordination barriers.

Finally, while cultural intelligence (CI) had a negligible direct effect on performance in the regression model, the demographics indicate that most respondents (over 90%) have had some exposure to cultural sensitivity training. However, the fact that 21% of respondents never attended training despite its availability highlights a perceived lack of importance or a potential misalignment in how CI initiatives are integrated into engineering workflows. This supports the academic argument that CI may function more as an indirect enabler – working through trust and communication – rather than a direct driver of technical project outcomes (Presbitero, 2021; Shan *et al.*, 2021).

14.3.2 Reliability analysis

The consistently high reliability scores shown in Table 14.2 indicate that the survey instrument was methodologically sound. Strong internal consistency improves the dependability

Table 14.2 Reliability analysis

Variable	Cronbach's alpha, α	Interpretation
Project performance (DV)	0.926	Very high internal consistency
Task distribution	0.879	Strong internal consistency
Work practice alignment	0.885	Strong internal consistency
Trust	0.897	Strong internal consistency
Cultural intelligence	0.943	Very high internal consistency
Time zone management	0.895	Strong internal consistency

Note: DV, dependent variable.

of the measurements, reduces random error and strengthens confidence in the study findings. The high Cronbach's alpha (α) values also support construct validity by indicating that the items adequately represent the variables being measured. As a result, the dataset provides a reliable basis for subsequent statistical analyses and for drawing credible conclusions.

14.3.3 Normality analysis

Table 14.3 shows that the skewness value (–0.536) and the kurtosis value (0.567) are within the range of –2 to + 2, indicating that the data distributions are approximately symmetrical and not excessively peaked or flat and thus normally distributed.

Histograms were generated for each variable to visually examine the distribution of the data, as shown in Figure 14.2. The resulting bell-shaped curves indicate that most observations cluster around the mean. This pattern suggests that the data approximate a normal distribution and satisfy the normality assumptions required for parametric analyses such as regression and analysis of variance (ANOVA).

Quantile plots were used to further assess data normality by comparing observed quantiles with those expected from a normal distribution. Across all variables, the data points closely followed the diagonal reference line, indicating a strong alignment between the observed distributions and the theoretical normal distribution. An example of the Q-Q plot is shown in Figure 14.3.

Table 14.3 Skewness and kurtosis

Descriptives			
		Statistic	Std. error
DV	Skewness	–0.536	0.178
	Kurtosis	0.567	0.354

Figure 14.2 Histogram of normality

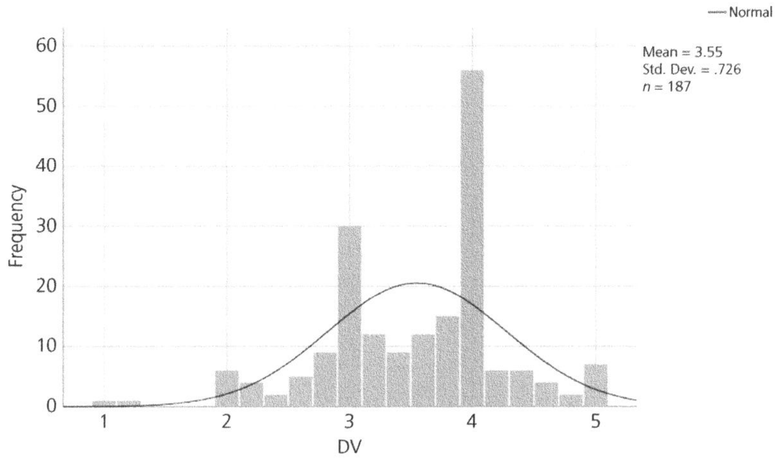

Figure 14.3 Normal Q-Q plot of dependent variable

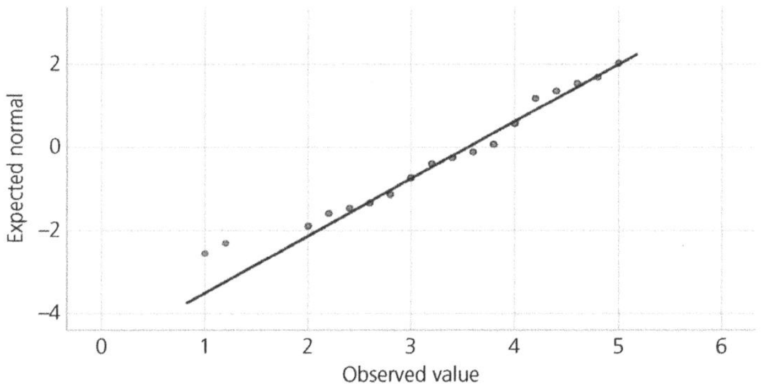

14.3.4 Descriptive analysis

Descriptive statistics summarise the central tendencies and dispersion of the key variables (project performance, task distribution, work practice alignment, cultural intelligence, trust and time zone management). As shown in Table 14.4, task distribution recorded a mean of 3.66 and a standard deviation (SD) of 0.63, indicating moderate agreement on the clarity and effectiveness of task assignment. Work practice alignment showed a mean of 3.63 (SD = 0.67), reflecting generally consistent views on workflow synchronisation across locations. Cultural intelligence had the highest mean of 3.79 (SD = 0.66), suggesting respondents perceive strong cultural awareness within their teams. Project performance averaged 3.55 (SD = 0.73), with comparatively higher variation in perceived outcomes. Trust and time zone management were reported as having means of 3.69 (SD = 0.70) and 3.64 (SD = 0.68),

Table 14.4 Descriptive statistics

	n	Minimum	Maximum	Mean	SD	Skewness		Kurtosis	
	Statistic	Statistic	Statistic	Statistic	Statistic	Statistic	Std. error	Statistic	Std. error
Project performance	187	1.00	5.00	3.5497	0.72634	−0.536	0.178	0.567	0.354
Task distribution	187	2.00	5.00	3.6588	0.63497	−0.108	0.178	0.027	0.354
Work practice alignment	187	1.60	5.00	3.6257	0.67409	−0.363	0.178	0.259	0.354
Cultural intelligence	187	2.00	5.00	3.7872	0.66028	−0.261	0.178	−0.145	0.354
Trust	187	2.00	5.00	3.6856	0.70448	−0.156	0.178	−0.275	0.354
Time zone management	187	2.00	5.00	3.6406	0.67589	−0.075	0.178	−0.233	0.354
Valid *n* (listwise)	187								

indicating moderate consensus and reinforcing their perceived importance for coordination and performance.

14.3.5 Correlation analysis

Correlation analysis in Table 14.5 showed statistically significant, positive associations between all independent variables and project performance ($p < 0.01$). Task distribution (TD) was moderately correlated with project performance (DV) ($r = 0.634$), indicating that clearer and more effective allocation is associated with better outcomes, consistent with evidence that structured allocation reduces redundancy and improves resource use (Hosseini *et al.*, 2015). Work practice alignment (WP) also showed a strong positive relationship with project performance ($r = 0.669$), supporting the view that standardised and synchronised workflows enhance consistency and efficiency (Morrison-Smith and Ruiz, 2020). Cultural intelligence (CI) demonstrated a moderate positive correlation ($r = 0.614$), suggesting that cultural adaptability is associated with improved cooperation and reduced friction in diverse teams (Presbitero, 2020). Trust (T) was strongly correlated with project performance

Table 14.5 Correlations

		Correlations					
		DV	TD	WP	CI	T	TZ
DV	Pearson correlation	1	0.634[a]	0.669[a]	0.614[a]	0.678[a]	0.693[a]
	Sig. (2-tailed)		0.000	0.000	0.000	0.000	0.000
	n	187	187	187	187	187	187
TD	Pearson correlation	0.634[a]	1	0.733[a]	0.569[a]	0.596[a]	0.535[a]
	Sig. (2-tailed)	0.000		0.000	0.000	0.000	0.000
	n	187	187	187	187	187	187
WP	Pearson correlation	0.669[a]	0.733[a]	1	0.603[a]	0.579[a]	0.584[a]
	Sig. (2-tailed)	0.000	0.000		0.000	0.000	0.000
	n	187	187	187	187	187	187
CI	Pearson correlation	0.614[a]	0.569[a]	0.603[a]	1	0.701[a]	0.587[a]
	Sig. (2-tailed)	0.000	0.000	0.000		0.000	0.000
	n	187	187	187	187	187	187
T	Pearson correlation	0.678[a]	0.596[a]	0.579[a]	0.701[a]	1	0.510[a]
	Sig. (2-tailed)	0.000	0.000	0.000	0.000		0.000
	n	187	187	187	187	187	187
TZ	Pearson correlation	0.693[a]	0.535[a]	0.584[a]	0.587[a]	0.510[a]	1
	Sig. (2-tailed)	0.000	0.000	0.000	0.000	0.000	
	n	187	187	187	187	187	187

[a]Correlation is significant at the 0.01 level (2-tailed).

($r = 0.678$), reinforcing its role in supporting transparency, information sharing, and collaboration. Time zone management (TZ) showed the strongest association with project performance ($r = 0.693$), underscoring the importance of mitigating temporal barriers to prevent workflow disruptions and delays (Meixner, 2018).

Intercorrelations between the predictors were also significant. Notably, task distribution and work practice alignment were highly correlated ($r = 0.733$, $p < 0.01$), indicating conceptual and practical overlap and suggesting that multicollinearity should be considered in subsequent regression analysis. Overall, the correlations indicate meaningful relationships but do not establish causality, which is assessed through regression modelling (Bryman, 2016; Field, 2024).

14.3.6 Multiple regression analysis

The R^2 value in Table 14.6 indicates that the five independent variables studied explain 66.3% of the variation in project performance. Nevertheless, 33.7% of the variables remain unexplained and warrant further investigation in future research.

Multiple regression results in Table 14.7 indicate that time zone management ($\beta = 0.358$, $p < 0.001$) and trust ($\beta = 0.314$, $p < 0.001$) are the strongest predictors of project performance.

Table 14.6 Model fit

R	R^2	Adjusted R^2	Std. error of the estimate
0.814[a]	0.663	0.654	0.42718

[a]Predictors: (constant), TZ, T, TD, CI, WP.

Table 14.7 Regression coefficients (standardised and unstandardised)

Coefficients[a]					
	Unstandardised coefficients		Standardised coefficient	t	Sig.
	B	Std. error	β		
(Constant)	−0.283	0.213		−1.329	0.186
TD	0.128	0.077	0.112	1.660	0.099
WP	0.209	0.075	0.194	2.795	0.006
CI	0.004	0.074	0.004	0.059	0.953
T	0.324	0.067	0.314	4.861	0.000
TZ	0.384	0.062	0.358	6.219	0.000

[a]Dependent variable: DV.

Work practice alignment also shows a significant positive effect ($\beta = 0.194$, $p = 0.006$). Task distribution has a positive but weaker effect ($\beta = 0.112$) and is not statistically significant at conventional thresholds ($p = 0.099$). Cultural intelligence shows a negligible and non-significant relationship with project performance ($\beta = 0.004$, $p = 0.953$).

The weaker effects for task distribution and cultural intelligence suggest the possibility of indirect pathways or interaction effects that the current model does not capture. Further research may examine potential mediators and employ designs that can assess change over time. Overall, the model clarifies the relative contribution of key GVET factors and highlights priority levers for improving project outcomes in distributed engineering environments.

14.3.7 Hypothesis testing

Correlation results indicate significant positive associations between all independent variables and project performance, supporting the hypothesised directions as shown in Table 14.8. Regression analysis was then used to determine which predictors retain a significant effect when modelled jointly and to assess hypothesis support.

H1 proposed that task distribution has a positive effect on project performance. Task distribution showed a positive but weak effect ($\beta = 0.112$). H1 is considered partially supported because, although it did not meet the strict 0.05 significance threshold in the multiple regression model ($p = 0.099$), it demonstrated a strong, statistically significant positive correlation with project performance in the bivariate analysis ($r = 0.634$, $p < 0.01$). The regression results still indicated a positive effect ($\beta = 0.128$), suggesting that task distribution contributes to project success. However, its impact may be secondary to stronger predictors, such as time zone management and trust. Furthermore, the high correlation between task distribution and work practice alignment ($r = 0.733$) suggests conceptual overlap, indicating that the influence of task allocation likely depends on integration with aligned workflows and governance mechanisms rather than acting as an isolated driver. H2 was supported, as work practice alignment significantly predicted project performance ($\beta = 0.194$, $p = 0.006$), reinforcing the value of standardised and integrated workflows across locations. H3 was not supported because cultural intelligence showed no meaningful direct effect on project performance ($\beta = 0.004$, $p = 0.953$), implying that its contribution may be indirect or contingent on mediators such as communication or trust. H4 was supported, with trust emerging as a strong predictor ($\beta = 0.314$, $p < 0.001$), highlighting its role in enabling collaboration, transparency and accountability. H5 was also supported, as time zone management was the strongest predictor of project performance ($\beta = 0.358$, $p < 0.001$), emphasising the importance of mitigating temporal barriers through effective scheduling and handover practices.

Overall, H2, H4 and H5 were supported, H1 was partially supported and H3 was not supported. The findings position work practice alignment, trust and time zone management as priority levers for improving GVET project outcomes, while task distribution and cultural intelligence warrant further investigation for indirect or interaction effects.

Table 14.8 Hypothesis testing

Hypothesis	Pearson correlation, *r*	Correlation significance	Standardised coefficient, *β*	Regression significance	Result
H1: task distribution designed to optimise competency, resource availability and cultural considerations will enhance project performance	0.634	0.000	0.112	0.099	Partially supported
H2: aligned work practices across global locations will result in improved project performance	0.669	0.000	0.194	0.006	Supported
H3: teams with enhanced cultural intelligence will experience reduced conflicts and higher project performance	0.614	0.000	0.004	0.953	Not supported
H4: trust among team members across locations significantly influences project performance	0.678	0.000	0.314	0.000	Supported
H5: effective time zone management will improve team morale and enhance project performance	0.693	0.000	0.358	0.000	Supported

14.4. Practical insights

Time zone management is the strongest predictor of project performance ($\beta = 0.358$, $p < 0.001$), indicating it should be treated as a strategic priority rather than a logistical afterthought. Organisations must move beyond ad hoc scheduling and invest in structured handover protocols, disciplined use of asynchronous tools and mandatory overlap windows for critical coordination. Failing to manage these temporal barriers directly translates into extended feedback loops, increased rework and dissatisfied stakeholders.

Trust serves as the fundamental engine of collaboration, ranking second among the most powerful predictors of success ($\beta = 0.314$, $p < 0.001$). Because trust is fragile in dispersed settings with limited face-to-face interaction, managers must be intentional in fostering transparency and interpersonal reliability. This requires a shift from micromanagement toward outcome-based performance management, predictable response behaviours and clear role clarity to ensure team members feel empowered and accountable.

Aligning work practices across global sites significantly improves efficiency by reducing the coordination friction that leads to errors and fragmented execution. However, the study suggests that alignment should not be rigid or headquarters-driven. Instead, organisations should define global minimum standards for workflows and digital tools while permitting enough contextual flexibility for local teams to adapt processes to their specific environments.

While task distribution is a central part of project planning, its direct impact on performance is statistically weaker ($\beta = 0.112$, $p = 0.099$) when not supported by other factors. This indicates that simply assigning the right task to the right location is insufficient; allocation decisions must be made jointly with workflow design and trust-building mechanisms. Effective distribution relies on an integrated approach where role clarity and dependencies are managed through aligned processes rather than as isolated planning exercises.

Cultural intelligence functions primarily as an indirect enabler rather than a direct driver of technical project outcomes. While cultural intelligence training remains highly valuable, its benefits are best realised when integrated into leadership development and collaboration practices rather than standalone programmes. By enhancing cultural adaptability, teams can more effectively navigate miscommunications and build the cohesion necessary to strengthen trust and communication openness.

Current project management standards, such as PMBOK, often remain oriented toward traditional, co-located assumptions and do not fully address the complexities of virtual execution. Organisations must evolve their governance frameworks to include GVET-specific practices that explicitly focus on temporal discipline, cross-location process integration and virtual trust-building. Ultimately, project success in this environment is driven less by technical allocation decisions and more by the quality of coordination and temporal discipline across the global network.

14.5. Conclusion and recommendations

This study provides a comprehensive evaluation of the factors influencing the performance of global virtual engineering teams (GVETs), supported by a robust dataset from 187 engineering professionals. The credibility of the findings is anchored in the high professional maturity of the sample: 85% of respondents are aged 36 or older and 59% have more than 10 years of experience in dispersed team environments. Furthermore, the sample's high educational attainment – 96% holding at least a Bachelor's degree – ensures that the insights reflect a technically sophisticated and strategically aware workforce.

The statistical analysis confirms that the proposed conceptual framework has strong explanatory power, with the five independent variables collectively predicting 66.3% of the variance in project performance ($R^2 = 0.663$). While correlation analysis showed that all variables were positively associated with success ($p < 0.01$), the multiple regression model revealed a clear hierarchy of influence.

- *Time zone management*: this emerged as the strongest predictor of project success ($\beta = 0.358, p < 0.001$), highlighting that managing temporal barriers is the most critical operational lever.
- *Trust*: following closely, trust was a powerful predictor ($\beta = 0.314, p < 0.001$), reinforcing its role as the foundation for collaboration and transparency in virtual settings.
- *Work practice alignment*: this factor also showed a significant positive effect ($\beta = 0.194$, $p = 0.006$), confirming that standardised workflows are essential for reducing rework and inefficiencies.
- *Task distribution*: while positively correlated, it demonstrated limited predictive strength and was not statistically significant in the joint model ($\beta = 0.112, p = 0.099$).
- *Cultural intelligence*: contrary to theoretical expectations, cultural intelligence showed a negligible direct effect on performance ($\beta = 0.004, p = 0.953$).

14.5.1 Theoretical and practical implications

The findings align with several key theories. The critical role of trust supports social exchange theory, which posits that reciprocal transparency and reliability are vital when physical interaction is limited. The impact of task distribution and work practice alignment aligns with the resource-based view and contingency theory, emphasising the need to match specialised human capital with structured yet adaptable processes.

Practically, the study suggests that project performance is driven less by technical allocation and more by coordination quality and temporal discipline. Organisations should treat time zone coordination as a strategic delivery priority rather than a logistical afterthought, implementing structured handover protocols and overlap windows for critical tasks. Additionally, trust-building must be intentional, moving away from micromanagement toward outcome-based performance measures.

14.5.2 Limitations and future research

The study is primarily limited by its geographical concentration in Malaysia (90%) and its cross-sectional design, which captures a single point in time and precludes definitive conclusions about long-term causality. Future research should adopt longitudinal designs to observe how trust and cultural intelligence develop across project lifecycles. Furthermore, given that cultural intelligence and task distribution showed limited direct effects, future studies should investigate their roles as mediators or moderators to uncover the indirect pathways through which they influence project outcomes.

REFERENCES

Alami A (2016) Global project management challenges. *PM World Journal* **5(1)**.

Barney J (1991) Firm resources and sustained competitive advantage. *Journal of Management* **17(1)**: 99–120.

Battistella C, Bortolotti T, Boscari S, Nonino F and Palombi G (2024) The impact of cultural dimensions on project management performance. *International Journal of Organizational Analysis* **32(1)**: 108–130.

Bryman A (2016) *Social Research Methods*, 5th edn. Oxford University Press, Oxford, UK.

Cochran WG (1977) *Sampling Techniques*, 3rd edn. Wiley, New York, NY, USA.

de Souza Santos RE and Ralph P (2022) A grounded theory of coordination in remote-first and hybrid software teams. In *Proceedings of the 44th International Conference on Software Engineering (ICSE), Pittsburgh, PA, USA*, pp. 25–35.

Field A (2024) *Discovering Statistics using IBM SPSS Statistics*, 6th edn. Sage, London, UK.

Hosseini MR, Chileshe N, Zuo J and Baroudi B (2015) Adopting global virtual engineering teams in AEC projects: a qualitative meta-analysis of innovation diffusion studies. *Construction Innovation* **15(2)**: 151–179.

Imtiaz S and Ikram N (2020) Framework for task allocation in global software development. *IEEE Access* **8**: 206235–206247.

Kiely G, Butler T and Finnegan P (2022) Global virtual teams coordination mechanisms: building theory from research in software development. *Behaviour & Information Technology* **41(9)**: 1952–1972.

Meixner TR (2018) *The Role of Trust in Building Effective Virtual Teams: A Mixed Methods Study in a Large Public Sector Organization*. Doctoral dissertation, Franklin University, Columbus, OH, USA.

Moosa L, Pearson H and Mthombeni M (2023) Invoking team trust to facilitate performance management in the context of virtual teams. *South African Journal of Business Management* **54(1)**: 1–12.

Morrison-Smith S and Ruiz J (2020) Challenges and barriers in virtual teams: a literature review. *SN Applied Sciences* **2(6)**: 1096.

Nahar N, Zhou S, Lewis G and Kästner C (2022) Collaboration challenges in building ML-enabled systems: communication, documentation, engineering, and process. In *Proceedings of the 44th International Conference on Software Engineering (ICSE), Pittsburgh, PA, USA*, pp. 413–425.

Patanakul P (2023) Contingency theory and its applications to complex project organizing. In *Research Handbook on Complex Project Organizing* (Winch G, Brunet M and Cao D (eds)). Edward Elgar, Cheltenham, UK, pp. 60–69.

PMI (Project Management Institute) (2025) *A Guide to the Project Management Body of Knowledge (PMBOK Guide)*, 8th edn. PMI, Newtown Square, PA, USA.

Presbitero A (2020) Task performance in global virtual team: examining the roles of perceived cultural dissimilarity and cultural intelligence of member and leader. *Personnel Review* **49(5)**: 1091–1105.

Presbitero A (2021) Communication accommodation within global virtual team: the influence of cultural intelligence and the impact on interpersonal process effectiveness. *Journal of International Management* **27(1)**: 100809.

Reiff J and Schlegel D (2022) Hybrid project management – a systematic literature review. *International Journal of Information Systems and Project Management* **10(2)**: 45–63.

Sagar SK, Oladinrin OT, Arif M and Rana MQ (2023) Interpretive structural model of trust factors in construction virtual project teams. *Construction Innovation* **23(1)**: 248–264.

Shan J, Konishi M, Pullin P and Lupina-Wegener A (2021) Effects of cultural intelligence on multicultural team effectiveness: the chain mediation role of common ingroup identity and communication quality. *Journal of Theoretical Social Psychology* **5(4)**: 519–529.

Swart K, Bond-Barnard T and Chugh R (2022) Challenges and critical success factors of digital communication, collaboration and knowledge sharing in project management virtual teams: a review. *International Journal of Information Systems and Project Management* **10(4)**: 84–103.

Taras V (2025) The barriers to global virtual team effectiveness: expectations vs. reality. *Organizational Dynamics* **54(3)**: 101144.

Watanuki HM and Moraes RDO (2022) Trust in virtual teams: a systematic review of information systems literature. *Journal of Information Systems and Technology Management* **19**: e202219006.

Yousef K (2024) Exploring the impact of cultural diversity in global projects: a comparative analysis of virtual and face-to-face teamwork. *International Journal of Cross Cultural Management* **24(2)**: 411–430.

Yousif HA, Gheni AY, Jusoh YY, Ilyana N and Shanmugam M (2022) Global virtual teams management system: project manager guide. *Bulletin of Electrical Engineering and Informatics* **11(3)**: 1642–1649.

Conclusion

Mohammed Reyasudin Basir Khan
ISBN 978-1-83742-803-8
https://doi.org/10.1108/978-1-83742-802-120261018
Emerald Publishing Limited: All rights reserved

Chapter 15
Institutionalising resilient and sustainable project ecosystems: leadership, teams, learning and governance in an era of disruption

Mohammed Reyasudin Basir Khan

Tun Razak Graduate School, Universiti Tun Abdul Razak (UNIRAZAK), Kuala Lumpur, Malaysia

15.1. From capability awareness to institutional transformation

Project environments are entering a phase characterised by structural volatility. Global supply chains remain vulnerable to geopolitical fragmentation (Sovacool and Ryu, 2025). Digital transformation is accelerating decision cycles while simultaneously introducing algorithmic risk (Dobrucali *et al.*, 2024). Sustainability expectations are expanding beyond compliance into active stakeholder scrutiny (Silvius *et al.*, 2017). These forces alter not only operational complexity but the risk architectures of projects themselves.

Traditional project management frameworks remain largely execution-oriented (PMI, 2025). They emphasise planning discipline, control systems and performance metrics. While these remain important, they do not adequately address systemic resilience. Resilience requires the structural embedding of adaptive capacity, ethical oversight, and learning and feedback loops (Lengnick-Hall *et al.*, 2011).

Institutional transformation, therefore, becomes necessary across four dimensions.

- Leadership must evolve from situational adaptation to governance-embedded adaptability.
- Team structures must support distributed intelligence across hybrid and artificial intelligence (AI)-enabled environments.
- Learning systems must function as long-term resilience infrastructure rather than episodic reflection.
- Communication must operate as legitimacy architecture under environmental, social and governance (ESG) accountability.

15.2. Institutionalising adaptive leadership

15.2.1 Beyond individual competence: leadership as governance design

Adaptive leadership has been widely discussed as a behavioural orientation (Uhl-Bien and Arena, 2018). However, in complex project ecosystems, adaptability must be institutionalised rather than personalised. Reliance on individual leader capability introduces fragility. When leadership transitions occur, adaptive capacity may erode unless embedded structurally.

Project governance literature increasingly recognises that leadership is intertwined with oversight systems (Müller and Lecoeuvre, 2014; Too and Weaver, 2014). Governance mechanisms shape decision rights, accountability pathways and stakeholder engagement protocols. Institutionalising adaptive leadership, therefore, requires redesigning governance architectures.

Three mechanisms are particularly critical.

- *Distributed decision authority frameworks*: projects operating under uncertainty benefit from decentralised decision-making rights that enable rapid responses at operational levels (Joslin and Müller, 2015). Centralised bottlenecks delay adaptation and amplify risk.
- *Strategic sensing routines*: dynamic capabilities theory emphasises the importance of sensing environmental shifts (Teece, 2007). Project governance boards must incorporate systematic environmental scanning mechanisms, including regulatory monitoring, stakeholder sentiment analysis and technological trend assessment.
- *Ethical oversight integration*: sustainability governance increasingly demands ethical scrutiny of project impacts (Martens and Carvalho, 2016). Leadership accountability must extend beyond cost and schedule performance toward social and environmental stewardship.

Institutionalising adaptive leadership transforms it from a personal skill to a structural expectation.

15.2.2 Leadership in AI-enabled project environments

Artificial intelligence and predictive analytics are reshaping project scheduling, risk modelling and resource allocation (Richey *et al.*, 2023). While digital tools enhance forecasting precision, they introduce new ethical and governance challenges. Algorithmic decision systems may embed bias, obscure accountability or prioritise efficiency over social considerations. Leadership must therefore integrate algorithmic oversight capacity.

Emerging scholarship highlights the importance of 'human-in-the-loop' governance in AI systems (Ransbotham *et al.*, 2017). Project leaders must

- validate algorithmic recommendations against contextual judgement
- ensure transparency in data sources and modelling assumptions
- establish escalation protocols for algorithmic anomalies
- integrate ESG parameters into optimisation models.

Adaptive leadership in AI contexts becomes a form of sociotechnical governance rather than merely interpersonal coordination.

15.2.3 Leadership and sustainability institutionalisation

Sustainability integration in projects has often been treated as an additional constraint (Silvius *et al.*, 2012). However, resilience literature suggests that sustainability enhances long-term stability by reducing environmental and reputational risk (Bansal and DesJardine, 2014).

Institutionalising sustainability-oriented leadership requires

- embedding ESG metrics within project performance dashboards
- linking executive incentives to sustainability indicators
- incorporating stakeholder impact assessments into stage-gate reviews
- aligning project portfolios with organisational sustainability strategy.

Leadership thus evolves into governance stewardship.

15.2.4 Leadership development pipelines for resilience

Organisations frequently promote project managers based on technical competence. However, resilience-oriented leadership requires capabilities in interpreting complexity, ethical reasoning and stakeholder negotiation (Hansen and Svejvig, 2022).

Development pipelines should therefore include

- scenario-based resilience simulations
- cross-cultural immersion experiences
- sustainability governance education
- data ethics and AI literacy training.

Leadership institutionalisation ensures continuity of adaptive capacity across project generations.

15.3. Resilient team architectures in hybrid and AI-integrated environments

If leadership provides integrative direction, teams provide executional intelligence. Yet contemporary teams are undergoing structural transformation.

15.3.1 From co-located teams to hybrid ecosystems

Digital platforms have normalised hybrid and geographically dispersed project teams (Marnewick and Marnewick, 2020). While flexibility increases access to talent, dispersion introduces coordination complexity.

Hybrid team resilience depends on

- explicit communication norms
- defined digital collaboration protocols

- temporal coordination across time zones
- psychological safety across virtual interfaces.

Research indicates that distributed teams are vulnerable to a decline in trust when informal relational interactions diminish (Gibson and Gibbs, 2006). Organisations must therefore deliberately design relational infrastructure rather than assume cohesion will emerge organically.

15.3.2 Cognitive diversity as risk mitigation

Diverse teams generate broader problem-solving repertoires (Edmondson, 2018). In volatile environments, cognitive diversity enhances adaptability. However, diversity without inclusion amplifies conflict.

Resilient team architectures therefore require

- structured deliberation mechanisms
- rotating facilitation roles
- transparent conflict resolution pathways
- inclusive leadership norms.

Diversity becomes strategic risk mitigation rather than symbolic compliance.

15.3.3 Human–AI collaboration models

Increasingly, project teams collaborate not only with human peers but with digital systems. AI-driven forecasting tools, automated reporting dashboards and predictive maintenance systems are becoming embedded within project workflows (Dobrucali *et al.*, 2024).

Human–AI collaboration requires

- role clarity between algorithmic recommendations and human discretion
- bias auditing procedures
- cross-functional data literacy
- transparent communication regarding model limitations.

Teams must develop ‘algorithmic trust calibration’ – neither blind acceptance nor reflexive rejection of digital outputs.

15.3.4 Psychological safety in digitally mediated environments

Psychological safety remains central to team performance (Edmondson, 2018). However, digital mediation alters interaction patterns. Misinterpretation risk increases in text-based communication. Delayed feedback may inhibit clarification.

Maintaining safety in hybrid teams requires

- regular synchronous engagement sessions
- norms encouraging clarification requests
- transparent escalation channels
- leadership modelling vulnerability.

Resilient teams do not eliminate conflict; they normalise structured dialogue in the face of uncertainty.

15.3.5 Team resilience as organisational asset

From a resource-based perspective, cohesive and adaptive teams constitute socially complex assets difficult for competitors to replicate (Barney, 1991). Trust density, shared cognition and collaborative norms accumulate through experience. Organisations should therefore treat team resilience as a strategic investment rather than an incidental by-product of project execution.

This requires

- longitudinal team capability tracking
- cross-project team continuity planning
- institutionalised reflection mechanisms
- incentive systems rewarding collaboration rather than individual heroics.

Team architecture thus becomes a structural dimension of organisational resilience.

15.4. Learning as strategic survival infrastructure

If leadership provides direction and teams provide distributed intelligence, learning provides continuity. In volatile environments, organisational survival depends less on initial planning precision and more on the capacity to absorb disruption, interpret signals and recalibrate behaviour. Learning, therefore, must transition from episodic post-project review to institutionalised resilience infrastructure.

15.4.1 From lessons learned to learning systems

Traditional project management methodologies encourage 'lessons learned' documentation at project closure (PMI, 2025). However, empirical studies suggest that lesson repositories are frequently underutilised, disconnected from decision-making processes and poorly integrated into future project planning (Love *et al.*, 2016; Williams, 2003). Documentation without institutional embedding produces archival accumulation rather than adaptive capability.

Organisational learning theory distinguishes between corrective and transformative learning (Argyris and Schön, 1996). In project environments, corrective learning (single-loop) improves task execution within existing frameworks. Transformative learning (double- and triple-loop) challenges underlying assumptions, risk models and governance structures.

Resilience requires progression beyond single-loop correction. Projects operating in digitally accelerated and sustainability-driven environments must routinely question foundational assumptions.

- Are risk assessment models incorporating climate-related uncertainties?
- Do procurement frameworks reflect supply chain geopolitical volatility?
- Are stakeholder engagement practices aligned with emerging ESG expectations?

Embedding structured reflection into stage-gate governance processes strengthens institutional memory and reduces repetition of systemic errors (Sense, 2011).

15.4.2 Learning under conditions of acceleration

Digital transformation compresses project cycles and shortens innovation timelines. While acceleration enhances responsiveness, it also increases the risk of institutional forgetting. When projects conclude rapidly and teams disperse, experiential knowledge dissipates. Recent scholarship on project-based organisations highlights the risk of 'knowledge evaporation' in temporary structures (Brady and Davies, 2004; Lindner and Wald, 2011). The mobility of skilled professionals across organisations further compounds memory erosion.

Institutionalising learning requires structural mechanisms such as

- cross-project knowledge integration forums
- AI-assisted knowledge mining from project documentation
- structured after-action reviews with executive participation
- digital knowledge repositories linked to decision dashboards.

Artificial intelligence offers opportunities for pattern recognition across historical project data (Richey *et al.*, 2023). However, automated analytics must be complemented by human interpretation to avoid over-reliance on quantitative proxies. Learning thus becomes a socio-technical infrastructure rather than an archival procedure.

15.4.3 Resilience, redundancy and adaptive capacity

The resilience literature emphasises redundancy and flexibility as buffers against disruption (Duchek, 2020; Lengnick-Hall *et al.*, 2011). Learning contributes to resilience by enabling rapid recalibration in the face of uncertainty.

Adaptive capacity depends on

- institutional memory depth
- error-reporting openness
- cross-functional knowledge flow
- strategic experimentation tolerance.

Projects that penalise disclosure or prioritise short-term performance metrics over reflection discourage learning behaviour. Conversely, governance systems that reward knowledge sharing and constructive error identification strengthen long-term resilience.

Organisational learning also intersects with sustainability integration. ESG expectations evolve continuously. Firms that embed sustainability learning routines, including stakeholder feedback loops and impact assessments, adapt more effectively to regulatory and reputational shifts (Bansal and DesJardine, 2014). Learning, therefore, operates not merely as performance enhancement but as survival infrastructure.

15.5. Communication and legitimacy in an era of ESG accountability

If learning strengthens internal resilience, communication sustains external legitimacy. In contemporary project ecosystems, legitimacy is increasingly fragile. Stakeholder activism, digital transparency and regulatory scrutiny amplify reputational risk.

15.5.1 Communication as governance architecture

Project governance literature recognises communication as central to oversight and accountability (Too and Weaver, 2014). However, communication must now be reconceptualised as institutional architecture rather than interpersonal exchange.

Governance legitimacy depends on

- transparent reporting structures
- consistent stakeholder engagement
- escalation clarity
- traceable decision rationales.

Digital platforms allow real-time dissemination of information. Yet they also expose inconsistencies. Misalignment between formal reports and operational realities can erode trust rapidly. Projects operating within sustainability-intensive sectors such as energy, infrastructure and construction face heightened scrutiny (Sovacool and Ryu, 2025). Communication strategies must integrate ESG disclosures, risk transparency and stakeholder impact reporting into core dashboards.

15.5.2 Stakeholder activism and social amplification

The proliferation of social media and digital advocacy networks increases the speed at which project controversies escalate. Reputational crises may emerge from perceived ethical lapses, environmental harm or labour disputes. Institutional legitimacy research emphasises that organisations must align actions with evolving societal norms (Suchman, 1995). In project contexts, legitimacy is maintained through proactive stakeholder dialogue rather than reactive crisis management.

Structured stakeholder engagement mechanisms include

- community consultation forums
- public impact reporting
- transparent grievance mechanisms
- independent oversight committees.

Communication thus functions as a reputational risk management infrastructure.

15.5.3 Cultural intelligence and cross-border governance

Global project environments involve cross-cultural coordination and regulatory diversity. Cultural intelligence enhances negotiation effectiveness, perceived authority, sensitivity and conflict management (Earley and Ang, 2003). However, cultural complexity now intersects with digital communication. Remote interactions reduce contextual cues, increasing the risk of misinterpretation.

Institutionalising cultural intelligence requires

- cross-cultural competence training
- inclusion metrics within governance dashboards
- culturally adaptive communication templates
- multilingual stakeholder reporting frameworks.

Failure to integrate cultural sensitivity into communication structures amplifies coordination risk and reputational vulnerability.

15.5.4 Transparency, data ethics and algorithmic accountability

As AI tools increasingly inform project decisions, transparency must extend to the algorithms themselves. Stakeholders may question how predictive models prioritise resource allocation or risk classification.

Emerging governance discourse emphasises algorithmic accountability and explainability (Ransbotham *et al.*, 2017). Project organisations must ensure

- documented model assumptions
- audit trails for automated decisions
- human override mechanisms
- ethical data governance policies.

Communication architecture, therefore, expands to include digital accountability.

15.6. Sustainability as structural project architecture

Sustainability is no longer peripheral to project strategy. It functions as a structural architecture, influencing investment decisions, stakeholder engagement and long-term viability.

15.6.1 From compliance to strategic alignment

Earlier project sustainability research emphasised compliance with environmental standards (Silvius *et al.*, 2012). More recent scholarship highlights the strategic integration of sustainability, linking it to competitive advantage (Martens and Carvalho, 2016).

Institutionalising sustainability requires

- portfolio-level sustainability mapping
- integration of UN Sustainable Development Goals (SDG) indicators into project metrics
- lifecycle environmental impact analysis
- cross-sector partnership frameworks.

SDG alignment strengthens legitimacy and positions projects within broader societal value creation frameworks.

15.6.2 Sustainability and resilience interdependence

Climate-related risks, resource scarcity and regulatory evolution introduce systemic vulnerabilities. Sustainable design enhances resilience by reducing exposure to environmental disruption. For example:

- Energy-efficient infrastructure reduces long-term operational risk.
- Localised supply chains mitigate geopolitical disruption.
- Circular economy integration reduces material volatility.

Thus, sustainability is not merely an ethical obligation but a resilience strategy.

15.7. Integrative transition toward future-oriented governance

Leadership, teams, learning and communication do not function as isolated organisational variables. They intersect within a broader sustainability architecture that shapes how projects are governed, interpreted and legitimised. The transition toward future-oriented governance, therefore, requires understanding these domains as interdependent components of a resilient project ecosystem.

Institutionalised adaptive leadership anchors this system. Leadership defines ethical boundaries, distributes decision authority and embeds accountability within governance structures. Without structural reinforcement, adaptability remains individual rather than organisational.

Team architecture operationalises governance. In hybrid and digitally mediated environments, resilience depends on deliberate design of collaboration routines, psychological safety and role clarity. Distributed expertise and human–AI interaction require coordination mechanisms that prevent fragmentation.

Learning infrastructure provides continuity. Institutionalised reflection and knowledge integration transform disruption into capability development. Learning ensures that projects accumulate strategic memory rather than replicate structural weaknesses.

Communication architecture sustains legitimacy. Transparent reporting, culturally intelligent engagement and ESG-aligned disclosure mechanisms regulate feedback loops across stakeholders. In contexts of digital transparency and activism, governance credibility depends on coherent communication systems.

These four domains converge within sustainability-oriented governance. Sustainability reframes project success beyond short-term performance, embedding environmental and social accountability into strategic decision-making. Resilience, therefore, emerges not from isolated optimisation but from systemic alignment across leadership, team capability, learning processes and communication structures.

Future-oriented governance is thus anticipatory rather than reactive, integrative rather than fragmented and ethically grounded rather than narrowly procedural. It represents a structural recalibration of project management toward long-term adaptive stability.

15.8. Digital futures and AI governance in project ecosystems

Digital transformation is not simply enhancing project tools; it is reshaping governance logic. Artificial intelligence, predictive analytics, digital twins and cloud-based collaboration platforms are increasingly embedded in planning, scheduling and monitoring systems (Dobrucali *et al.*, 2024; Richey *et al.*, 2023). However, technological sophistication alone does not guarantee resilience. The institutional challenge lies in governing digital integration responsibly.

15.8.1 From digital efficiency to digital governance

Early digital transformation initiatives in project management focused primarily on efficiency gains such as automating reporting, improving forecasting accuracy and reducing coordination delays (Kerzner, 2025). While efficiency remains valuable, digital integration now intersects with ethical oversight, cybersecurity risk and algorithmic transparency.

Three governance challenges are particularly salient.

- *Algorithmic opacity*: predictive scheduling and risk modelling systems may operate as 'black boxes', limiting interpretability (Ransbotham *et al.*, 2017). Without explainability, accountability becomes unclear.
- *Data governance and privacy*: large-scale data integration across stakeholders raises confidentiality and regulatory compliance concerns.
- *Automation bias*: overreliance on algorithmic outputs may suppress human judgement, particularly under time pressure.

Project governance boards must therefore incorporate digital literacy into oversight structures. Leadership accountability now includes validating digital systems and ensuring alignment with ethical standards.

15.8.2 AI–human decision integration

AI tools can enhance forecasting precision but they cannot replace contextual reasoning. Research in sociotechnical systems emphasises that optimal performance arises from complementary human–machine collaboration rather than substitution (Jarrahi, 2018).

In project environments, AI may assist in

- predicting schedule slippage patterns
- detecting anomaly signals in cost variance
- identifying supply chain vulnerabilities
- simulating environmental impact scenarios.

However, final decisions must integrate

- stakeholder sensitivity
- ethical trade-offs
- cultural implications
- long-term strategic positioning.

Institutionalising 'human-in-the-loop' governance ensures that digital acceleration does not undermine legitimacy or resilience.

15.8.3 Cybersecurity and systemic vulnerability

Digital interdependence increases exposure to cyber threats. Infrastructure projects, particularly those involving critical utilities, are vulnerable to cyber disruption (Sovacool and Ryu, 2025). Cybersecurity must therefore be integrated into risk governance frameworks rather than treated as an IT function.

Resilient digital governance requires

- redundant data storage systems
- cyber risk simulation exercises
- incident response protocols embedded in project planning
- executive oversight of digital risk exposure.

Digital transformation thus introduces both opportunity and systemic vulnerability. Institutional maturity determines whether digitalisation strengthens or destabilises resilience.

15.9. Global collaboration and geopolitical volatility

Projects increasingly operate across borders, regulatory regimes and cultural contexts. While global collaboration expands the innovation potential, it amplifies institutional complexity.

15.9.1 Geopolitical fragmentation and supply chain risk

Recent disruptions such as pandemics, trade tensions and regional conflicts have exposed the fragility of globally dispersed supply chains (Sovacool and Ryu, 2025). Project resilience now requires supply diversification, local sourcing strategies and scenario-based risk modelling.

Institutional responses include

- multi-tier supplier visibility dashboards
- geopolitical risk assessment integration in procurement decisions
- strategic redundancy in critical component sourcing
- flexible contractual structures allowing rapid adjustment.

Traditional risk registers often underestimate geopolitical volatility. Integrating macro-environmental intelligence into project governance strengthens anticipatory capacity.

15.9.2 Cross-border regulatory complexity

Infrastructure, energy and technology projects face varying regulatory standards across jurisdictions. ESG disclosure expectations differ internationally. Failure to align compliance frameworks can delay implementation or damage legitimacy.

Project organisations must therefore develop regulatory mapping capabilities and cross-border governance expertise. Cultural intelligence intersects with legal literacy in managing stakeholder expectations across regions. Global project ecosystems require hybrid governance models that reconcile local norms with global corporate standards.

15.10. Toward a resilient institutional architecture

The preceding analysis suggests that resilience cannot be reduced to risk mitigation. It emerges from systemic integration across leadership, teams, learning and communication. This section synthesises these elements into a forward-oriented institutional architecture.

15.10.1 Structural characteristics of resilient project ecosystems

Resilient project ecosystems exhibit

- adaptive governance structures with distributed authority
- hybrid human–AI collaboration protocols
- institutionalised learning systems embedded in stage-gate processes
- transparent ESG-aligned communication dashboards
- cross-cultural coordination mechanisms
- cybersecurity-integrated risk frameworks.

These characteristics operate interdependently. Weakness in one domain amplifies vulnerability in other domains.

15.10.2 Resilience as design principle

Resilience must be incorporated at the design stage rather than appended during a crisis. This involves

- embedding sustainability indicators within the scope definition
- integrating digital governance protocols during system selection
- designing team structures for diversity and adaptability
- establishing learning checkpoints at each project phase.

Resilience becomes architectural rather than reactive.

The interdependence of leadership, team architecture, learning infrastructure and governance mechanisms can be conceptualised as an embedded resilience system rather than as discrete organisational variables. Figure 15.1 presents the future-oriented resilient project ecosystem model, illustrating how institutionalised adaptive leadership anchors four mutually reinforcing capability domains – hybrid team architecture, continuous learning infrastructure, digital and AI governance, and communication aligned with ESG legitimacy – within a broader context shaped by geopolitical volatility, digital acceleration, sustainability accountability and stakeholder activism. The model emphasises that resilience and sustainable value creation emerge not from isolated optimisation of individual components but from systemic alignment across governance structures, cultural norms and adaptive capacities.

15.11. Toward a future research and policy agenda

The evolution of project management into institutional resilience architecture opens several research avenues:

- measuring adaptive leadership maturity in project governance
- quantifying trust density in hybrid teams
- evaluating AI–human collaboration effectiveness
- developing sustainability capability maturity models
- assessing legitimacy risk under digital transparency
- modelling cross-border governance complexity.

Figure 15.1 Future-oriented resilient project ecosystem model

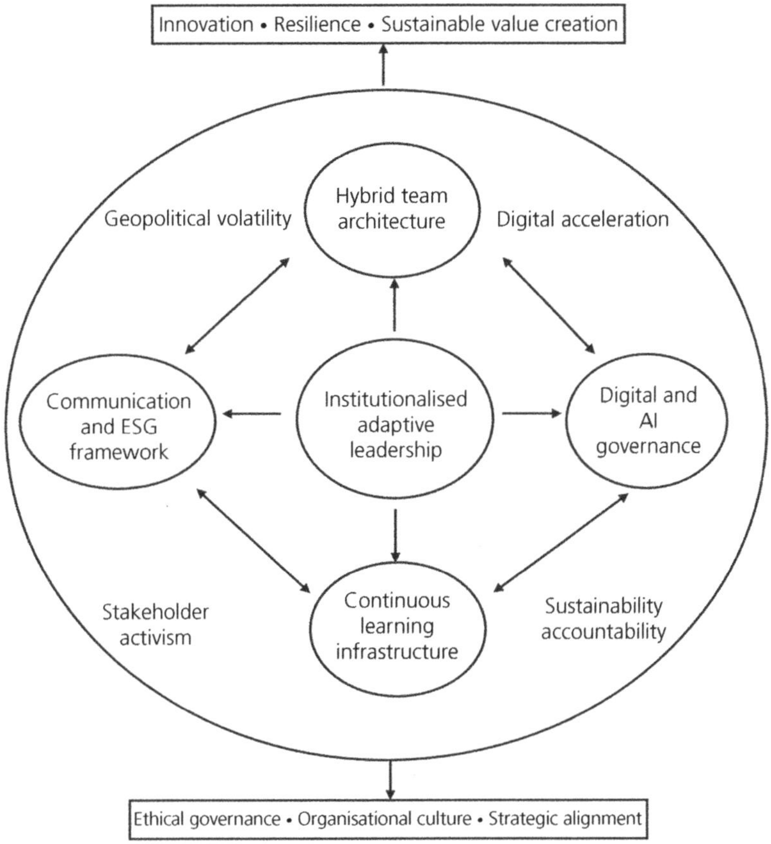

Policy implications include:

- national infrastructure resilience standards
- ESG reporting harmonisation frameworks
- professional certification updates incorporating AI ethics
- cross-sector resilience partnerships aligned with SDG 17.

Project management education must also evolve to include:

- systems and complexity modelling
- data ethics and digital governance
- sustainability integration frameworks
- cross-cultural negotiation simulation.

15.12. Executive and policy implications

The institutionalisation of resilient project ecosystems carries significant implications for executives, policy makers and professional bodies. While earlier sections addressed structural design, this section translates those insights into actionable governance directions.

15.12.1 Implications for executive leadership

Senior executives must reposition project management from an operational function to a strategic governance capability. This requires the following.

- *Elevating project governance to board-level oversight*: boards should review resilience metrics alongside financial performance indicators (Hansen and Svejvig, 2022).
- *Embedding ESG metrics into portfolio dashboards*: sustainability performance should be monitored systematically rather than episodically (Martens and Carvalho, 2016).
- *Investing in leadership pipeline transformation*: leadership development programmes must include complexity interpretation, digital ethics and stakeholder negotiation competencies.
- *Institutionalising AI governance protocols*: algorithmic systems require documented accountability and human oversight structures (Ransbotham *et al.*, 2017).

Executives must recognise that resilience is cumulative. It cannot be purchased through tools alone; it is constructed through consistent governance reinforcement.

15.12.2 Implications for policy makers and professional bodies

National and transnational institutions increasingly rely on large-scale infrastructure, digital transformation and sustainability-oriented projects. Policy makers should therefore

- develop resilience-oriented procurement standards
- harmonise ESG reporting frameworks across jurisdictions
- encourage public–private partnerships aligned with SDG objectives
- incorporate cybersecurity risk evaluation into project approval processes.

Professional bodies such as PMI and IPMA may consider

- updating certification frameworks to integrate AI governance
- incorporating sustainability metrics into competence baselines
- developing resilience maturity assessment tools.

Institutional reform at professional and regulatory levels reinforces the transformation of project management into a societal capability.

15.13. Future research directions

The evolving institutional landscape creates fertile ground for scholarly inquiry.

15.13.1 Leadership and governance

Future research should examine

- metrics for adaptive governance maturity
- the interaction between board oversight structures and project resilience outcomes
- comparative analyses of distributed against centralised decision architectures.

Longitudinal studies may reveal how leadership institutionalisation influences portfolio-level stability.

15.13.2 Hybrid team performance

Research opportunities include

- measuring trust density in digital environments
- evaluating the impact of AI-assisted coordination on psychological safety
- analysing diversity integration mechanisms in multinational project ecosystems.

Hybrid team dynamics represent a frontier in project scholarship.

15.13.3 Learning analytics and knowledge sustainability

Emerging questions include

- how AI-enabled analytics influence organisational learning quality
- whether accelerated project cycles undermine institutional memory
- how triple-loop learning can be operationalised within Agile frameworks.

Bridging knowledge-based theory with empirical project data remains a priority.

15.13.4 Sustainability and legitimacy

Future inquiry should explore

- the relationship between SDG alignment and stakeholder trust
- the financial performance implications of ESG-integrated project portfolios
- legitimacy risk modelling under digital transparency conditions.

Sustainability governance in project contexts remains under-theorised relative to its strategic importance.

15.14. Concluding synthesis: project management as societal infrastructure

This volume began by reframing project management as an integrated capability system. This concluding chapter extends that argument by demonstrating that such capabilities must be institutionalised within governance architectures that can endure structural disruption.

Leadership, when embedded within oversight systems, becomes governance stewardship. Teams, when designed for diversity and hybrid collaboration, become adaptive intelligence networks. Learning, when institutionalised, becomes resilience infrastructure. Communication, when aligned with ESG transparency, becomes a legitimacy architecture. Digital acceleration, geopolitical volatility and sustainability accountability are not temporary disturbances. They are defining structural conditions of the twenty-first century. Project management must therefore transition from a discipline of control to a discipline of institutional resilience.

In this evolution, project ecosystems serve not only organisational objectives but societal stability. Infrastructure, energy systems, healthcare programmes, digital transformation initiatives and sustainability transitions all depend on resilient project governance. Thus, project management becomes more than a methodology; it becomes a form of societal infrastructure enabling coordinated adaptation under uncertainty.

REFERENCES

Argyris C and Schön DA (1996) *Organizational Learning II: Theory, Method and Practice*. Addison-Wesley, Reading, MA, USA.

Bansal P and DesJardine MR (2014) Business sustainability: it is about time. *Strategic Organization* **12(1)**: 70–78.

Barney J (1991) Firm resources and sustained competitive advantage. *Journal of Management*, **17(1)**: 99–120.

Brady T and Davies A (2004) Building project capabilities. *Organization Studies* **25(9)**: 1601–1621.

Dobrucali E, Demirkesen S, Sadikoglu E, Zhang C and Damci A (2024) Investigating the impact of emerging technologies on construction safety performance. *Engineering, Construction and Architectural Management* **31(3)**: 1322–1347.

Duchek S (2020) Organizational resilience: a capability-based conceptualization. *Business Research* **13(1)**: 215–246.

Earley PC and Ang S (2003) *Cultural Intelligence: Individual Interactions Across Cultures*. Stanford University Press, Stanford, CA, USA.

Edmondson AC (2018) *The Fearless Organization: Creating Psychological Safety in the Workplace for Learning, Innovation, and Growth*. Wiley, Hoboken, NJ, USA.

Gibson CB and Gibbs JL (2006) Unpacking the concept of virtuality: the effects of geographic dispersion, electronic dependence, dynamic structure, and national diversity on team innovation. *Administrative Science Quarterly* **51(3)**: 451–495.

Hansen LK and Svejvig P (2022) Seven decades of project portfolio management research (1950–2019) and perspectives for the future. *Project Management Journal* **53(3)**: 277–294.

Jarrahi MH (2018) Artificial intelligence and the future of work: human–AI symbiosis in organizational decision making. *Business Horizons* **61(4)**: 577–586.

Joslin R and Müller R (2015) Relationships between a project management methodology and project success in different project governance contexts. *International Journal of Project Management* **33(6)**: 1377–1392.

Kerzner H (2025) *Project Management: A Systems Approach to Planning, Scheduling, and Controlling*. Wiley, Hoboken, NJ, USA.

Lengnick-Hall CA, Beck TE and Lengnick-Hall ML (2011) Developing a capacity for organizational resilience through strategic human resource management. *Human Resource Management Review* **21(3)**: 243–255.

Lindner F and Wald A (2011) Success factors of knowledge management in temporary organizations. *International Journal of Project Management* **29(7)**: 877–888.

Love PED, Teo P, Davidson M, Cumming S and Morrison J (2016) Building absorptive capacity in an alliance: process improvement through lessons learned. *International Journal of Project Management* **34(7)**: 1123–1137.

Martens ML and Carvalho MM (2016) Sustainability and success variables in the project management context: an expert panel. *Project Management Journal* **47(6)**: 24–43.

Marnewick C and Marnewick AL (2020) Technology readiness: a precursor for Industry 4.0. *Journal of Contemporary Management* **17(1)**: 129–149.

Müller R and Lecoeuvre L (2014) Operationalizing governance categories of projects. *International Journal of Project Management* **32(8)**: 1346–1357.

PMI (Project Management Institute) (2025) *A Guide to the Project Management Body of Knowledge (PMBOK Guide)*, 8th edn. PMI, Newtown Square, PA, USA.

Ransbotham S, Kiron D, Gerbert P and Reeves M (2017) Reshaping business with artificial intelligence: closing the gap between ambition and action. *MIT Sloan Management Review* **59(1)**: 1–12.

Richey RG Jr, Chowdhury S, Davis-Sramek B, Giannakis M and Dwivedi YK (2023) Artificial intelligence in logistics and supply chain management: a primer and roadmap for research. *Journal of Business Logistics* **44(4)**: 532–549.

Sense AJ (2011) The project workplace for organizational learning development. *International Journal of Project Management* **29(8)**: 986–993.

Silvius G, Schipper R and Planko J (2012) *Sustainability in Project Management*. Gower, Farnham, UK.

Silvius AG, Kampinga M, Paniagua S and Mooi H (2017) Considering sustainability in project management decision making: an investigation using Q-methodology. *International Journal of Project Management* **35(6)**: 1133–1150.

Suchman MC (1995) Managing legitimacy: strategic and institutional approaches. *Academy of Management Review* **20(3)**: 571–610.

Sovacool BK and Ryu H (2025) Beyond economies of scale: learning from construction cost overrun risks and time delays in global energy infrastructure projects. *Energy Research & Social Science* **123**: 104057.

Teece DJ (2007) Explicating dynamic capabilities: the nature and microfoundations of (sustainable) enterprise performance. *Strategic Management Journal* **28(13)**: 1319–1350.

Too EG and Weaver P (2014) The management of project management: a conceptual framework for governance. *International Journal of Project Management* **32(8)**: 1382–1394.

Uhl-Bien M and Arena M (2018) Leadership for organizational adaptability: a theoretical synthesis and integrative framework. *The Leadership Quarterly* **29(1)**: 89–104.

Williams T (2003) Learning from projects. *Journal of the Operational Research Society* **54(5)**: 443–451.

Advancing Leadership and Collaboration in Project Management: Culture, Knowledge and Team Dynamics

Mohammed Reyasudin Basir Khan
ISBN 978-1-83742-803-8
https://doi.org/10.1108/978-1-83742-802-120261019
Emerald Publishing Limited: All rights reserved

Index

Printed and bound by CPI Group (UK) Ltd, Croydon, CR0 4YY

14/10/2026

14981999-0006